Applied Chemistry: A Textbook for Engineers and Technologists

Applied Chemistry: A Textbook for Engineers and Technologists

H. D. Gesser

University of Manitoba
Winnipeg, Manitoba, Canada

Kluwer Academic / Plenum Publishers
New York, Boston, Dordrecht, London, Moscow

Library of Congress Cataloging-in-Publication Data

Gesser, Hyman D., 1929–
Applied chemistry: a textbook for engineers and technologists/H.D. Gesser.
p. cm.
Includes bibliographical references and index.
ISBN 0-306-46553-1
1. Chemistry, Technical. I. Title.

TP515 .G48 2001
660—dc21

2001029579

ISBN 0-306-46553-1

233 Spring Street, New York, N.Y. 10013

http://www.wkap.nl/

10 9 8 7 6 5 4 3 2 1

A C.I.P. record for this book is available from the Library of Congress

Printed in the United States of America

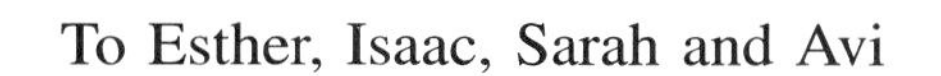

To Esther, Isaac, Sarah and Avi

Preface

This book is the result of teaching a one semester course in Applied Chemistry (Chemistry 224) to second year engineering students for over 15 years. The contents of the course evolved as the interests and needs of both the students and Engineering Faculty changed. All the students had at least one semester of Introductory Chemistry and it has been assumed in this text that the students have been exposed to Thermodynamics, Chemical Kinetics, Solution Equilibrium, and Organic Chemistry. These topics must be discussed either before starting the Applied subjects or developed as required if the students are not familiar with these prerequisites.

Engineering students often ask "Why is another Chemistry course required for Non-Chemical Engineers?"

There are many answers to this question but foremost is that the Professional Engineer must know when to consult a Chemist and be able to communicate with him. When this is not done the consequences can be a disaster due to faulty design, poor choice of materials or inadequate safety factors.

Examples of blunders abound and only a few will be described in an attempt to convince the student to take the subject matter seriously.

The Challenger space shuttle disaster which occurred in January 1986 was attributed to the cold overnight weather which had hardened the O-rings on the booster rockets while the space craft sat on the launch-pad. During flight the O-ring seals failed, causing fuel to leak out and ignite. The use of a material with a lower glass transition temperature (T_g) could have prevented the disaster.

A similar problem may exist in automatic transmissions used in vehicles. The use of silicone rubber O-rings instead of neoprene may add to the cost of the transmission but this would be more than compensated for by an improved and more reliable performance at $-40°C$ where neoprene begins to harden; whereas the silicone rubber is still flexible.

A new asphalt product from Europe incorporates the slow release of calcium chloride ($CaCl_2$) to prevent icing on the roads and bridges. Predictably, this would have little use in Winnipeg, Canada, where $-40°C$ is not uncommon in winter.

The heavy water plant at Glace Bay, Nova Scotia was designed to extract D_2O from sea water. The corrosion of the plant eventually delayed production and the redesign and use of more appropriate materials added millions to the cost of the plant.

A chemistry colleague examined his refrigerator which failed after less than 10 years of use. He noted that a compressor coil made of copper was soldered to an expansion tube made of iron. Condensing water had corroded the—guess what?—iron tube. Was this an example of designed obsolescence or sheer stupidity. One wonders, since the savings by using iron instead of copper is a few cents and when the company is a well known prominent world manufacturer of electrical appliances and equipment.

With the energy problems now facing our industry and the resulting economic problem that results, the engineer will be required to make judgements which can alter the cost benefit ratio for his employer. One must realize that perpetual motion is impossible even though the US Supreme Court has ruled that a patent should be granted for a devise which the Patent Office considered to be a Perpetual Motion Machine. An example of this type of proposal appeared in a local newspaper which described an invention for a car which ran on water. This is accomplished by a battery which is initially used to electrolyze water to produce H_2 and O_2 that is then fed into a fuel cell which drives an electrical motor which propels the car. While the car is moving, an alternator driven by the automobile's motion, charges the battery. Thus, the only consumable item is water. This is an excellent example of perpetual motion.

A similar invention of an automobile powered by an air engine has been described. A compressed air cylinder powers an engine which drives the automobile. A compressor which is run by the moving car recompresses the gas into a second cylinder which is used when the first cylinder is empty. Such perpetual motion systems will abound and the public must be made aware of the pitfalls.

Have you heard of the Magnatron? Using 17 oz of deuterium (from heavy water) and 1.5 oz of gallium will allow you to drive an engine 110,000 miles at a cost of $110. Are you sceptical? You should be, because it is an example of the well-known Computer GIGO Principle (meaning garbage in = garbage out).

An engineer responsible for the application of a thin film of a liquid adhesive to a plastic was experiencing problems. Bubbles were being formed which disrupted the even smooth adhesive coat. The answer was found in the dissolved gases since air at high pressure was used to force the adhesive out of the spreading nozzle. The engineer did not believe that the air was actually soluble in the hexane used to dissolve the glue. When helium was used instead of air, no bubbles formed because of the lower solubility of He compared to O_2 and N_2 in the solvent. Everything is soluble in all solvents, only the extent of solution varies from non-detectable (by present methods of measurement) to completely soluble. The same principle applies to the permeability of one substance through another.

An aluminum tank car exploded when the broken dome's door hinge was being welded. The tank car, which had been used to carry fertilizer (aqueous ammonium nitrate and urea), was washed and cleaned with water—so why had it exploded? Dilute ammonium hydroxide is more corrosive to aluminum than the concentrated solution. Hence, the reaction

$$3NH_4OH + Al \rightarrow Al(OH)_3 + 3NH_3 + 1.5H_2$$

produces hydrogen which exploded when the welding arc ignited the H_2/O_2 mixture. The broad explosive range of hydrogen in air makes it a dangerous gas when confined.

Batteries are often used as a back-up power source for relays and, hence, stand idle for long periods. To keep them ready for use they are continuously charged. However, they are known to explode occasionally when they are switched into service because of the excess hydrogen produced due to overcharging. This can be avoided by either catalyzing the recombination of the H_2 and O_2 to form water

$$2H_2 + O_2 \rightarrow 2H_2O$$

by a nickel, platinum or palladium catalyst in the battery caps, or by keeping the charging current equal to the inherent discharge rate which is about 1% per month for the lead-acid battery.

It has recently been shown that the flaming disaster of the Hindenburg Zeppelin in 1937, in which 36 lives were lost, may have been caused by static electricity igniting the outer fabric. This was shown to contain an iron oxide pigment and reflecting powdered aluminum. Such a combination, known as a Thermite mixture, results in the highly exothermic Gouldshmidt reaction (first reported in 1898)

$$Fe_2O_3 + 2Al \rightarrow Al_2O_3 + 2Fe \qquad \Delta H^0 = -852 \text{ kJ/mol of } Fe_2O_3$$

In the early days of the railway, rails were welded with the molten iron formed in this reaction. The combination of powdered aluminum and a metal oxide has been used as a rocket fuel and evidence has been obtained to indicate that after the disaster the Germans replaced the aluminum by bronze which does not react with metal oxides. Thus, the bad reputation hydrogen has had as a result of the accident is undeserved and the resulting limiting use of the airship was due to faulty chemistry and could have been avoided.

The original design and structure of the Statue of Liberty, built about 100 years ago, took into account the need to avoid using different metals in direct contact with each other. However, the salt sea spray penetrated the structure and corroded the iron frame which supported the outer copper shell. Chloride ions catalyzed the corrosion of iron. The use of brass in a steam line valve resulted in corrosion and the formation of a green solid product. The architect was apparently unaware of the standard practice to use amines such as morpholine as a corrosion inhibitor for steam lines. Amines react with copper in the brass at high temperatures in the presence of oxygen to form copper-amine complexes similar to the dark blue copper ammonium complex, $Cu(NH_3)_4^{2+}$.

Numbers are a fundamental component of measurements and of the physical properties of materials. However, numbers without units are meaningless. Few quantities do not have units, e.g., specific gravity of a substance is the ratio of the mass of a substance to the mass of an equal volume of water at 4°C. Another unitless quantity is Reynolds Number, $R_e = \rho v l/\eta$ where ρ is the density; v is the velocity; η is the viscosity of the fluid and l is the length or diameter of a body or internal breath of a pipe. The ratio $\eta/\rho = \mu$ the kinematic viscosity with units of l^2/t. $R = vl/\mu$ and has no units if the units of v, l, and μ are consistent.

To ignore units is to invite disaster. Two examples will illustrate the hazards of the careless- or non-use of units. During the transition from Imperial to SI (metric) units in Canada, an Air Canada commercial jet (Boeing 767) on a trans Canada flight (No 143) from Montreal to Edmonton on July 23, 1983 ran out of fuel over Winnipeg.

Fortunately the pilot was able to glide the airplane to an abandoned airfield (Gimli, MB) used for training pilots during World War II. The cause of the near disaster was a mix-up in the two types of units involved for loading the fuel and the use of a unitless conversion factor. (See Appendix A for a detailed account of this error).

The second example of an error in units cost the USA (NASA) $94,000,000. A Mars climate probe missed its target orbit of 150 km from the Mars' surface and approached to within 60 km and burned up. The error was due to the different units used by two contractors and which was not inter-converted by the NASA systems engineering staff. This book uses various sets of units and the equivalences are given in Appendix A. This is designed to keep the student constantly aware of the need to watch and be aware of units.

The above examples show how what may be a simple design or system can fail due to insufficient knowledge of chemistry. This textbook is not intended to solve all the problems you might encounter during your career. It will, however, give you the vocabulary and basis on which you can build your expertise in engineering.

The exercises presented at the end of each chapter are intended to test the students' understanding of the material and to extend the topics beyond their initial levels.

The author is indebted to the office staff in the Chemistry Department of the University of Manitoba who took pencilled scrawls and converted them into legible and meaningful text. These include Cheryl Armstrong, Tricia Lewis and Debbie Dobson. I also wish to thank my colleagues and friends who contributed by critical discussions over coffee. I also wish to express my thanks to Roberta Wover who gave me many helpful comments on reading the manuscript and checking the exercises and Web sites. Mark Matousek having survived Chem. 224 several years ago, applied some of his acquired drawing skills to many of the illustrations shown. Nevertheless, I must accept full responsibility for any errors or omissions, and I would be very grateful if these would be brought to my attention.

Some general references are listed below:

Kirk & Othmer, Encyclopedia of Chemical Technology, 4th Ed., 30 Vol., J. Wiley & Sons, New York (1995).

Ullmann's Encyclopedia of Industrial Chemistry, 26 Vol., VCH, Germany (1992).

Encyclopedia of Physical Science and Technology. 15 Vol., + Year Books, Academic Press, Orlando, Florida (1987).

V. Hopp and I. Hennig, Handbook of Applied Chemistry, Hemisphere Publ. Co., Washington (1983).

McGraw-Hill Encyclopedia of Science and Technology. 15 Vol., + Year Books, Ndew York (1982).

W. Steedman, R.B. Snadden and I. H. Anderson, Chemistry for the Engineering and Applied Sciences, 2nd Ed., Pergamon Press, Oxford (1980).

I. P. Muklyonov, Editor, Chemical Technology, 3rd Ed., 2 Vol., Mir Publ., Moscow (1979), in English.

R. M. E. Diamant, Applied Chemistry for Engineers, 3rd Ed., Pitman, London (1972).

G. R. Palin, Chemistry for Technologists, Pergamon Press, Oxford, (1972).

Chemical Technology: An Encyclopedic Treatment, 7 Vol., Barnes and Noble Inc. New York (1972).

F. G. Butler and G. R. Cowie, A Manual of Applied Chemistry for Engineers, Oliver and Boyd, London, (1965).

L. A. Munro, Chemistry in Engineering, Prentice Hall, Englewood Cliffs, New Jersey (1964).

E. Cartwell, Chemistry for Engineers — An Introductory Course, 2nd Ed. Butterworths, London, (1964).
E. S. Gyngell, Applied Chemistry for Engineers, 3rd Ed., Edward Arnold, London (1960).
Thorpe's Dictionary of Applied Chemistry, 4th Ed., 11 Vol., Longmans, Green, London, (1957)

The World Wide Web is an excellent source of technical information though it is important to recognize that discretion must be exercised in selecting and using the information since the material presented is not always accurate or up-to-date. Some selected web sites are added to the Further Readings lists at the end of each chapter.

Contents

Abbreviations Used in this Text

AAGR	Average annual growth rate
AAS	Atomic absorption spectrometry
ABS	Acrylonitrile-butadiene-styrene polymer
AEM	Anion exchange membrane
AFC	Atomic fluorescence spectrometry
AFR	Air fuel ratio
AGR	Advanced gas reactor
ANFO	Ammonium nitrate fuel oil
ASTM	American Society for Testing and Materials
bbl	Barrel for oil, see Appendix A
BC	Bimetallic corrosion
BET	Brunauer–Emmett–Teller
BLEVE	Boiling liquid expanding vapor explosion
BOD	Biochemical oxygen demand
BP	Boiling point
BWR	Boiling water reactor
CANDU	Canadian deuterium uranium reactor
CASING	Crosslinking by activated species of inert gases
CC	Crevice corrosion
CEM	Cation exchange membrane
CN	Cetane number
CNG	Compressed natural gas
COD	Chemical oxygen demand
CPVC	Chlorinated polyvinylchloride
CR	Compression ratio
DC	Direct current
DNA	Deoxyribonucleic acid
DP	Degree of polymerization
DR	Drag Reducer
DR	Distribution ratio
DTA	Differential thermal analysis
ECE	Economic Commission for Europe
ECM	Electrochemical machining

EDS	Exxon Donner Solvent
EHL	Elastohydrodynamic lubrication
EIS	Electrochemical impedance spectroscopy
ENM	Electrochemical noise method
EO	Extreme pressure (lubrication)
ER	Electrorheological fluid
ETBE	Ethyl tert butyl ether
EV	Expected value
EV	Electric vehicle
FAC	Free available chlorine
FEP	Hexafluoropropylene + PTFE
FP	Flash point
GAC	Granulated activated carbon
GBC	Grain boundary corrosion
GNP	Gross National Product
GR	Gas cooled reactor
hv	photon
HAR	High aspect ratio
HDI	Hexane diisocyanate
HLW	High level waste
HMN	Heptamethylnonane
HRI	Hydrocarbon Research Inc
HWR	Heavy water reactor
ICAPS	Inductively coupled argon plasma spectrometry
ICE	Internal combustion engine
Is	Specific impulse
LEL	Lower explosion limit
LH_2	Liquid hydrogen
LNG	Liquified natural gas
LWR	Light water reactor
M-85	Methanol with 15% gasoline
MDF	Macro defect free
MeV	Million electron volts
MI	Machinability index
MMT	Methylcyclopentadiene Manganese II tricarbonyl
MON	Motor octane number
MPC	Maximum permissible concentration
MPN	Most probable number
MTBE	Methyl tertiary butyl ether
MW	Megawatt, 10^6 W
NMOG	Non-methane organic gases
NTP	Normal conditions of temperature and pressure, 25°C and 1 atm pressure.
OB	Oxygen balance
OECD	Organization for Economic Cooperation and Development
ON	Octane number (average of MON + RON)
OPEC	Organization of Petroleum Exporting Countries
OTEC	Ocean thermal energy conversion

PAH	Polynuclear aromatic hydrocarbons
PAN	Polyacrylonitrile
PC	Pitting corrosion
PCB	Polychlorinated biphenyl
PF	Phenol-formaldehyde
pH	$pH = -\log_{10} [H^+]$, neutral water has pH = 7
PMMA	Polymethylmethacrylate
ppm	parts per million, $\mu g/g$ or mL/m^3
PS	Polystyrene
PTFE	Polytetrafluoroethylene
PV	Pressure volume (product)
PVAc	Polyvinylacetate
PVC	Polyvinyl chloride
PWR	Pressurized water reactor
Quad Q,	Unit of energy, see Appendix A
Rad	Radiation absorbed dose
RBE	Relative biological effectiveness
RDX	Cyclonite or hexogen
Rem	Roentgen equivalent to man
rpm	Revolutions per minute
RO	Reverse osmosis
RON	Research octane number
SAN	Styrene-acrylonitrile copolymer
SCC	Stress corrosion cracking
SCE	Saturated calomel electrode
SHE	Standard hydrogen electrode
SI	Spark ignition
SIT	Spontaneous ignition temperature
SNG	Synthetic natural gas
SOAP	Spectrographic oil analysis program
SP	Smoke point
SRC	Solvent refined coal
SSPP	Solar sea power plants
STP	Standard condition of temperature and pressure, 0°C and 1 atm pressure
TDC	Top dead center
TDI	Toluene diisocyanate
TEL	Tetraethyl lead
TGA	Thermal gravimetric analysis
THM	Trihalomethanes
TLV	Threshold limit value
TML	Tetramethyl lead
TNG	Trinitroglycerol
TNT	Trinitrotolluene
TOE	Tons of oil equivalent (energy)
TW	Terawatts, 10^{12} W
UC	Uniform corrosion
UEL	Upper explosion limit

UFFI	Urea-formaldehyde foam insulation
UN	United Nations
UV	Ultraviolet light, $\lambda < 380$ nm
VCI	Vapor corrosion inhibitors
VI	Viscosity index
VOC	Volatile organic compounds
VOD	Velocity of detonation
WHO	World Health Organization

Acknowledgments

Table 1-1, Table 1-2, International Energy Annual. Table 1.3, Fig. 1.16, Royal Society London and H. Tabor, Non-convecting solar pond, *Phil. Trans. Roy. Soc. London* **A295** 423 (1980). Table 1.4, G. L. Wick and J. D. Isaacs, Salinity Power, (1975), Institute of Marine Resources #75-9, University of California, La Jolla, CA USA. Fig. 1.3, Physics in Canada and J. Jovanovich, Is Nuclear Power Essential? Sept. (1988). Fig. 1.8, Embassy of France (Ottawa). Fig. 1.9, B de Jong, Net radiation by a horizontal surface at the earth, Delft University Press (1973). Fig. 1.11, Goodyear-Akron OH. Fig. 1.10, A. D. Walt, IEEE Spectrum **8** 51 (Copyright © 1971 IEEE). Fig. 1.14, American Wind Energy Association, Washington DC, USA. Fig. 1.15, A. Lavi and C. Zener, IEEE Spectrum **10** 23 (Copyright © 1973 IEEE). Fig. 1.17, Embassy of Israel (Ottawa). Fig. 1.18, Scripps Oceanographic Institute, University of California, La Jolla CA, and A. Fisher, Energy from the sea, Popular Science, May (1975). Fig. 1.20, Reprinted with permission from R. S. Norman, Science **186** 351 (1974), (Copyright © 1974, AAAS, American Association for the Advancement of Science). Table 2.1, International Energy Annual. Table 2.2, A.S.T.M. West Conshohocken PA. Table 2.4, Ontario Hydro Research, Toronto, ON. Fig. 2.3. National Peat Board, Ireland. Fig. 3.1, Restek Inc. Bellefonte PA. Fig. 3.2, Fig. 3.5, Natural Resources Canada. Reproduced with the permission of the Minister of Public Works and Government Services Canada (1977). Table 3.2, International Enery Annual. Table 3.5, Table 3.6, Fig. 3.6, Royal Society London and R. C. Neavel, Exxon Donor Solvent Liquefaction Process, *Phil. Trans. R. Soc.* London **A300** 141 (1981). Fig. 3.7, *Royal Society London* and B. K. Schmid and D. M. Jackson, The SCR-II Process, *Phil. Trans. R. Soc. London* **A300** 129 (1981). Fig. 3.8, S. H. Moss and W. G. Schlinger, Coal Gasification, Lubrication **66** 25 (1980) Texaco Inc. White Plains NY, Courtesy of Equilon Enterprises LLC, Shell and Texaco working together. Fig. 3.9, Royal Society London and J. C. Hoogendoorn, Motor Fuels and Chemicals from Coal via Sasol Synthetic Route, *Phil. Trans. R. Soc. London* **A300** 104 (1981). Fig. 3.10. Sasol Ltd, Sasolburg, South Africa. Fig. 4.5, Reprinted with permission from C. F. Kettering, On the power and efficiency of internal combustion engines, Industrial and Engineering Chemistry **36** 1079, Copyright (1944) American Chemical Society. Fig. 4.6, Reprinted with permission from R. Lobinski et al., Organolead in Wine, *Nature* **370** 24 (1994). Copyright (1994) Macmillan Magazines Ltd. Table 4.7, Reprinted with permission from C. K. Westbrook and W. J. Pitz, The Chemical Kinetics of Engine Knock, Energy & Technology Review, Feb./Mar. (1991) (The US Government retain non-exclusive, royalty-free copyrights). University of California, Lawrence Livermore National Laboratory, Livermore CA. Table 4.8,

International Energy Outlook, Paris (1994). Table 4.9, Biomass Energy Institute Inc., Bio-Joule (1983). Table 5.1, Reprinted with permission from: T. Y. Chang et al., Alternative Transportation Fuels and Air Quality, *Environ. Sci. Technology* **25** 1190, Copyright (1991) American Chemical Society and Ford Motor Co. Fig. 6.1A, Shell International, London. Fig. 6.1B, Tokyo Gas Co., Japan. Fig. 6.2, Tokyo Gas Co., Japan. Fig. 6.8, L. Belkbir et al. Comparative Study of the Formation-Decomposition Mechanisms and Kinetics in $LaNi_5$, and Magnesium Reversible Hydrides, *Int. J. Hydrogen Energy*, **6** 285 (1981), with permission from The International Association of Hydrogen Energy. Table 6.2, Reprinted from Shell Petroleum Handbook, 6th Edition © 1983 with permission from Elsevier Science. Fig. 6.5, Shali Fu, Wuhan, China. Table 6.8, Linde/Union Carbide Corp. Table 7.1, International Atomic Energy Agency (I.A.E.A.). Table 7.4, A.E.C.L. Table 7.13, A.E.C.B. Table 7.14, A.E.C.L. Fig. 7.1, Based on data from I.A.E.A. Fig. 7.5, Atomic Industrial Forum/Nuclear Energy Institute. Fig. 7.6, Fig. 7.7, Fig. 7.8, Fig 7.9, A.E.C.L. Fig. 7.10, H. Inhaber, Energy Risk Assessment, With permission from Gordon & Breach Sci. Publ. NY (1982) © OPA (Overseas Publishers Association NV). Fig. 7.11, E.P.R.I. (Electric Power Research Institute) Journal (1976). Fig. 8.4, Fig. 8.5, Fig. 8.7, Fig. 8.8, Texaco Inc. *Lubrication* **67** #1 (1981). Fig. 8.2, Fig. 8.3, Fig. 8.9, J. Hickman and K. Middleton, Some Surface Chemical Aspects of Lubrication, Advancement of Science, June (1970) British Association for the Advancement of Science, London. Table 9.3, Fig. 9.2, Fig. 9.3, Fig. 9.4, Fig. 9.5, Fig. 9.6, Y. Sugie et al., Characteristics in Electrochemical Machining for Various Steels, Denki Kagaku, **46** 147 (1978) with permission from the Electrochemical Society of Japan. Table 9.8, In part from E.P.R.I. Journal, Oct. (1976). Chapter 10, Cartoon, Science Dimension **14** 12 (1983) N.R.C. (Ottawa). Fig. 10.3, Y. Waseda and K. T. Aust, Review. Corrosion Behavior of Metallic Glasses, *J. Material Sci.* **16** 2338 (1981) With permission from Chapman & Hall Publ. Table 11.5, Table 11.6, Table 11.7, Fig. 11.9, with permission from General Electric, Schenectady, NY. Fig. 11.4, W. Watt, Production and Properties of High Modulus Carbon Fibres. *Proc. Roy Soc. London* **A319** 8 (1970), British Crown Copyright/Defence Evaluation and Research Agency, reproduced with permission of the Controller, Her (Britannic) Majesty's Stationery Office and the Royal Society of London. Fig. 11.10, Cadillac Plastics. Fig. 12.2, Witco Corp. Fig. 15.3 and Fig. 15.4, H. V. Thurman, Introductory Oceanography, 8th Edition, © 1990, Reprinted by permission of Prentice-Hall Inc., Upper Saddle River, NJ, 07458. Fig 15.8, Delta Engineering (Ottawa). Fig. 15.1, Fig. 16.1, Mir Publ. Moscow. Table 16.2, Fig. 16.2, J. D. Birchall, A. J. Howard and K. Kendall, New cements—inorganic plastics of the future, Chemistry in Britain, Dec. (1982) The Royal Society of Chemistry. Table 16.3, Fig. 16.6, Corning Glass Works. Fig. 16.5, Reprinted with permission from American Machinist (1997)—A Penton Publication. Fig. 16.8, Michelin North America (Canada) Inc. Fig. B.4, Fig. B.5, Fig. B.6, Union Carbide Corp. Table B.1, C.R.C. Handbook of Chemistry and Physics, 71st Edition. (1990–91).

Applied Chemistry: A Textbook for Engineers and Technologists

1

Energy: An Overview

1.1. INTRODUCTION

World energy pundits have long proclaimed that the fossil fuels in the Earth's crust are limited and will be exhausted some day. This argument is similar to the accepted pronouncements of cosmologists that "the entropy of the Universe is increasing towards a maximum" or that "the sun will one day burn itself out." The world energy crisis of 1973 was precipitated when OPEC curtailed oil production and fixed their own price for oil for the first time, and within a year the price of oil went from about \$1.20/bbl to about \$10/bbl. This was aggravated by the increase in oil imports to the USA, which ceased to be self-sufficient in oil in the late 1960s. The present world price for oil is about \$30/bbl (after reaching a maximum of about \$40/bbl and dropping to less than \$15/bbl) and it is expected to eventually increase again. The price of oil will depend on demand as well as the financial needs of the oil producers. The successful development of alternate energy sources, e.g., fusion, could bring the price down to \$5/bbl. Since energy is an integral part of every function and product from food (which requires fertilizer) to plastics which are petroleum based, to steel or other metals which require energy for extraction, beneficiation, reduction and fabrication, worldwide inflation can be directly attributed to the rising price of oil.

An illustration of the importance of energy to the economy of a country is shown in Fig. 1.1, where the annual Gross National Product (GNP) of a country is plotted against the total annual energy consumed by that country per unit population. Some small anomalies are apparent but in general the higher the GNP* of a country the larger is its per capita energy consumption. Energy is essential to progress and there is no substitute for energy. Society's use of energy has continuously increased but sources have invariably changed with time. This is illustrated in Fig. 1.2.

It is interesting to note (see Fig. 1.3) that the per capita use of commercial energy for UK and USA has been essentially constant for 100 years whereas that for Germany, Russia, and Japan showed an exponential growth (doubling time of 12 years) towards the constant US/UK values. The effect of the world's population growth on energy usage is obvious.

Energy can conveniently be classified into renewable and nonrenewable sources as shown in Fig. 1.4. Such a division is quite arbitrary and is based on a time scale which distinguishes hundreds of years from millions. The world's reserve supply of oil is

*The Gross National Product (GNP) is the sum total of the market value of goods and services produced per annum for final consumption, capital investment or for government use.

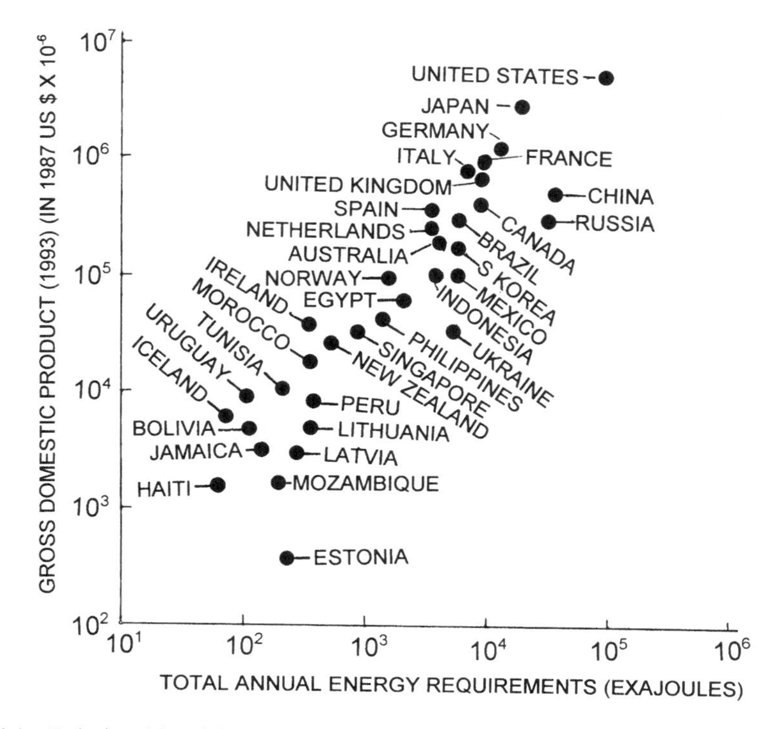

FIGURE 1.1. Relationship of GNP to energy available per capita in 1994 for various countries.

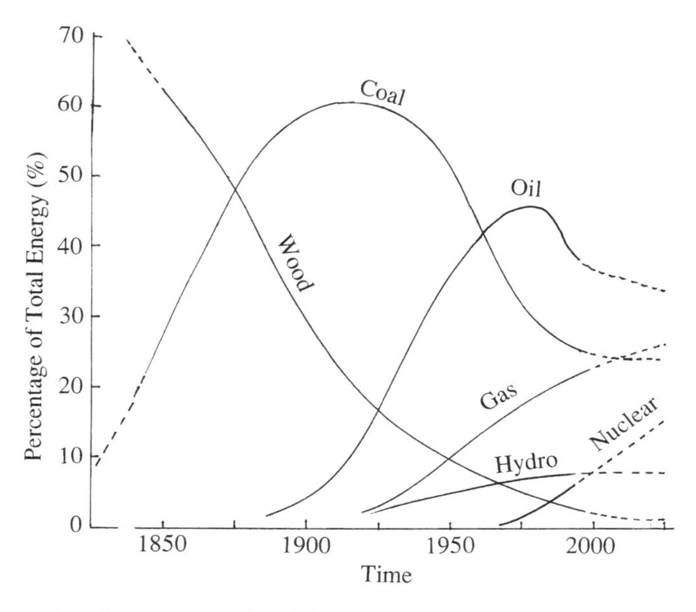

FIGURE 1.2. The changing patterns of world energy consumption as a percentage of total usage.

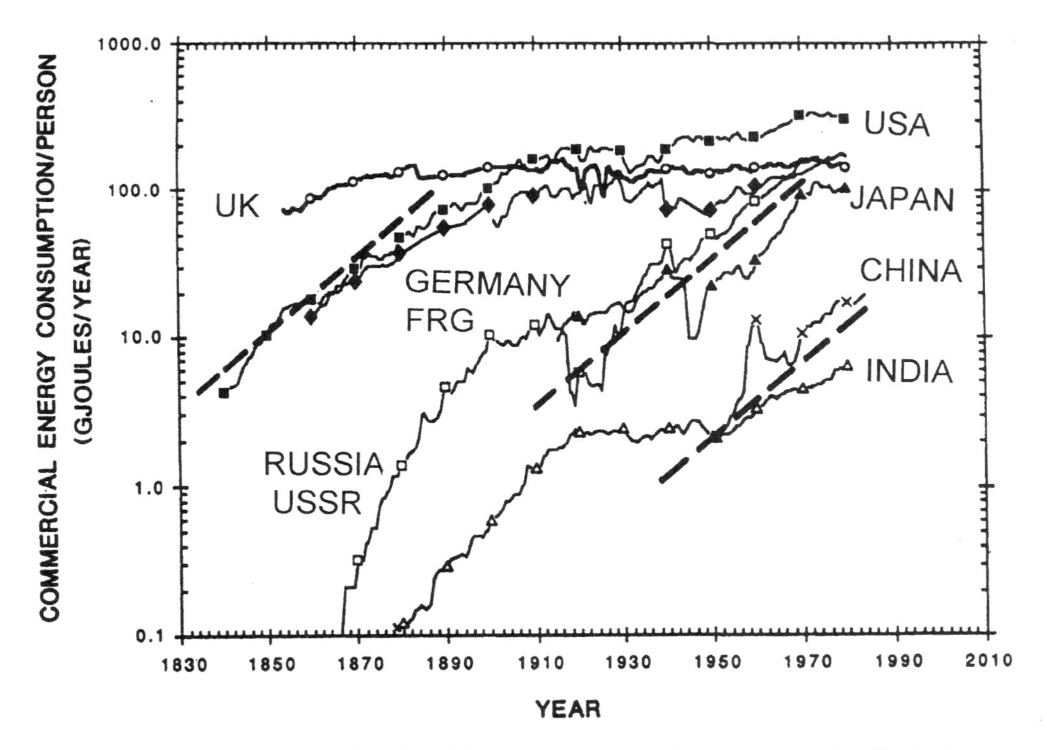

FIGURE 1.3. The commercial (industrial) energy consumption per capita in Gigajoules per person per year is plotted (note logarithmic scale) as a function of time.

estimated at about 5.8×10^3 Quads and that of natural gas at about 5.1×10^3 Quads (see Table 1.1). Oil and gas production has still been slowly rising with oil production expected to peak in 2035. Gas production is expected to peak by 2050 and so will last slightly longer, assuming that more oil and gas resources are made available. Coal is the major fossil fuel on earth and consists of over 75% of the available fossil fuel energy (see Table 1.1). Recent world energy consumption is given in Table 1.2.

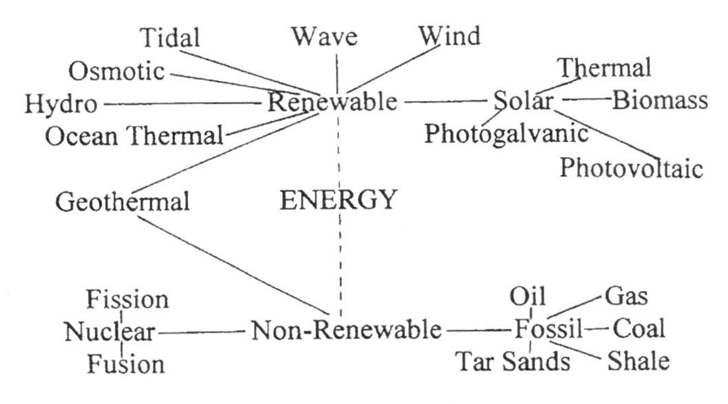

FIGURE 1.4. A classification of energy sources.

TABLE 1.1
World Reserves of Fossil Fuels in Quads (1998)

Area	Natural Gas	Oil	Coal
North America	297	395	7900
Latin America	221	430	660
Western Europe	166	110	2755
East Europe, CIS	1950	355	8000
Africa	350	425	1870
Middle East	1723	3770	5
Australia, Far East	430	290	8910
Total	5137	5775	30,100

Conservative considerations of our energy consumption predict that coal will supply 1/4 to 1/3 of the world's energy requirements by the year 2050. Its use can be relied upon as an energy source for about another 200 years; however, other considerations (such as the greenhouse effect and acid rain) may restrict the uncontrolled use of fossil fuel in general and coal in particular.

The increase in use of fossil fuel during the past few decades has resulted in a steady increase in the CO_2 concentration in the atmosphere. This is shown in Fig. 1.5. In 1850 the concentration of CO_2 was about 290 ppm and by the year 2025 the estimated concentration will be about double present values (350 ppm) if fossil fuels are burned at the present rate of 5 Gton of C/year. By the year 2025 the world's energy demands will have increased to over 800 Quads from 250 Quads in 1980. If a large fraction of this energy is fossil fuel, i.e., coal, then the annual increase in the concentration of CO_2 in the atmosphere is calculated to be greater than 10 ppm.

The CO_2 in the atmosphere is believed to have an adverse effect on the World's climate balance. The atmosphere allows the solar visible and near ultraviolet rays to penetrate to the earth where they are absorbed and degraded into thermal energy, emitting infrared radiation which is partially absorbed by the CO_2, water vapor and other gases such as CH_4 in the atmosphere (see Fig. 1.6).

There is at present a thermal balance between the constant energy reaching the earth and the energy lost by radiation. The increase in CO_2 in the atmosphere causes

TABLE 1.2
World Consumption of Energy by Sources in 1986, 1995, and 1997 (Quads)

	1986	%	1995	%	1997	%
Oil	126.6	40.4	141.1	38.9	148.7	39.3
Coal	86.3	27.5	77.5	21.4	92.8	24.6
Natural gas	62.3	19.9	93.1	25.7	83.9	22.2
Hydroelectric energy	21.3	6.8	25.7	7.2	26.6	7.1
Nuclear energy	16.3	5.2	23.2	6.4	24.0	6.3
Renewable sources	0.7	0.2	1.6	0.4	1.8	0.5
Total	313.5	100	362.2	100	377.8	100

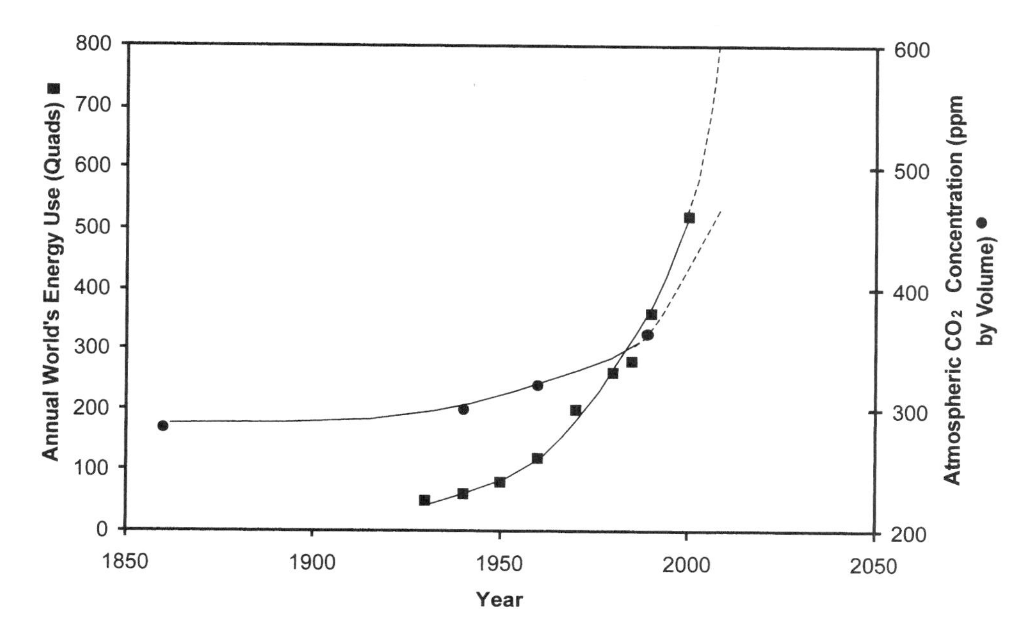

FIGURE 1.5. The concentration of carbon dioxide in the earth's atmosphere as a function of time (●) and a plot of the total annual world's energy use against time (■). The extrapolated values are speculative and subject to large errors.

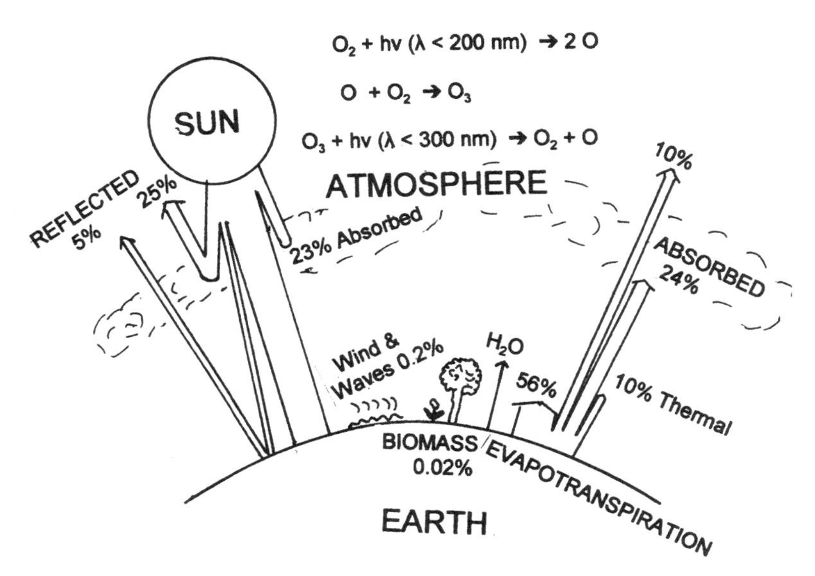

FIGURE 1.6. The Greenhouse Effect—A Schematic Representation. The CO_2 and other greenhouse gases, such as CH_4, absorb infrared radiation and become vibrationally excited. The excitation energy is then degraded into heat (kinetic energy).

an increase in the absorption of the radiated infrared from the earth (black body radiation) and a rise in the thermal energy or temperature of the atmosphere. This is called the *greenhouse effect*. Temperature effects are difficult to calculate and estimates of temperature changes vary considerably, although most agree that a few degrees rise in the atmospheric temperature (e.g., 3°C by the year 2025) could create deserts out of the prairies and convert the temperate zones into tropics, melt the polar ice caps and flood coastal areas. For example, for a 1°C rise in the earth's temperature the yield of wheat would be expected to drop by 20%, though rice yields might rise by 10%. If the average temperature of the oceans increased by 1°C, the expansion would cause a rise in sea level of about 60 cm (assuming no melting of glacier ice).

One uncertain factor in the modelling and predictions is that there is a lack in a material balance for CO_2, i.e., some CO_2 is unaccounted for indicating that some CO_2 sinks (i.e., systems which hold or consumed CO_2) have not been identified. The oceans and forests (biomass growth) consume most of the CO_2 and it is possible that these sinks for CO_2 may become saturated or on the other hand some new sinks may become available. With such uncertainties it is obvious that reliable predictions cannot be made. However, the climate changes which will occur as the CO_2 concentration increases are real and a threat to world survival. Recent measurements by satellites of the temperature of the upper atmosphere over a ten-year period have indicated no overall increase in temperature. This measurement has yet to be confirmed.

Coal is considered the "ugly duckling" of fossil fuel as it contains many impurities which are released into the atmosphere when it is burned. An important impurity is sulfur which introduces SO_2 and SO_3 into the atmosphere, resulting in acid rain that can actually change the pH of lakes sufficiently to destroy the aquatic life. The acid rain is also responsible for the destruction of the forests in Europe and the eastern parts of Canada and USA. The clean conversion of coal to other fuels may circumvent the pollution problems but would not overcome the greenhouse effect since CO_2 ultimately enters the atmosphere.

Thus, the depletion of fossil fuels may not be soon enough and tremendous efforts are being made in the search for viable economic alternatives such as nuclear energy, or renewable energy such as solar, wind, tidal, and others.

1.2. RENEWABLE ENERGY SOURCES

Ultimate sources of renewable energy are the earth, which gives rise to geothermal energy, the moon, which is responsible for tidal power, and the sun, which is the final cause of all other—hydro, wind, wave, thermal, and solar photodevices. A brief discussion of each source is essential for an overall appreciation of the difficulties we are facing and possible solutions to our energy requirements.

1.3. GEOTHERMAL

Thermal energy from within the earth's crust is classified as geothermal energy. At depths greater than 10 km the temperature of the magma is above 1000°C and is a potential source not yet fully exploited. The temperature of the earth's core is about

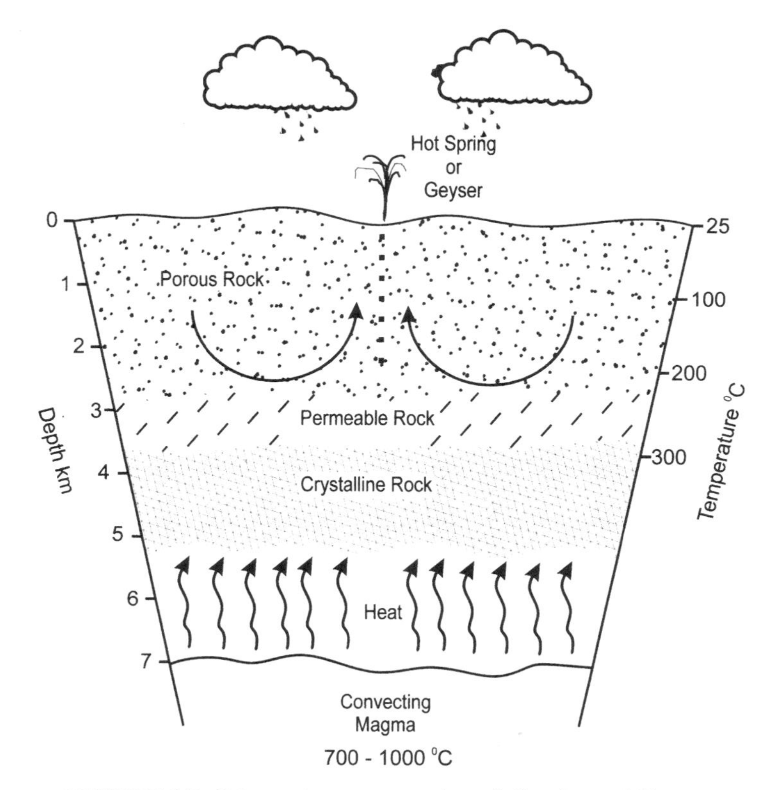

FIGURE 1.7. Schematic representation of Geothermal Energy.

4000°C. Drilling to depths of 7.5 km is presently possible and may some day reach 15 to 20 km. The surface source of thermal energy is due to the decay of natural radioactive elements and to the frictional dissipation of energy due to the movement of plate tectonics. The heat is usually transmitted to subsurface water which is often transformed into steam that can force water to the surface. Old Faithful at Yellowstone National Park, WY, USA, is an example of a geyser erupting 50 m every hour for 5 min (see Fig. 1.7). Geothermal energy is exploited near San Francisco where a 565 MW power plant is run on geothermal steam and 15 MW of thermal energy from hot water reservoirs is used for heating and industrial heat processes. At present there are five geothermal plants in operation in Mexico with a total output of more than 500 MW. Similar uses of geothermal energy have been developed in Italy, Japan, Iceland, USSR, and New Zealand, and it is rapidly being exploited in many other parts of the world.

It has been estimated that the geothermal energy in the outer 10 km of the earth is approximately 10^{23} kJ or about 2000 times the thermal energy of the total world coal resources. However, only a small fraction of this energy would be feasible for commercial utilization. Estimates of geothermal energy presently in use and converted into electrical power is about 2×10^3 MW. The greater use of geothermal energy could help save much of our energy needs and would reduce the rate of increase of CO_2 in the atmosphere.

1.4. TIDAL POWER

Tidal power is believed to have been used by the Anglo Saxons in about 1050. Tidal power is a remarkable source of hydroelectrical energy. The French Rance River power plant in the Gulf of St. Malo in Brittany consists of 24 power units, each of 10 MW. A dam equipped with special reversible turbines allows the power to be generated by the tidal flow in both directions (see Fig. 1.8).

Several tidal projects have been in the planning stages for many years and include the Bay of Fundy (Canada–USA), the Severn Barrage (Great Britain), and San Jose Gulf (Argentina). Though tidal power is reliable, it is not continuous, and some energy storage system would make it much more practical. However, as the price of fossil fuels rises, the economics of tidal power becomes more favorable.

1.5. SOLAR ENERGY

The Sun is approximately 4.6×10^9 years old and will continue in its present state for another 5×10^9 years. The Sun produces about 4×10^{23} kJ/sec of radiant energy, of which about 5×10^{21} kJ/year reaches the outer atmosphere of the earth. This is about 15,000 times more than man's present use of energy on earth. We are fortunate however that only a small fraction of this energy actually reaches the earth's surface

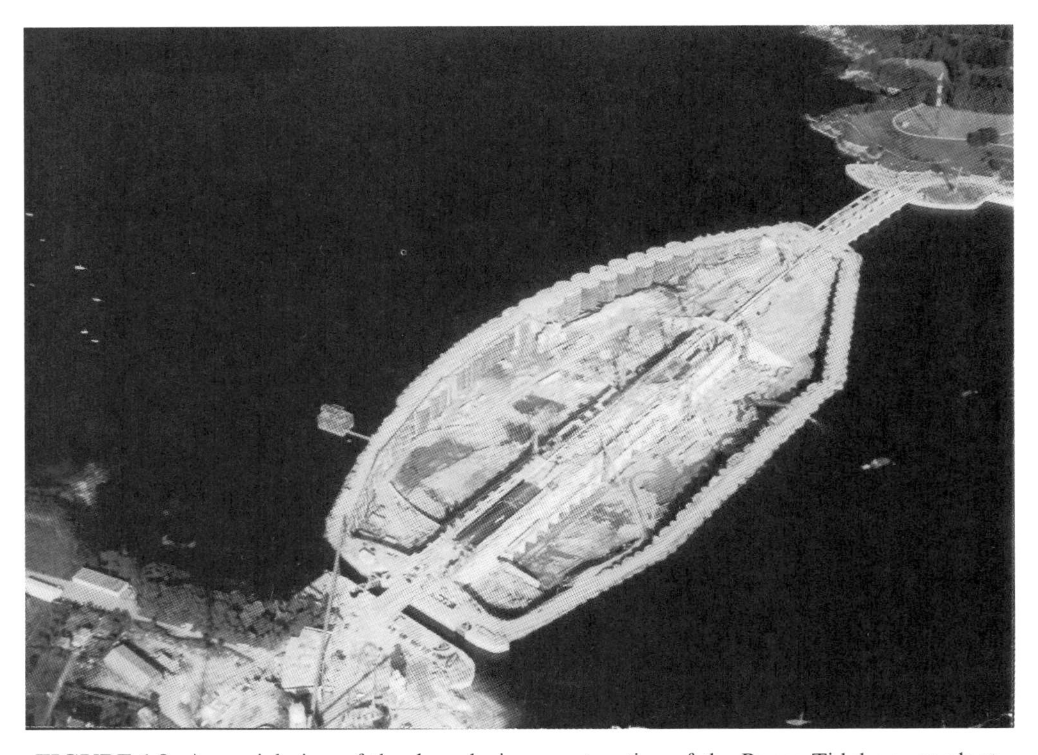

FIGURE 1.8. An aerial view of the dam during construction of the Rance Tidal power plant.

(see Fig. 1.6). About 30% is reflected back into space from clouds, ice, and snow, about 23% is absorbed by O_2, O_3, H_2O, and upper atmosphere gases and dust, and about 47% is absorbed at or near the earth's surface and is responsible for heating and supporting life on earth. Of the energy absorbed by the earth about 56% is used to evaporate water from the sea and plants (evapotranspiration). Another 10% is dissipated as sensible heat flux. The remainder is radiated back into space, about 10% into the upper atmosphere and about 24% is absorbed by our atmosphere. A small but important fraction of the Sun's energy, about 0.2%, is consumed in producing winds and ocean waves. An even smaller fraction, 0.02%, is absorbed by plants in the process of photosynthesis of which about 0.5% of the fixed carbon is consumed as nutrient energy by the earth's 6×10^9 people. The variation in solar intensity reaching the Earth due to its elliptical orbit about the Sun is only 3.3%. The production of fixed carbon by photosynthesis is about 10 times present world consumption of energy by human society. Thus, solar energy is sufficient for man's present and future needs on earth. The main difficulty is in the collection and storage of this energy.

Solar energy can be utilized directly in flat bed collectors for heating and hot water, or concentrated by parabolic mirrors to generate temperatures over 2000°C. The thermal storage of solar energy is best accomplished with materials of high heat capacity such as rocks, water, or salts such as Glauber's salt, which undergo phase changes, e.g.,

$$Na_2SO_4 \cdot 10H_2O_{(s)} \rightarrow Na_2SO_{4(s)} + 10H_2O_{(l)} \qquad \Delta H = 81.5\ \text{kJ/mol of salt} \tag{1.1}$$

(The transition temperature for Glauber's salt is 32.383°C).

Solar energy can also be directly converted into electrical energy by photovoltaic and photogalvanic cells, or transformed into gaseous fuels such as hydrogen by the photoelectrolysis or photocatalytic decomposition of water.

The Sun consists of about 80% hydrogen, 20% helium, and about 1% carbon, nitrogen, and oxygen. The fusion of hydrogen into helium, which accounts for the energy liberated, can occur several ways. Two probable mechanisms are:

The Bethe mechanism: (1939)	Q(MeV)
$^{12}C + {}^1H \rightarrow {}^{13}N + \gamma$	1.94
$^{13}N \rightarrow {}^{13}C + e^+ + \nu\ (t_{1/2} = 9.9\ \text{min})$	1.20
$^{13}C + {}^1H \rightarrow {}^{14}N + \gamma$	7.55
$^{14}N + {}^1H \rightarrow {}^{15}O + \gamma$	7.29
$^{15}O \rightarrow {}^{15}N + e^+ + \nu\ (t_{1/2} = 2.2\ \text{min})$	1.74
$^{15}N + {}^1H \rightarrow {}^{12}C + {}^4He$	4.96
$4\,{}^1H \rightarrow {}^4He + 2e^+ + 2\nu$	24.68
$2e^+ + 2e^- \rightarrow 2\gamma(+2 \times 1.02)$	2.04
Total energy	26.72

The Salpeter mechanism: (1953)	Q(MeV)
$^1H + {}^1H \rightarrow {}^2H + e^+ + \nu$	0.42
$^2H + {}^1H \rightarrow {}^3He + \gamma$	5.49
$^3He + {}^3He \rightarrow {}^4He + 2\,{}^1H$	12.86
$4\,{}^1H \rightarrow {}^4He + 2e^+ + 2\nu$	24.68

Both reactions occur, though the Bethe mechanism requires a higher temperature and therefore predominates in the central regions of large stars.

The solar constant is 2.0 cal/cm^2 min or 1370 W/m^2 above the earth's atmosphere and about 1.1 kW/m^2 normal to the Sun's beam at the equator. At other latitudes this value is reduced due to the filtering effect of the longer atmospheric path. Two maps shown in Fig. 1.9 indicate the world's distribution of solar energy in July and January at sea level. The spectral distribution is given in Fig. 1.10 where the black body radiation from the earth is also shown.

Ideal sites for solar energy collection are desert areas such as one in northern Chile which has low rainfall (1 mm/year) and 364 days/year of bright sunshine. The Chile site (160 $\times$ 450 km^2) receives about 5 $\times$ 10^{17} kJ/year (1 kJ/m^2/hr $\times$ 60 min/h $\times$ 8 h/day $\times$ 365 day/year $\times$ 72,000 km^2 $\times$ 10^{6}m^2/km^2). This is about a third of the world's use of energy in 1995. Thus, theoretically the desert areas or nonarid lands could be used to supply the world with all its energy requirements and there is no doubt that before the next century has passed solar energy will probably dominate a large portion of the world's energy sources.

Figure 1.4 shows the subclassification of solar energy into thermal, biomass, photovoltaic and photogalvanic. The most familiar aspect of solar energy is the formation of biomass or the conversion of carbon dioxide, water and sunlight into cellulose or food, fuel and fiber. Thus, wood was man's major fuel about 200 years ago to be displaced by coal, the modified plants of previous geological ages. Wood is a renewable energy source but it is not replenished quickly enough to be an important fuel today. A cord of wood is 128 ft^3 (8′ $\times$ 4′ $\times$ 4′) of stacked firewood. It is not recognized as a legal measure. A cord contains about 72 ft^3 of solid wood or about 4300 lb to which must be added about 700 lb of bark or a total of about 5000 lb, varying with the wood and its moisture content. The thermal energy of wood is from 8000 to 9000 Btu/lb. It has been argued that biomass used for fuel is not practicable because it displaces land which could be used for agriculture—a most essential requirement of man whose nutritional demands are continuously increasing. This objection is not valid if the desert is used, as in the case of the Jojoba bean which produces an oil that has remarkable properties, including a cure for baldness, lubrication, and fuel.

Plants which produce hydrocarbons directly are well known—the best example is the rubber tree which produces an aqueous emulsion of latex—a polymer of isoprene (mol. wt. 2 $\times$ 10^6 D) (see Fig. 1.11). The annual harvest of rubber in Malaysia was 200 lb/acre/year before World War II but by improving plant breeding and agricultural practices, the production has increased to 10 times this value.

Melvin Calvin, Nobel prize winner in Chemistry in 1961 for his work on the mechanism of photosynthesis, has been one of the principal workers in the search for plants which produce more suitable hydrocarbons, e.g., a latex with a mol. wt. of 2000 Daltons which can be used as a substitute for oil. One plant he has studied, Euphorbia (*E. Lathyris*) yields, on semiarid land, an emulsion which can be converted into an oil at about 15 bbl/acre. Another tree, Capaiba, from the Amazon Basin, produces an oil (not an aqueous emulsion) directly from a hole drilled in the trunk about 1 m from the ground. The yield is approximately 25 L in 2 to 3 hours every 6 months. This oil is a

C_{15} terpene (tri-isoprene) which has been used in a diesel truck (directly from tree to tank) without processing.

Recent studies have shown that oils extracted from plants such as peanuts, sunflowers, maize, soya beans, olives, palm, corn, rapeseed and which are commonly classed as vegetable oils in the food industry, can be used as a renewable fuel. These oils are composed primarily of triglycerides of long chain fatty acids. When used directly as a diesel fuel they tend to be too viscous, clog the jet orifices, and deposit carbon and gum in the engine. Some improvement is obtained by diluting the oil with alcohol or regular diesel fuel or by converting the triglycerides into the methyl or ethyl esters. This is done in two steps: (1) hydrolysis and (2) esterification. The methyl or ethyl ester produced is more volatile and less viscous but is still too expensive to burn as a fuel.

The energy ratio for biomass energy, i.e., the energy yield/consumed energy for growth and processing is variable and usually between 3 and 10. Plant breeding and genetic engineering should greatly improve this ratio.

Grain, sugar cane, and other crops containing carbohydrates can be harvested for the starch and sugar which can be fermented to ethyl alcohol. The residue which is depleted in carbohydrates but richer in protein is still a valuable feed stock.

Thus, the Energy Farm, where a regular crop can be utilized as a fuel, is obviously a requirement if stored solar energy is to replace dwindling fossil fuels.

1.6. PHOTOVOLTAIC CELLS

The direct conversion of solar energy into electrical energy is accomplished by certain solid substances, usually semiconductors, which absorb visible and near ultraviolet (UV) light, and by means of charge separation within the solid lattice a voltage is established. This generates a current during the continuous exposure of the cell to sunlight. Typical solar cells are made of silicon, gallium arsenide, cadmium sulfide, or cadmium selenide. The main hindrance to widespread use of solar cells is their high cost, which is at present about $400/$m^2$ of amorphous silicon (13% efficient) $50/$m^2$ would make such solar cells economic and practical. The reduction in cost to some composite cells of CdS/Cu_2S have been reported. Even lower costs may be expected as a result of the major efforts being made to develop inexpensive methods of forming the polycrystalline or amorphous materials by electroplating, chemical vapor deposition, spray painting, and other processes to dispense with the expensive single crystal wafers normally used. A typical photovoltaic cell is shown in Fig. 1.12.

A 6×9 m^2 panel of solar cells operating at 10% efficiency with a peak output capacity of 5 kW at midday would yield an average of 1 kW over the year — more than the electrical energy requirements of an average home if electrical storage was utilized to supply energy for cloudy and rainy days and during the night. More recently, a solar powered airplane crossed the English Channel using photovoltaic cells to power an electrical motor. This clearly demonstrates the potential power of solar energy.

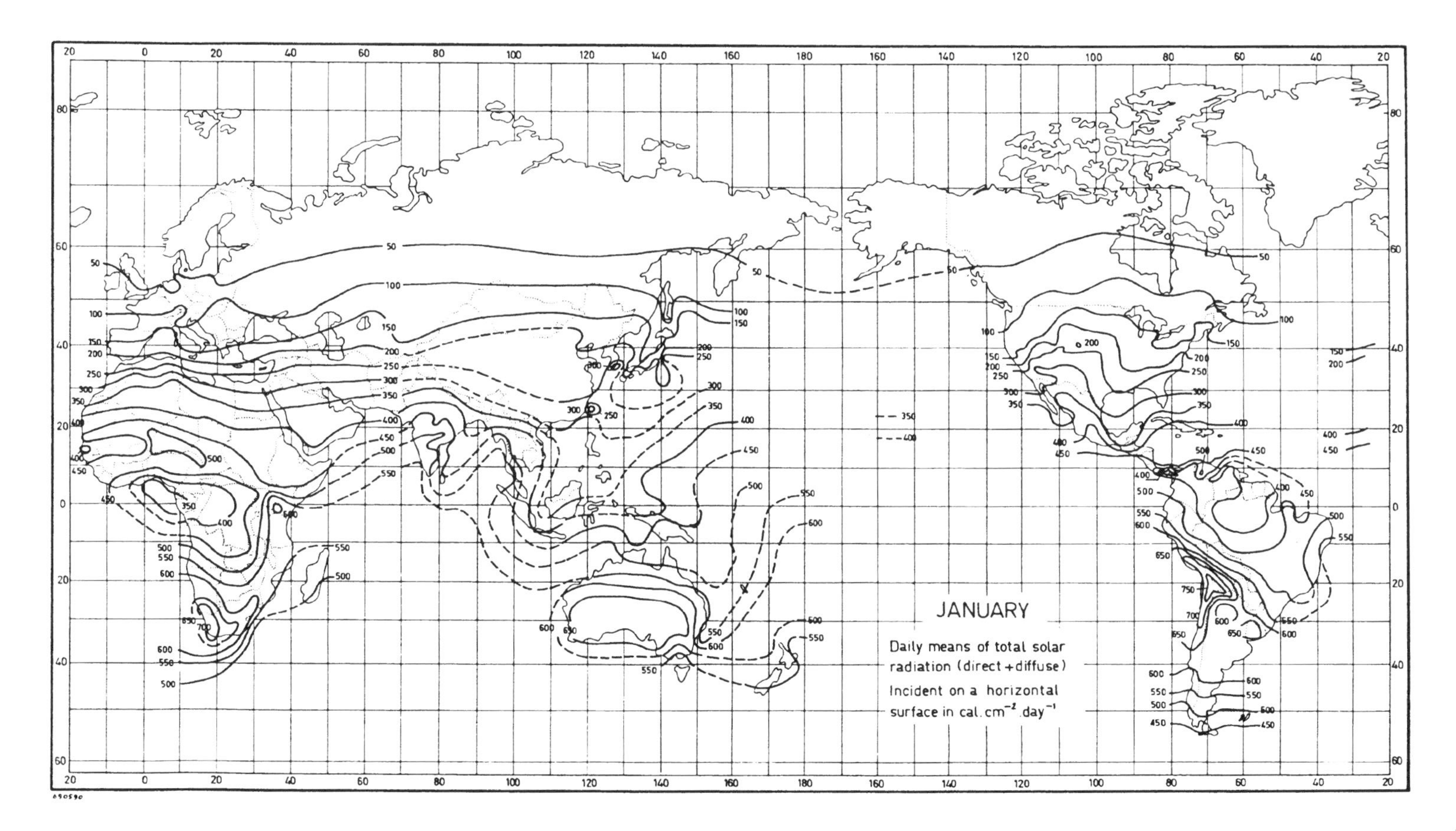
JANUARY
Daily means of total solar radiation (direct + diffuse)
Incident on a horizontal surface in cal. cm⁻² .day⁻¹

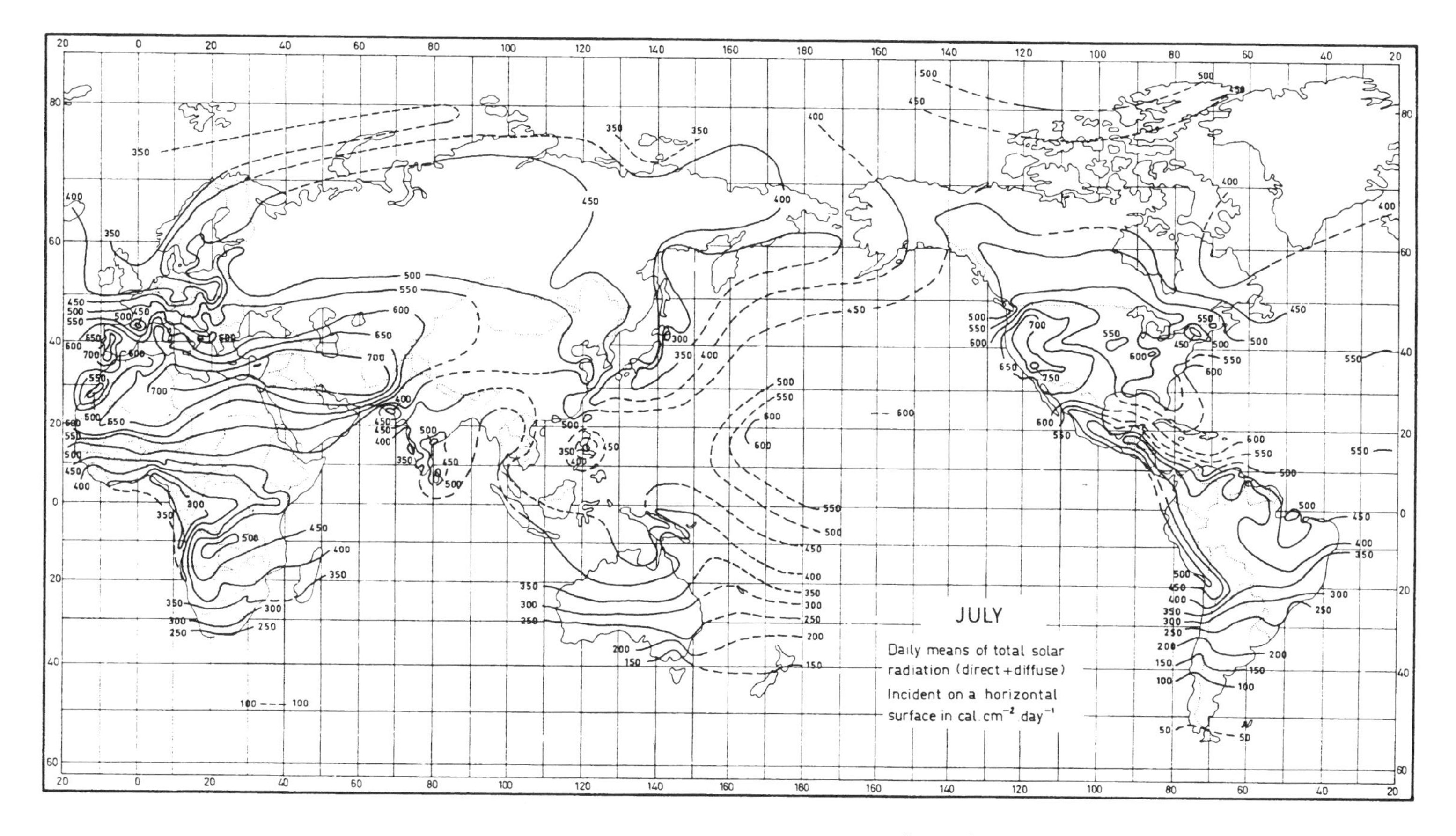

FIGURE 1.9. The contours of constant mean daily insolation (range from 50 to 750 cal cm^{-2} day^{-1}) for the months of January and July and show the insolation levels and their variation from summer to winter for locations throughout the world.

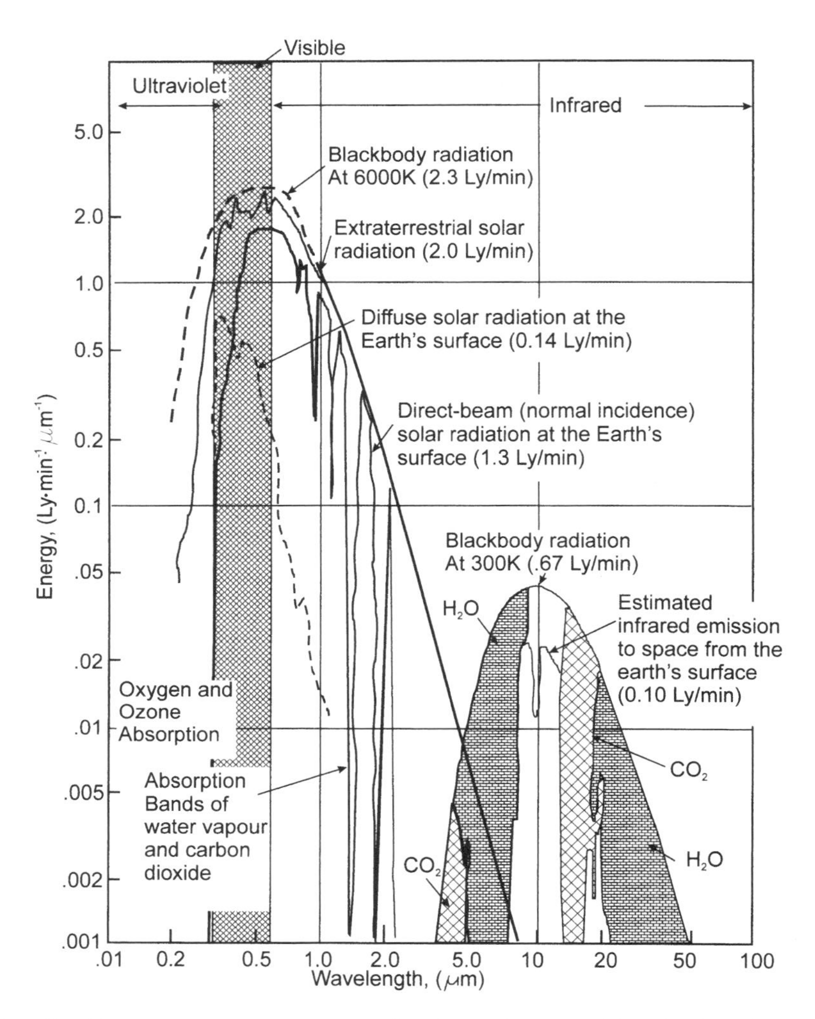

FIGURE 1.10. Electromagnetic spectrums of solar and terrestrial radiation. Note that the Langley (Ly) is a measure of solar energy density and is equal to 1 cal/cm^2 of irradiated surface.

1.7. PHOTOGALVANIC CELLS

Cells in which the solar radiation initiates a photochemical reaction, which can revert to its original components via a redox reaction to generate an electrochemical voltage, are called *photogalvanic cells.* This is to be distinguished from a solar rechargeable battery where light decomposes the electrolyte which can be stored and recombined to form electrical energy via an electrochemical cell, e.g.,

$$FeBr_3 + h\nu \rightarrow FeBr_2 + 1/2Br_2$$
$$FeBr_2 + 1/2Br_2 \rightarrow FeBr_3 \quad \mathscr{E}^0 = 0.316\ \mathrm{V} \tag{1.2}$$

FIGURE 1.11. A tapper at Goodyear's Dolok Merangir rubber plantation on the Indonesian island of Sumatra uses an extension knife to draw latex from a rubber tree. The bark of a rubber tree is cut up part of the year and down the rest to allow the tree to replenish itself. The rubber is sold by Goodyear to other manufacturers for making such diverse products as surgical gloves, balloons, overshoes, and carpet backing.

Photogalvanic cells usually consist of electrodes which are semiconductors and a solution which can undergo a redox reaction. The band gap of the semiconductor must match the energy of the redox reaction before the cell can function. Light absorbed by the electrodes promotes electrons from the valence band to the conduction band where they migrate to the surface (in *n*-type semiconductors) where reaction with the

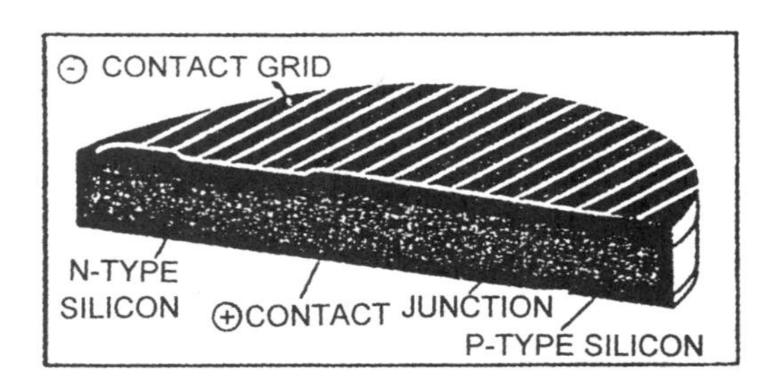

FIGURE 1.12. Typical solar cell 100 mm in diameter.

electrolyte can occur. This is shown in Fig. 1.13 for the system in which two photochemically-active semiconductor electrodes are used, one in which the *p*-type oxidizes Fe(II) to Fe(III).

$$Fe^{2+} \rightarrow Fe^{3+} + e^{-} \quad (1.3)$$

The reverse reaction occurs at the other *n*-type electrode. Many such cells have been prepared but the efficiency is very low due to the limited surface area of the electrodes. More recently porous transparent semiconductor electrodes have been made which can increase efficiency by some orders of magnitudes and it remains to be seen if these systems are stable over long periods. Such cells when shorted can be used for the photoelectrolysis of water or the production of hydrogen, but more will be said about this later.

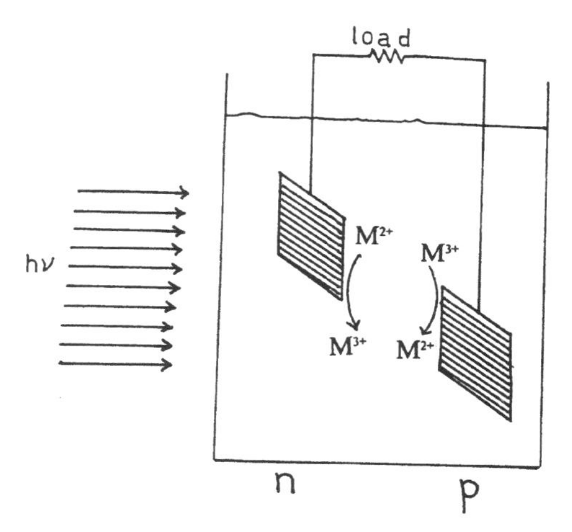

FIGURE 1.13. A photogalvanic cell using *n*-type and *p*-type semiconductor electrodes and a regenerating redox system ($M^{3+} \rightleftharpoons M^{2+} + e^{-}$) to carry the current.

1.8. WIND ENERGY

Wind first powered sailing ships in Egypt about 2500 B.C. and windmills in Persia about 650 A.D. The use of windmills for the grinding of grain was well established in the Low countries (Holland and Belgium) by 1430 where they are still used to this day.

The maximum power available from a horizontal axis windmill is given by

$$P_{max} = \tfrac{1}{2}\rho A V^3 \tag{1.4}$$

where ρ = density of air, A = cross-sectional area of the windmill disk, and V = air velocity.

However, to maintain a continuous air flow past the windmill implies that the extractable power, P_{ext}, must be less than P_{max} and it can be shown that $P_{ext} = (16/27)P_{max}$ or 59.3% is the optimum extractable power available and usually 70% of this value is realized practically. The energy which is usually extracted by an electrical generator or turbine can be stored in a bank of batteries for future use. Several very large windmills have been built to generate electrical energy but costs are still too high to make them commonplace. The vertical axis windmill is much simpler than the horizontal axis type but large units have not as yet been tested.

Wind farms have been successfully operating in California where many small windmills are located on exposed terrain (see Fig. 1.14). The continuous wind at about 17 mph is sufficient to make the generation of electrical energy a viable project. Three

FIGURE 1.14. A wind-farm in Tehachapi Pass, CA with an overall installed capacity of 650 MW, enough electricity for more than 140,000 homes.

regions: Altamont Pass east of San Francisco, Telachapi south of Bakersfield, and San Gorgonis near Palm Springs, east of Los Angeles produced 30% of the world's wind-generated electricity in 1995. The wind-generated electricity cost 7.5 cents per kWh in 1993. By 2005 the estimated cost is expected to be less than half this value.

1.9. HYDRO POWER

Hydro power relies on the conversion of potential energy into kinetic energy which is used to turn an electrical generator and turbines. The development of highly efficient turbines has increased the output of power stations and allowed for their installation in many new areas. The main disadvantage in hydro power is that the transmission of electrical energy over long distances results in losses which effectively place a limit on such distances. Also, it is not convenient, because of environmental factors, to store the potential energy or the electrical energy. Several methods have been used, such as storing water behind a dam, pumping the water into another reservoir, compressing air in a large cavern, or electrolyzing water and transporting the hydrogen in a pipeline. This latter alternative has an interesting by-product, namely heavy water which is essential for the CANDU nuclear reactor. During the electrolysis of water, the lighter isotope ^{1}H is liberated as H_2 more readily than the heavier ^{2}H or deuterium D_2 gas. Hence the heavy water accumulates in the electrolyte and can eventually be purified by distillation.

1.10. OCEAN THERMAL

The oceans cover over 70% of the earth's surface and are continually absorbing solar radiation. The penetration of the solar energy is only 3% at 100 m. This results in a temperature gradient (shown in Fig. 1.15) which can be used to generate electrical

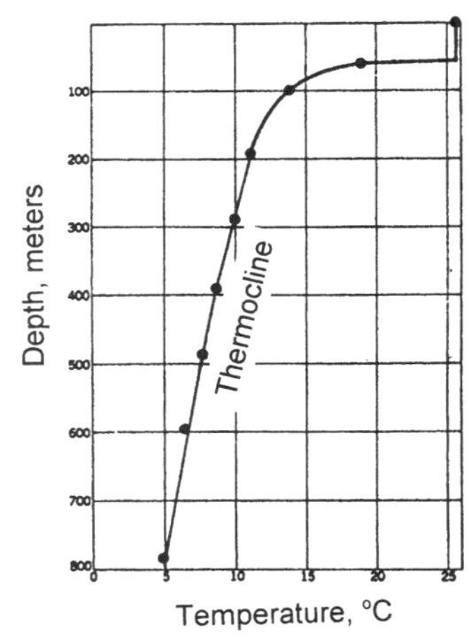

FIGURE 1.15. Typical ocean temperature profile at various depths. The surface layer, which is about 200 m deep obtains its heat from the sun and stays at about 25°C. The cold water at the lower depths comes from the Arctic region and can be as low as 5°C.

TABLE 1.3
Estimated Annual Yield of a Solar Pond

Area	1 km^2
1. Insolation	2000 GWh(t)
2. Pond heat yield	400 GWh(t)
Equivalent fuel oil	43,000 t
3. Power yield	33 GWh(e)
Source temperature	87°C
Sink temperature	30°C
H-X drops	10° total
Carnot effect	13.2%
Turbine factor	0.6
Overall thermal energy	ca. 8%
Equivalent cont. power	0.4 MW
(58% load factor)	4.7×10^4 m^3
4. Desalinated water	(12,900 m^3/day)

energy. A variety of schemes have been proposed and some experimental units have been tested as early as 1929 off Cuba and more recently near Hawaii. The main disadvantages to Ocean Thermal Energy Conversion (OTEC) or Solar Sea Power Plants (SSPP) are: (1) the need to operate turbines with a very low temperature gradient of about 15°C to 20°C, (2) the corrosive nature of seawater, and (3) the usually large distances from shore that the plant has to be located. The Carnot thermodynamic efficiency* of the heat engine is only about 3% but by using ammonia or a low boiling organic compound as the boiler fluid it is possible to approach maximum conversion efficiency.

It is possible to magnify the solar thermal gradient by using nonconvecting solar ponds. Much of the recent development work has been done by Tabor in Israel where Dead Sea brine is used to establish a density gradient in a pond approx. 1 m deep. The bottom of the pond is dark and absorbs the solar energy, heating the more dense lower layer. With temperature differences of more than 50°C it is possible to achieve a Carnot efficiency of over 13%. A schematic diagram of a solar pond is shown in Fig. 1.16 with an estimate of the operating parameters shown in Table 1.3. The diffusion of salt tends to destroy the density gradient and hence will allow convection to upset the temperature gradient. To maintain the density gradient, salt-free water is added to the surface and salt-enriched water is fed into the bottom layer. The flash chamber can also yield desalinated water as a by-product as indicated in Fig. 1.16 and Table 1.3. The electrical energy is obtained from a Rankine cycle turbine which was developed for such solar energy conversion and which runs on an organic vapor.

The problems associated with solar ponds involve: (1) Surface mixing due to winds which tend to create convecting zones near the surface. This can be reduced by adding a floating netting grid to the surface. A floating plastic sheet cannot be used because it acts as a collector of dust and becomes opaque in a few days. (2) Ecological factors

*The Carnot efficiency is the theoretical maximum efficiency by which a heat engine can do work when operating between two temperatures. The greater the difference in temperatures the greater is the efficiency by which heat can be converted into work.

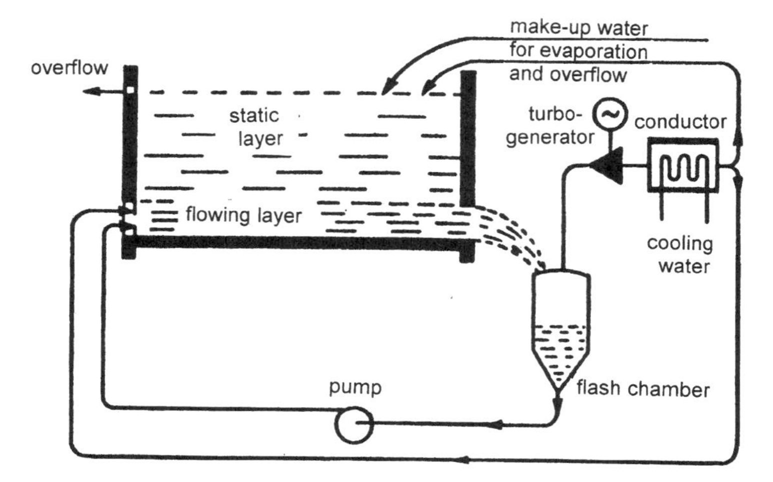

FIGURE 1.16. Schematic representation of the "falling pond" method of extracting heat from the bottom of a pond.

must be considered since brine may be slowly lost through the bottom of the pond to the aquifer. Hence the location of the pond in flat sterile land is preferred.

Solar ponds are presently being tested in Israel where plans are in progress for a large unit located at the Dead Sea (see Fig. 1.17).

1.11. WAVE ENERGY

Wave energy has been described as liquid solar energy since it originates from the sun, which causes wind, which in turn forms waves. Wave energy has enormous potential and presents a tremendous challenge to the engineer. Its presence on the high seas is almost continuous. It has been estimated that the power available along a kilometer of shore can be over 20 MW. Several devices have been tested and usually operate air turbines which generate electricity. The power take-off and mooring are still problems to be solved. An example of a simple wave energy device is shown in Fig. 1.18. A one-way flap allows a head of water to be stored over 5 to 100 waves. When the full head is reached the water is released and allowed to drive the turbine as it empties.

As a renewable energy source, wave power is an untapped source which could help alleviate the rising cost of energy and its continued development must be encouraged.

1.12. OSMOTIC POWER

As a river empties into the sea it is possible to extract hydro power by converting gravitational potential energy into kinetic energy—mechanical energy and electrical energy. At the same time, the difference in salt concentration between the river and the

FIGURE 1.17. Solar pond in Israel.

sea results in a difference in chemical potential or Free Energy (ΔG) which can be exploited by means of the osmotic pressure.

The osmotic pressure of a solution is the pressure exerted by the solution on a semipermeable membrane which separates the solution from pure solvent. The semipermeable membrane allows only the solvent to pass freely through it while opposing the transport of solute. This is illustrated in Fig. 1.19. The solvent tends to dilute the solution and passes through the membrane until equilibrium is established, at which point a pressure differential, π, exists between solution and solvent. This pressure differential is called the *osmotic pressure*.

At equilibrium the osmotic pressure, π, of a solution relative to the pure solvent is given by a relation which resembles the Ideal Gas Law ($PV = nRT$) where

$$\pi V = nRT \qquad \text{or} \qquad \pi = (n/V)RT = CRT \tag{1.5}$$

where V = volume of solution, R = Ideal Gas Constant (0.0821 L atm/°K mol), C = concentration of solute (molar mass/liter), n = mols of dissolved solute, ions, etc., T = absolute temperature (273° + 10°) = 283 K.

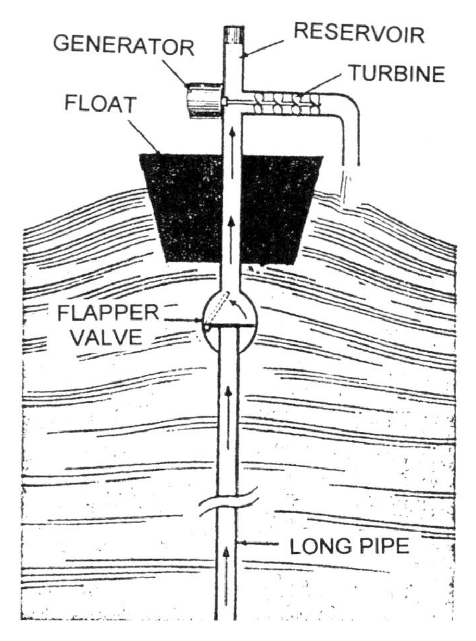

FIGURE 1.18. Wave-powered generators convert kinetic energy of waves to electricity. Scripps wave pump amplifies effective height of wave by trapping water in reservoir on each downward plunge of the buoy; water in long standpipe surges through one-way flapper valve because it oscillates out of phase with up-and-down motion of waves. When enough water is trapped (after about nine waves), it is released to spin the turbine-generator.

The average molar concentration of salt in seawater is about 0.5 M and since it may be assumed that the salt is primarily NaCl, then the total molar concentration of solute is 1.0 M (i.e., 0.5 M Na^+ + 0.5 M Cl^-). Hence

$$\pi = 1.0\,\text{mol/L} \times 0.0821\ \text{L, atm/K, mol} \times 283\,\text{K} = 23\,\text{atm}$$

This osmotic pressure represents a hydrostatic head of water of about 700 ft (30 ft/atm × 23 atm) or over 200 m. A schematic diagram of osmotic power is illustrated in

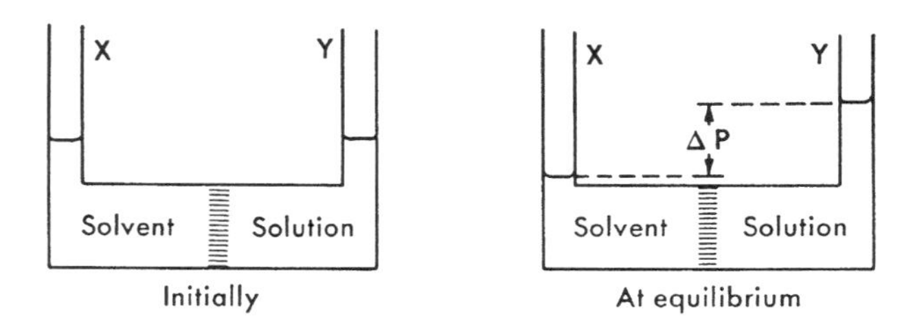

FIGURE 1.19. Development of osmotic pressure is illustrated by the difference between an intial state and the equilibrium state. Solvent, but not solute, passes through the semipermeable membrane, tending to dilute the solution and thereby allowing a differential pressure, ΔP, to develop. At equilibrium the differential hydrostatic pressure is equal to the osmotic pressure, π.

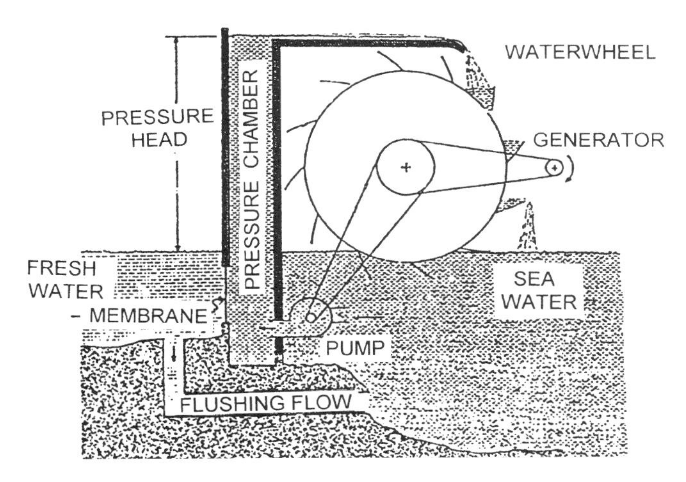

FIGURE 1.20. Diagram of an osmotic salination energy converter used to extract power from the natural flow of freshwater into the sea. The river, or part of its flow, must pass through the membrane as it tends to dilute the salt in the seawater.

Fig. 1.20 where seawater is pumped into a pressure chamber at a constant rate depending on the flow of the river. The river passes through the membrane, diluting the seawater and creating the hydrostatic head which can then turn a water wheel and generate electricity. The membrane area must be enormous to accommodate the permeability of the river. An estimate of power output is 0.5 MW/m^3 of input flow resulting in an amortized cost of about 5 cents/kWh. Lower costs are expected if the salt concentration gradient is higher, e.g., seawater ($\pi = 23$ atm) emptying into the Dead Sea ($\pi = 500$ atm) to produce electrical power from the gravitational as well as the chemical potential. The potential power from various rivers is listed in Table 1.4.

TABLE 1.4
Potential Power due to Salinity Gradients

Source	Flow rate (m^3/s)	Power (watts)
$\pi = 23$ atm		
Amazon River (Brazil)	2×10^5	4.4×10^{11}
La Plata-Parana River (Argentina)	8×10^4	1.7×10^{11}
Congo River (Congo/Angola)	5.7×10^4	1.2×10^{11}
Yangtze River (China)	2.2×10^4	4.8×10^{10}
Ganges River (Bangladesh)	2×10^4	4.4×10^{10}
Mississippi River (USA)	1.8×10^4	4.0×10^{10}
USA waste water to oceans	500	1.1×10^9
Global run-off	1.1×10^6	2.6×10^{12}
$\pi = 500$ atm		
Salt Lake		5.6×10^9
Dead Sea		1.8×10^9

Only with the developments in membrane technology as a result of work in reverse osmosis will osmotic power become a significant factor in world energy supply.

It has been estimated by the Scripps Institution of Oceanography that world power needs in the year 2000 will be about 33 million megawatts (33 TW). The seas can provide all this and more: wave energy 2.5 TW, tidal power 2.7 TW, current power 5 TW, osmotic power 1400 TW, OTEC 40,000 TW.

With such optimistic projections and present day technology it may soon be possible to rely on our renewable energy resources with confidence and assurance so long as we continue to minimize the environmental effects.

EXERCISES

1. Pick a country the name of which starts with the same letter of the alphabet as your own family or given name and obtain the latest data for Fig. 1.1. Give the source.
2. Explain why geothermal energy is classed as both a renewable and nonrenewable energy source. Do you agree with this explanation?
3. Complete Table 1.1 with references to uranium available as a nuclear fuel, 860 Quads from the earth and 10^5 Q from seawater.
4. It has been argued that the CO_2 absorption band in the atmosphere is almost saturated and if more CO_2 is produced there will be no additional absorption. Explain why this is incorrect.
5. From Figs. 1.6 and 1.10 it would appear that water is more important than CO_2 in governing the Greenhouse Effect. Comment on this.
6. Explain why burning wood and burning coal are not equivalent as far as the Greenhouse Effect is concerned.
7. The normal energy requirements of a house in Winnipeg in winter is 1 million Btu/day or 10^6 kJ/day. (a) What weight of Glauber salt would be required to store solar energy for one month of winter use? (b) If the optimum conditions for using the Glauber salt is a 35% by weight (Na_2SO_4) solution in water, what volume of solution would be required? (Note: Density of GS solution is 1.29 g/mL.
8. Canada produces over 700 million pounds of vegetable oils per annum. If this were to replace all Canadian petroleum oil used, how long would it last?
9. If a typical home has 1200 ft^2 of floor space and a sloping A-type roof at 45° to 60° for optimum solar collection, the roof area facing the sun would be approx. 850 ft^2. Calculate the energy per day available as heat and as electricity via photovoltaic cells (assume 5% efficiency) for both summer and winter.
10. What would be the rise in sea level if the average temperature of the oceans increased by 1°C at 20°C. The density of water at 20°C is 0.99823 g/mL and at 21°C the value is 0.99802 g/mL. Assume no ice melts and that the ocean area does not increase. The average depth of the ocean is 3865 m.
11. Reverse electrodialysis is a method of extracting electrical energy directly from the flow of a fresh water river into the sea (salt water). Ion-exchange membranes are used to separate the flow of fresh and salt water. Draw a diagram of the system and explain how it works. See Section 15.5.

FURTHER READING

M. H. Halmann and M. Steinberg, *Greenhouse Gas Carbon Dioxide Mitigation*, CRC Press, (1999).
D. L. Klass, *Biomass for Renewable Energy, Fuels, and Chemicals*, Academic Press, New York, (1998).
A. Bejan, P. Vadasz, and D. G. Kroger, *Energy and the Environment*, Kluwer Academic, New York, (1999).
A. Bisio and S. Boots, *Encyclopedia of Energy and the Environment*, 2 Vol., Wiley, New York, (1996).
R. Campbell-Howe, Editor, 21st National Passive Solar Conference Proceedings, *American Solar Energy*, Boulder, Colorado, (1996).
P. Takahashi & A. Trenka, Editors, Ocean Thermal Energy Conversion, *UNESCO Energy Engin. Ser.* Wiley, New York, (1996).
S. L. Sah, *Renewable and Novel Energy Sources*, State Mutual Book and Periodical Service Ltd., New York, (1995).
R. Nansen, Sun Power, *The Global Solution for the Coming Energy Crisis*, Ocean Press, Seattle, Washington, (1995).
P. Gipe, *Wind Energy Comes of Age*, Wiley, New York, (1995).
W. A. Duffield, *Trapping the Earth for Heat*, US Geology Survey, Washington DC, (1994).
W. Avery and C. Wu, *Ocean Thermal Energy Conversion*, John Hopkkins Appl. Phys. Lab., Baltimore, Maryland, (1994).
M. Heimann, *The Global Carbon Cycle*, Springer, New York, (1994).
R. Golob and E. Brus, *The Almanac of Renewable Energy: The Complete Guide to Emerging Energy Technologies*, Holt, New York, (1994).
T. F. Markvart, Editor, *Solar Electricity*, Wiley, New York (1994).
J. J. Kraushaar and R. A. Ristinen, *Energy and Problems of a Technical Society*, 2nd Ed., Wiley, New York, (1993).
M. Potts, *The Independent Home: Living Well with Power from the Sun, Wind and Water*, Chelsea Green Pub., White River Junction, VT, (1993).
J. Carless, *Renewable Energy: A Concise Guide to Green Alternatives*, Walker, New York, (1993).
R. J. Seymour, Editor, Ocean Energy Recovery: The State of the Art, *Am Soc. Civil Eng.*, New York, (1992).
F. P. W. Winteringham, *Energy Use and the Environment*, Lewis Pub., London, (1992).
C. J. Winter, R. L. Sezmann, and L. L. Vant-Hull, Editors, *Solar Power Plants*, Springer-Verlag, New York, (1991).
Handbook of Unusual Energies, Gordon Press, New York, (1991).
J. Priest, Energy, *Principles, Problems Alternatives*, 4th Ed., Addison-Wesley, Reading, Massachusetts, (1991).
Developments in Tidal Energy, *Am. Soc. Civil Eng.*, New York, (1990).
Scientific American, *Managing Planet Earth*, Sept., (1989).
R. L. Loftness, *Energy Handbook*, 2nd Ed., Van Nostrand, New York, (1984).
R. Shaw, *Wave Energy*, Wiley, New York, (1982).
H. Inhaber, *Energy Risk Assessment*, Gordon and Breach, New York, (1982).
Energy—A special report in the public interest, National Geographic Feb. (1981).
S. P. Parker, Editor, *Encyclopedia of Energy*, 2nd Ed., McGraw-Hill, New York, (1981).
J. O'M. Bockris, *Energy Options—Real Economics and the Solar-Hydrogen System*, Taylor and Francis Ltd., London, (1980).
W. Sassin, Energy., *Sci. Am.* **243**(3) Sept. p. 118, (1980).
Energy in Transition, 1985–2010. Final Report of the Committee on Nuclear and Alternative Energy Systems, NRC., Nat'l. Acad. Sci. Washington DC (1979), W. H. Freeman, San Francisco, California, (1980).
Energy and Environment: Readings from Scientific American, W. H. Freeman, San Francisco, California, (1980).
R. Stobaugh and D. Yergin, Editors, *Energy Future. Report of the Energy Project at Harvard Business School*, Ballantine Books, New York, (1979).
H. Messel, Editor, *Energy for Survival*, Pergamon Press, New York, (1979).
B. Sorensen, *Renewable Energy*, Academic Press, New York, (1979).
Energy—The Fuel of Life—by the Editors of Encyclopedia Britannica, Bantum Books, New York (1979).
L. C. Ruedisili and M. W. Firebaugh, *Perspectives on Energy—Issues, Ideas and Environmental Dilemmas*, 2nd Ed., Oxford Univ. Press, New York, (1978).

D. A. Tillman, *Wood as an Energy Resource*, Academic Press, New York, (1978).
D. M. Considine, Editor, *Energy Technology Handbook*, McGraw-Hill, New York, (1977).
D. N. Lapedes, Editor, *Encyclopedia of Energy*, McGraw-Hill, New York, (1976).
N. D. Morgan, Editor, *Energy and Man: Technical and Social Aspects of Energy*, IEEE Press, New York, (1975).
A. Fisher, *Energy from the Sea, Popular Science*, May p. 68, June p. 78, July, (1975).
W. C. Reynolds, *Energy—from Nature to Man*, McGraw-Hill, New York, (1974).
Energy and Power (full issue), *Sci. Am.* Sept., (1971).
R. H. Oort, The Energy Cycle of the Earth, *Sci. Am.* **223**(3) p. 54, Sept., (1970).
F. Daniels, *Direct Use of the Sun's Energy*, Ballantine Books, New York, (1964).
A. M. Zaren and O. D. Erway, *Introduction to the Utilization of Solar Energy*, McGraw-Hill, New York, (1963).
P. C. Putnam, *Energy in the Future*, Van Nostrand, (1956).
F. Daniels, Editor, *Solar Energy Research*, University of Wisconsin (1955).
World Energy Resource-- International Energy Annual, http://www.eia.doe.gov/
Energy Conservation DataBook, Japan,
http://www.eccj.or.jp/databook/1998e/index.html
Energy Sources, http:/ /www.yahoo.com/science/energy/
Categories: Alternate Energy, Biomass, Fly Wheels, Fuel Cells, Geo Thermal,
Hydropower, Nuclear, Renewable, Solar, Wind,
Ocean Wave Energy Company, http://www.owec.com/
Ozone/Greenhouse, http://www.epa.gov./globalwarming/index.html
Ozone secretariate, http://www.unep.org/ozone/
American Bioenergy Association, http://www.biomass.org/
Solar Energy Technology, http://www.sunwize.com/
Solar energy kits and systems, http://buelsolar.com/
International Solar Energy Society, http://www.ises.org/
Fossil Fuels Association, http://www.fossilfuels.org
Onta Fossil Fuels Industry, http://www.fossil-fuels.com/
Government views of fossil energy, http://www.fe.doe.gov/
Electric Power Research Institute, http://www.epri.com/
Hydroquebec, http://www.hydroquebec.com/hydroelectricity/index.html
Tidal Power, http://www.tidalelectric.com/
Wave Energy, http://www.waveenergy.dk/apparater/engelsk—klassificering.html
Wave Power Patents, http://www.waveenergy.dk/apparater/fly.htm
American Wind Energy Association, http://www.awea.org

2

Solid Fuels

2.1. INTRODUCTION

Fuels are conveniently classified as solids, liquids, and gaseous fuels. Solid fuels include peat, wood, and coal and can encompass solid rocket fuels as well as metals. The earliest fuels used by man were nonfossil fuels of wood and oil from plants and fats from animals. The windmill and water wheels were other sources of energy.

Sources of power have changed with the years and will continue to change as shown in Fig. 1.2. In 1992 approx. 7% of the world's power was supplied by hydroelectric plants and the remaining 92.5% from fuels. Natural gas provides 22% of the total power, petroleum, 40%, and coal, 25% (of which 7% is derived from hard coal and 18% from the soft coals) and nuclear fuels provide about 7%. Oil, which has displaced coal as the major fuel, will soon be replaced by natural gas which in turn will eventually be replaced by nuclear energy.

The origin of coal is not known with certainty. One popular theory claims that coal originated about 250 million years ago as a result of the decay of vegetation primarily from land and swamps and not of marine origin. Bacterial action undoubtedly helped with the reduction process. The first step following the exclusion of oxygen was the formation of peat—a slimy mass of rotting organic matter and debris. Under the pressure of sediments the peat became dehydrated and hard, forming low grade coal, called *lignite*. Under further pressure and time the reactions of condensation and consolidation (50-fold decrease in volume) converted the lignite into a higher grade coal—bituminous coal. A highest grade coal—anthracite—has the highest percentage of carbon. A simplified flow description for coal formation is shown in Fig. 2.1. The three component ternary phase diagram for the C, H, and O content of the various grades of coal is shown in Fig. 2.2 where comparison is made with cellulose and lignin,* the general precursor to coal. The simple weight percentage point of cellulose is indicated by a filled triangle (▲) and the atom percentage is shown as a filled square (■). The bond-equivalent points are meant to account for the bonding (valency) of the elements, namely 4 for C, 2 for O and 1 for H. The atom percentage multiplied by the bond factor and normalized (to 100% for all the elements) is called the *bond-equivalent percentage* and is shown as filled circles (●). The values for CH_4, CO_2, and H_2O are also shown in Fig. 2.2 as open circles (○). The direction from cellulose to anthracite shows clearly that the loss of water and oxygen must occur during the coalification process.

* Wood consists of about 25% lignin which acts to bind the cellulose fibers together.

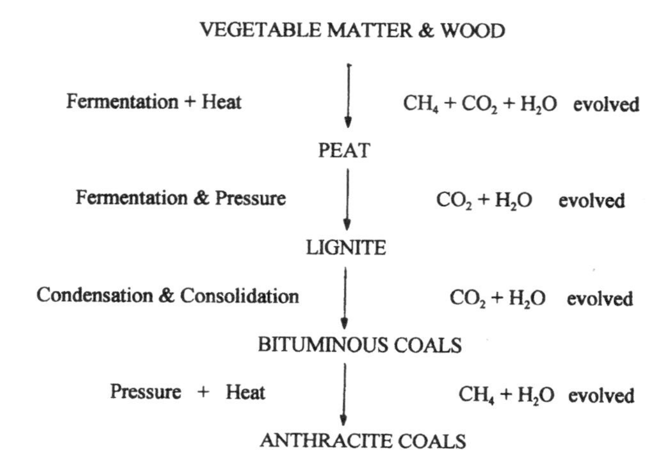

FIGURE 2.1. Possible route for the formation of coal from plant matter.

Wood was obviously man's first fuel, followed by animal fats and vegetable oil. There is evidence that candles were used during the first Minoan civilization about 3000 B.C. Coal was used by the Chinese about 100 B.C. and the "black stone" was reported to be used by Greek smiths about 250 B.C. The Romans in Great Britain also used coal. Marco Polo describes the mining of "black stone" in his travels 1271–1298 A.D.

Coal is primarily known as a fuel but it is also a valuable chemical. It can be reacted with lime, CaO, at high temperatures (electric arc) to form calcium carbide,

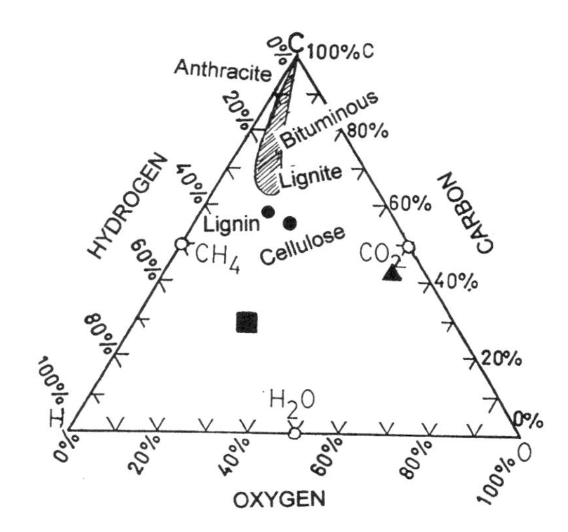

FIGURE 2.2. The 3-phase ternary bond-equivalent diagram for C–O–H showing the transition from cellulose to anthracite. Cellulose $(C_6H_{10}O_5)_x$ is also shown as ▲ for weight % and as ■ for atom % in the diagram. The other points are bond-equivalent %.

TABLE 2.1
World Coal Resources and Use (Dec. 1996)

	Recoverable reserves (10^9 tons)			Uses (10^6 tons)	
	Hard[a]	Soft[a]	Total	Production	Consumption
Russia	54	119	173	331	284
USA	126	149	275	1087	1028
China	68.6	57.6	126.2	1690	1532
Germany	26.4	47.4	73.8	286	277
Australia	52.1	47.5	99.6	276	135
UK	1.1	0.5	1.6	62	78
Canada	5.0	4.5	9.5	87	62
Poland	13.4	2.4	15.8	232	195
South Africa	61.0	—	61.0	230	169
India	80.2	2.2	82.4	352	342
Rest of the world	77.5	92.7	170.2	922	1167
Total	565.3	522.8	1088.1	5625	5269

[a]Hard = anthracite and bituminous; soft = lignite and subbituminous.

CaC_2, which when hydrolyzed forms acetylene (C_2H_2):

$$3C + CaO \rightarrow CaC_2 + CO \tag{2.1}$$

$$CaC_2 + 2H_2O \rightarrow Ca(OH)_2 + C_2H_2 \tag{2.2}$$

Coal is also used to prepare active carbon which is used for the purification of air and water and in numerous industrial processes.

The distribution of coal in the world is given in Table 2.1. The recoverable reserves consist of the coal which can be economically mined with presently known technology and conditions. The proven reserves are the coal known to be present, within $\pm 20\%$, by extensive drilling and experience and is about twice the recoverable reserves. The estimated total world coal resources are about 8 times larger and is based on geologically favorable formation within the earth's crust and on previous experience within the countries concerned. It excludes under sea or under ice sites.

Of the world's reserves, the distribution between hard and soft coal is about equal on a global basis. There is enough coal to last the world for 200 to 1000 years depending on the rate of usage and the rate of exploration.

Coal in 1947 was at a record high price of $4.16 per ton at the mine. More than 600 Mtonne were mined that year in the USA, but already oil was beginning to displace it. The old steam engines, which used a ton of coal for every 4 miles hauling a heavy freight, consumed a quarter of the coal production (125 Mtonne). By 1960 when the railroad had almost completely converted from coal to diesel, the rail industry used only 2 Mtonne of coal.

2.2. WOOD AND CHARCOAL

Wood, a renewable source, is not an important industrial fuel today. However, its use continues in some rural areas where it is often supplemented with liquid propane. In some underdeveloped countries wood is still the principal source of energy. Dry wood contains from 1% to 12% moisture whereas green wood contains from 26% to 50% water. The resinous woods, like pine or cedar, yield about 18.5 MJ/kg of air-dried wood or, allowing for the moisture content, about 21 MJ/kg on a dry weight basis. Hardwoods have a heating value of about 19.4 MJ/kg. The energy available in present forest stocks is estimated to be equivalent to about 270×10^9 tonnes of coal or about 2/3 of the equivalent oil reserves.

Wood-burning fireplaces have become a popular form of heating in the past few years. The normal open hearth fireplace is not an efficient producer of heat since its draft sends most of the hot air up the chimney creating a partial vacuum pulling cold air into the home. The recent introduction of glass doors to close in the fire and the introduction of outside air to the fire for the combustion process has improved the efficiency of the fireplace, especially with forced air circulation around the fire chamber.

Open fires, however, are a source of pollution since the smoke produced contains large quantities of polynuclear aromatic hydrocarbons such as benzo(a)pyrene, a carcinogen also found in cigarette smoke. In some communities it has been necessary to restrict the burning of wood because of the resulting air pollution.

2.3. PEAT

Peat is formed when dead vegetation is saturated with water which prevents the action of aerobic bacteria. Thus, most of the carbon of the cellulosic matter is retained, and with ageing, peat is formed. It accumulates at an average rate of 0.7 mm/year or world wide at 210 Mtonne of carbon. Canada (40%) and Russia (36%) have more than 3/4 of the world's peat land (320 Mha or 150×10^3 Mtonne of carbon). In Russia peat deposits occupy about 1/10 of the total country's terrain. It is a spongy watery mass when first obtained from the peat bog. Six tonnes of dry peat yield about one tonne of fuel. A commercial grade of peat contains about 25% water. Air-dried peat has a heating value of about 16.3 MJ/kg. Peat is rich in bitumens, carbohydrates, and humic acids and as a chemical source it can yield waxes, paraffins, resins, and oils. Peat also is a source of pharmaceutical and curative preparations as well as a livestock feed supplement.

Peat is not used in North America as a fuel to any great extent, but in Europe it has been employed in domestic heating for centuries. Peat is often harvested by massive machinery for industrial use (Fig. 2.3). The machine deposits the cut and macerated peat in long furrows, which are then cross-cut into blocks and conveyed to dry storage or dryers. Ireland harvests 4 Mtonne annually as fuel for generating 20% of the nations electricity. In Finland about 3 Mtonne of peat, with an energy content of 30×10^9 MJ, is used annually. This is about 6% of the total fossil fuel energy used. Peat as a primary fuel was used in 15 power plants in 1983 to generate 950 MW. Peat is also used to generate power for electricity in Russia.

FIGURE 2.3. This machine cuts and macerates the peat and then sends it along the extended arm which deposits it in rows. The peat is cut into blocks by the discs as the equipment moves forward.

The pyrolysis of peat in the absence of oxygen is being studied in New Zealand, where the oil–wax product is then hydrogenated to form the equivalent to a common crude oil. New Zealand has enough peat to produce 400 Mbbl of oil.

The growth of peatlands consumes CO_2 and the harvesting of peat and its combustion as a fuel has a 2-fold adverse effect on the CO_2 balance in our atmosphere, first by releasing CO_2 into the atmosphere and second by removing it as a sink for CO_2.

2.4. COAL

Coal, the generic term applied to solid fossil fuels, ranges from lignite, which is basically a matured or modified peat, to meta-anthracite, which is more than 98% carbon. The qualities of different coals are classified in different ways either depending on the chemical composition, the heating value, or even the ash content and its fusion temperature. The International Classification of Hard Coals by Type (1956) has been widely accepted. The criteria used include volatile matter and calorific value as well as the swelling and caking properties of the coal. The Economic Commission for Europe has proposed its own classification and codification system. In North America the common classification of coals is by the ASTM method. The rank classification system* is based on the application and commercial use of coal, namely in combustion for electric power generation and in the preparation of coke for the metallurgical industry.

* The ASTM (American Society for Testing and Materials) has provided detailed definitions (D121-85) and tests (D-388-88) for rank classification.

2.5. ANALYSIS OF COAL

The elemental analysis of coal, i.e., its C, H, O, N, S, and ash residue may be important to a chemist who wishes to use coal as a chemical or source of carbon, but to an engineer who wants to burn the coal in a heat or power generating plant or a coking oven other parameters are more important—most notably its heat of combustion, moisture level, volatile matter, carbon and sulfur content as well as the ash. This is called the *proximate analysis* and it is determined as follows:

Moisture Content A sample of coal is ground to pass 20/60 mesh and weighed. It is dried in an oven at 110°C and reweighed. The loss in weight represents the moisture content of the coal.

Volatile Content The coal sample is heated in an inert atmosphere up to 900°C. The loss in weight varies with the temperature since some of the coal is decomposed into oils and tars which volatilize at various temperatures.

Ash Content A coal sample is heated in a muffle furnace at 900°C in the presence of air to combust the coal leaving the ash residue which is heated to constant weight. If the sample of coal is first freed of volatile matter then the loss in weight represents the fixed carbon in the coal.

Heat Content This is determined in a bomb calorimeter where a dry sample of coal is burned in an excess of oxygen and the heat evolved is measured. This is often referred to as the caloric value or the heat content of the coal. This includes the combustion of the volatile and tar components as well as the fixed carbon and is therefore related to the actual heat generating value of the coal. The heat of combustion of pure carbon is 32.8 MJ/kg.

2.6. ASTM CLASSIFICATION

An abbreviated version of the ASTM classification of coal by rank is shown in Table 2.2. The highest rank—meta anthracite—contains the highest percentage of carbon and the lowest amount of volatile matter. The four classes are divided into 13 groups according to carbon content and heating value. Thus, in the bituminous class there may be gas coals, coking coals, and steam coals. The proximate and ultimate analysis of typical coals are given in Table 2.3. It is possible to calculate the heat content of coal from its ultimate analysis. From the standard heats of formation of CO_2 (-393.5 kJ/mol), H_2O (-285.8 kJ/mol), and SO_2 (-296.8 kJ/mol) it is possible to derive the formula

$$Q = (1/100)[32.76\text{C} + 142.9(\text{H} - \text{O}/8) + 9.3\text{S}] \tag{2.3}$$

where Q is the exothermic heat evolved in the combustion of the coal in units of kJ/g and where C, H, O, and S represent the weight percentage of the element in the coal. Oxygen, which is assumed to be present as H_2O, is normally determined by difference. The assumption concerning oxygen is not too unreasonable since more than half of the oxygen in coal is normally present as —OH. If the ash content is significant then a

TABLE 2.2
Classification of Coals by Rank[a]

Class/group	Fixed carbon, C (%)	Volatile matter, V (%)	Calorific value, ΔH (MJ/kg)
Anthracite			
Meta-anthracite	$98 \leqslant C$	$V < 2$	
Anthracite	$92 < C < 98$	$2 < V < 8$	
Semianthracite	$86 < C < 92$	$8 < V < 14$	
Bituminous			
Low volatile bituminous	$78 < C < 86$	$14 < V < 22$	
Medium volatile bituminous	$69 < C < 78$	$22 < V < 31$	
High volatile A bituminous	$C < 69$	$31 < V$	$32.6 < \Delta H$
High volatile B bituminous			$30.2 < \Delta H < 32.6$
High volatile C bituminous[b]			$26.7 < \Delta H < 30.2$
			$24.4 < \Delta H < 26.7$
Subbituminous			
Subbituminous A coal			$24.4 < \Delta H < 26.7$
Subbituminous B coal			$22.1 < \Delta H < 24.4$
Subbituminous C coal			$19.3 < \Delta H < 22.1$
Lignitic			
Lignitic A			$14.7 < \Delta H < 19.3$
Lignite B			$\Delta H < 14.7$

[a]Based on dry, moisture and mineral free (ash free basis).
[b]Variable values depending on the agglomerating properties of the coal when freed of volatile matter.

TABLE 2.3
Proximate and Ultimate Analysis of Five Canadian Coals (Moisture Free)

	BC	NS	NS	Sask	Sask
Proximate analysis					
Volatile carbon	26.1	34.6	35.4	43.5	41.4
Fixed carbon	58.6	49.8	61.7	43.1	46.1
Ash	15.3	15.6	2.9	13.4	12.5
Ultimate analysis					
Carbon	74.4	66.0	84.7	61.1	66.1
Hydrogen	4.3	4.5	5.6	3.6	2.2
Sulfur	0.8	4.9	1.3	1.1	0.6
Nitrogen	1.2	1.4	1.3	1.0	1.3
Ash	15.3	15.6	2.8	13.4	12.5
Oxygen[a]	4.0	7.6	4.3	19.8	17.3
Calorific value[b] MJ/kg	28.6	23.8	33.9	17.3	18.3

[a]Determined by difference.
[b]Coal sample as received.

correction must be made for it. The nitrogen content of coal is usually small and often ignored. Since $\Delta H_f^0(NO_2) = 33.2$ kJ/mol the added term to the equation would be -2.4 N. Several other formulas have been proposed which are correct for some of the structural aspects of coal.

2.7. ASH

The combustion of coal results in the formation of an ash—the noncombustible component of coal—part of which is carried off with the combustion products as very fine particulate powder called *fly ash*. The heavy ash remaining in the combustion chamber is called *bottom ash* or *boiler slag*. The continuous mining process produces a coal containing about 28% ash. Coal used to generate steam usually has about 15% ash.

The ash is an undesirable component of the coal and is usually reduced during the cleaning process that coal is normally subjected to after being shipped from the mine. This cleaning process removes the rocks, clay, and minerals which invariably mix with the coal.

Some applications such as chain-grate stokers require a minimum of 7–10% ash to protect the metal parts of the furnace. One of the most important characteristics of the bottom ash is its melting point (or fusion temperature) which determines the ease with which it is removed from the furnace. The melting point of these inherent impurities in a coal can be represented by a three component phase diagram of oxides such as Fe_2O_3, Al_2O_3, and SiO_2 or by minerals such as $Al_2O_3 \cdot 2SiO_2$ (clay), SiO_2 (quartz), and $MgO \cdot Al_2O_3 \cdot 2 \cdot 5SiO_2$ (feldspar) where lines join the common melting points of the various mixtures. The choice of the components depends on the coal and the impurities in the ash. A high fusion ash melts above 1316°C whereas a low fusion ash melts below 1093°C. In some instances, iron oxide or sand (SiO_2) is added to the coal to increase the melting point of the ash so as to favor clinker formation which results in easy removal. Ash with a low fusion temperature can form an undesirable glassy coat of the furnace and grill. In the case where steam is generated the deposition of this adherent glassy deposit on the fireside of the steel boiler tubes causes the ash to insulate the steel tubes from the heat and thus reduce the efficiency of steam generation.

Fly ash is usually precipitated by a Cottrell electrostatic precipitator and/or collected in filter bags. The composition of the fly ash usually differs from the bottom ash due to thermal fractionation of the oxides. This is illustrated in Table 2.4 where the composition of the bottom ash is compared with various fractions of the fly ash which is classified into 5 groups:

(a) the alkali and alkaline earth metals
(b) refractory metals
(c) transition metals
(d) halogens
(e) the volatile elements.

In 1995 the U.S. utilities produced 65 Mt/y of fly ash. The 21 coal-fired thermal generating plants in Canada produced 16.6 GW consuming 37.8 Mt of coal of which

TABLE 2.4
Trace Element Analysis of Coal and its Various Combustion (ash) Fractions[a] ($\mu g/g$)

	Element		Coal	Bottom ash	Precipitated[b] ash	Bag[c] ash	Stack[d] ash
A.	Barium	Ba	150	1200	1600	1540	1700
	Calcium	Ca	4000	49,000	34,000	21,000	23,000
	Cesium	Cs	0.8	10	7	6	8
	Magnesium	Mg	420	4700	5400	440	4800
	Potassium	K	1220	16,000	13,000	11,000	13,300
	Rubidium	Rb	23	75	160	90	90
	Sodium	Na	500	7100	4600	3400	4000
	Strontium	Sr	120	1200	1200	1000	1100
B.	Aluminum	Al	11,000	100,000	112,000	103,000	103,000
	Beryllium	Be	0.9	—	8	10	10
	Dysprosium	Dy	1.0	6	10	5	6
	Europium	Eu	0.3	3	3	3	3
	Hafnium	Hf	0.7	—	9	3	7
	Lanthanum	La	5.0	70	60	50	50
	Lutetium	Lu	0.2	1	2	0.6	2
	Samarium	Sm	1.0	10	12	9	12
	Tantalum	Ta	0.3	2	3	1	3
	Terbium	Tb	0.3	4	4	0.8	3
	Thorium	Th	1.4	20	15	10	12
	Uranium	U	0.7	—	9	8	9
	Ytterbium	Yb	0.4	4	5	3	4
C.	Chromium	Cr	12	160	140	220	370
	Cobalt	Co	3	20	40	35	60
	Copper	Cu	6	20	84	142	170
	Iron	Fe	6300	140,000	90,000	70,000	70,000
	Manganese	Mn	24	360	270	250	510
	Nickel	Ni	4	70	35	200	340
	Scandium	Sc	3	—	40	20	20
	Silver	Ag	2	50	40	7	40
	Titanium	Ti	560	5000	6500	6000	6000
	Vanadium	V	20	130	180	240	240
D.	Bromine	Br	14	Nil	6	30	12,000
	Chlorine	Cl	1200	Nil	Nil	800	900,000
	Fluorine	F	80	—	100	400	33,000
	Iodine	I	1	Nil	Nil	Nil	1300
E.	Antimony	Sb	0.5	3	6	11	30
	Arsenic	As	12	35	90	230	300
	Boron	B	16	—	220	2000	2600
	Cadmium	Cd	0.3	3	2	4	20
	Gallium	Ga	10	30	60	140	140
	Lead	Pb	6	70	50	160	200
	Mercury	Hg	0.4	0.2	0.5	0.5	500
	Selenium	Sc	3	2	14	14	300
	Zinc	Zn	27	—	330	380	1100

[a] Values rounded off—average of several coals.
[b] Electrostatic precipitators 99% efficient.
[c] Filter bags.

TABLE 2.5
Chemical Analysis of Fly Ash (Saskatchewan)

Component	Content (%)	Component	Content ($\mu g/g$)
Na (as Na_2O)	3.02	As	<1
K (as K_2O)	0.48	B	291
Ca (as CaO)	10.87	Cd	2
Mg (as MgO)	1.09	Cr	20
Al (as Al_2O_3)	21.02	Cu	26
Si (as SiO_2)	57.64	Co	13
Fe (as Fe_2O_3)	2.81	Ga	107
Ti (as TiO_2)	0.76	Pb	36
P (as P_2O_5)	0.09	Mn	292
S (as SO_3)	<0.1	Hg	<0.5
		Mo	<50
		Ni	32
		Ag	2.9
		V	72
		Zn	48
		Se	<0.5

10% was ash. The fly ash produced was 2.5 Mt. The composition of the fly ashes varies considerably. One particular fly ash from Saskatchewan (see Table 2.5) was recently tested for the extraction and recovery of gallium which can be converted to GaAs. In the near future GaAs will be replacing silicon as the semiconductor of the electronic industry.

Another potential use of some fly ash is as a source of alumina (Al_2O_3), replacing bauxite which is being exhausted at an ever increasing rate.

2.8. COAL AND ITS ENVIRONMENT

Though coal is cheap and plentiful it is a "dirty" fuel which contaminates our environment and contributes to the CO_2 imbalance in our atmosphere as well as oxides of nitrogen from the combustion process. The principal contaminant in coal is sulfur which burns to form sulfur dioxide (SO_2) which is oxidized to sulfur trioxide (SO_3) in the atmosphere. In the commercial production of sulfuric acid, sulfur (S) is burned to form SO_2.

$$S + O_2 \rightarrow SO_2 \tag{2.4}$$

The SO_2 can dissolve in water to form sulfurous acid (H_2SO_3). The SO_2 is also catalytically converted to SO_3 using V_2O_5 as a catalyst:

$$2SO_2 + O_2 \xrightarrow{V_2O_5} 2SO_3 \tag{2.5}$$

The SO_3 is then treated with water or sulfuric acid solution to form H_2SO_4.

Similar reactions can occur in the atmosphere with the resulting formation of acid (H_2SO_4) rain which can fall considerable distances from the source. Thus the acid rain falling in Norway and Sweden primarily originated from coal burning in the Ruhr valley and the UK. Similarly the acid rain reaching Ontario comes primarily from the US iron and steel centers and industrial Ohio Valley though it is claimed that a large part of the SO_2 is also coming from the nickel smelters in Sudbury, Ontario. A joint US–Canadian study has determined the sources and recommended appropriate solutions to prevent the lakes from becoming too acidic to support aquatic life (fish).

Acid rain is not the only environmental contaminant from coal burning. The average concentration of mercury in coal is about 0.3 μg/g (i.e., less than 1 ppm). A 755 MW steam turbine driven power-station burns approx. 7100 tonne of coal per day. This corresponds to about 2.5 kg/day of mercury being sent up the stack. With present North America estimates of coal consumption at about 10^{10} tonne/year, about 3000 tonne of mercury is put into the environment. This is about 4 times the natural source.*

Coal also contains uranium and thorium and their radioactive decay products. Though as much as 98% of the fly ash is precipitated, the 2% remaining escapes up the stack. For a power plant burning 2×10^4 tonne/day if there is approx. 10% of fly ash then approx. 40 tonne/day fly ash escapes up the stack. Measurements have shown that this contains about 500 μCi ^{226}Ra per day. The fly ash behaves like ordinary smoke and its dispersion follows standard equations. If we assume a wind speed of 1 m/s, a stack of 120 m high will result in a maximum concentration at ground level at 400 m from the stack of 9×10^{-14} μCi ^{226}Ra per cm^3 air. The maximum permissible concentration (MPC) for ^{226}Ra in air is 10^{-11} μCi/cm^3. Thus the concentration of ^{226}Ra is 2 orders of magnitude lower than the MPC, but since fly ash contains other radionuclides (^{230}U, ^{210}Pb, etc.) the long-term effects on those living close to or downwind from coal burning power plants must be carefully monitored.

The use of fly ash in concrete does not solve the disposal problem but only shifts it to another locale. If cement is composed of 30% fly ash, it has been estimated that the radon diffusing out of the concrete (porosity = 5%) into a room (10 m $\times$ 10 m $\times$ 4 m) would be about 10^{-9} μCi/cm^3 — 100 times lower than the MPC. However, with higher porosity concrete and lower ventilation rates the margin of safety decreases and it means that concrete containing fly ash should not be used in structures for habitation though it would be permissible for use in foundations, bridges and roads.

A recent coal fired power station located between Los Angeles and Las Vegas has shown that it is possible to burn coal with as much as 3.5% sulfur without contaminating the environment with SO_2 and NO_x. This is done using two existing technologies, coal gasification and combined cycle generation, i.e., generating electricity simultaneously from turbines running on gas and steam. The plant pulverizes 1000 tonnes of coal per day, which is converted to synthesis gas by reaction with water and oxygen. The ash and minerals fuse and are removed whereas the H_2S is removed after the gas is cooled by generating steam. Water is then added to the clean gas to reduce NO_x formation upon combustion and more steam is generated to drive the turbines. The cost of power for a 600 MW plant is estimated to be 4 to 5 cents per kWh.

* The average concentration of mercury in rock is about 0.1 μg/g and since about 10^{10} tonne of rock is weathered annually it is estimated that about 800 tonne of Hg is put into the atmosphere each year. This mercury is distributed uniformly around the earth and presents little environmental danger. It is the nonuniform or concentrated dumping of mercury that is dangerous and must be avoided.

2.9. FLUIDIZED BED COMBUSTION

The fluidized bed reactor is about 60 years old but only in recent years has its application to coal combustion taken on commercial significance. The fluidized bed is the dispersion of a solid, usually in powder form, by a gas, under flow conditions such that the solid takes on the properties of a gas. Such reactors can be designed to operate continuously. Thermal conduction (heat transfer) in such systems can be high and, as a result, in the case of coal and air, the combustion can occur at much lower temperatures than in the fixed bed system. Thus, the addition of limestone ($CaCO_3$) or dolomite ($CaCO_3 \cdot MgCO_3$) to the fluidized bed system can result in the reduction of oxides of sulfur and oxides of nitrogen.

This is accomplished by the reactions which can occur in the combustion bed.

$$CaCO_3 \rightarrow CaO + CO_2 \tag{2.6}$$

$$CaO + SO_x \rightarrow CaSO_{(x+1)} \tag{2.7}$$

The lower combustion temperature keeps the equilibrium for the NO_x reaction toward the $O_2 + N_2$. See Exercise 18.

$$O_2 + N_2 \rightleftharpoons 2NO \tag{2.8}$$

This would mean that high sulfur coal could eventually be used in combustion processes without contaminating the environment.

Unfortunately the effect of increased CO_2 levels in the atmosphere will still contribute to the greenhouse effect. Only the combustion of biomass, H_2, or renewable energy sources do not enhance the greenhouse effect.

Recent work by the Coal Utilization Research Laboratory in Leatherhead, England, has shown that a slurry of 68% coal in water burned very well in a pressurized fluidized bed reactor. However, much more work remains to be done before such mixtures can be pumped through pipelines and burned directly in specially designed reactors.

2.10. COKE

Coke is produced when a bituminous coal is devolatilized by heating (in the absence of air) to temperatures ranging from 900 to 1200°C. As the temperature is slowly raised, physical and chemical changes take place. Adsorbed water is lost at temperatures up to 250°C. Some CO and CO_2 is liberated up to 300°C with some pyrogenic water. Above 350°C the coal becomes plastic, and begins to decompose between 500 and 550°C to form gases and tars. The product at this point is called *semicoke* which still off-gases up to 700°C where the hot coke acts as a catalyst for the decomposition of the volatile products. Above 700°C the coke is hardened and agglomerated. The by-product consists of coke-oven gas (55–60% H_2, 20–30% CH_4, 5–8% CO, 2–3% heavy hydrocarbons, 3–5% N_2, and 1–3% CO_2, as well as traces of O_2) having a calorific value of about 17 MJ/m^3.

Coal tar is also produced during the coking process. It contains over 300 substances which usually include 5–10% naphthalene, 4–6% phenanthrene, 1–2% carbazole, 0.5–1.5% anthracene, <0.5% phenol, ~1% cresol and ~1% pyridene compounds and other aromatic hydrocarbons. About 50–60% of the tar consists of high molecular weight hydrocarbons.

The principal use of coke is in the blast furnace for steel making, the manufacture of calcium carbide (CaC_2), and other metallurgical processes. The annual world production of coke has been at about 4×10^8 tonnes for the past 10 years. US production of coke has decreased from 46 Mtonne in 1980 to 22 Mtonne in 2000 as steel production moved to Asia and the far east. The coals used for coke production are usually a blend of two or more coals consisting of a high-volatile coal blended with low-volatile coals in a ratio of from 90:10 to 60:40. Such blends increase the rate of coking and produce a better product. The large sized fused coke (≈75 mm) is used in blast furnaces whereas the smaller sized (≈20 mm), called *breeze*, is used in boiler-firing, iron ore sintering, electric smelting and other applications where a purer grade of carbon is required.

The ash content of a metallurgical grade of coke must be less than 10% with a sulfur content <1.5% and a volatile component <1%. The calorific value of such coke is between 31.4 and 33.5 MJ/kg. One tonne of coal normally produces from 650 to 750 kg of coke.

The development of the coal gasification process allows for the removal of contaminants which make the coal a "dirty fuel." If a new sink for CO_2 can be invented, the need to go fully nuclear could be delayed.

EXERCISES

1. Discuss the statement "Coal is a dirty fuel."
2. What is the difference between a primary and secondary fuel?
3. What does charcoal and coke have in common? Compare their properties.
4. Describe a laboratory experiment by which wood could be converted into coal.
5. Explain why the proximate analysis of a coal is useful.
6. Explain why chloride salts should be washed from coal before it is used in combustion processes.
7. Wet coal can be air dried at room temperature. How does this differ from the moisture content of a coal?
8. Explain why the components in an ash from a coal has a different relative concentration compared to the initial coal.
9. Two recent formulas have been proposed to evaluate the heat of combustion of coals:

$$-\Delta H_c(\text{kJ/g}) = 357.8\text{C} + 1135.7\text{H} + 59.4\text{N} + 111.9\text{S} - 84.5\text{O} \tag{2.9}$$

Lloyd and Bavenport

$$-\Delta H_c(\text{kJ/g}) = 351\text{C} + 116.1\text{H} + 62\text{N} + 104\text{S} - 110\text{S} \tag{2.10}$$

Boie's formula

TABLE 2.A
Elemental Analysis of Some Solid Fuels

	wt.%					$-\Delta H_c$ (kJ/g) measured
	C	H	N	S	O	
Oil shales						
(A) Australia	84.6	11.5	0.56	0.30	3.0	43.70
(B) France	85.1	11.4	1.12	0.32	2.06	43.19
(C) Sweden	85.0	9.0	0.71	1.72	3.6	40.58
Coal						
(D) South Africa	84.3	5.9	2.50	0.70	6.6	35.60
(E) Athabasca tar	82.5	10.0	0.47	4.86	1.7	41.14
Sand						
(F) Syncrude oil						
(a)	87.8	10.1	0.34	0.12	1.6	42.54
(b)	79.2	5.1	1.30	1.30	13.1	32.36

Compare the accuracy of these formulas with Eq. (2.3) using the data given in Table 2.A.

10. The three component terniary phase diagram of a bottom ash is shown in Fig. 2.A. (a) How much sand (SiO_2) must be added to each tonne of coal (type A) (point x) if the melting temperature of the ash is to be raised from 1000°C (point x) to 1200°C (point y)? The ash content of the coal is 10% and the oxides in the ash are graphed as weight %. (b) A second coal (type B) has 15% bottom ash with a composition as shown in Fig. 2.A (point z). What

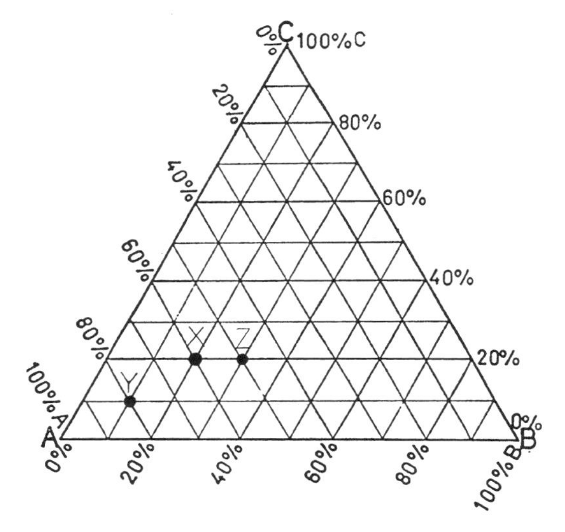

FIGURE 2A. A = SiO_2, B = Al_2O_3, C = Fe_2O_3. Ash at point X has a melting point of 1000°C and at point Y the melting point is 1200°C.

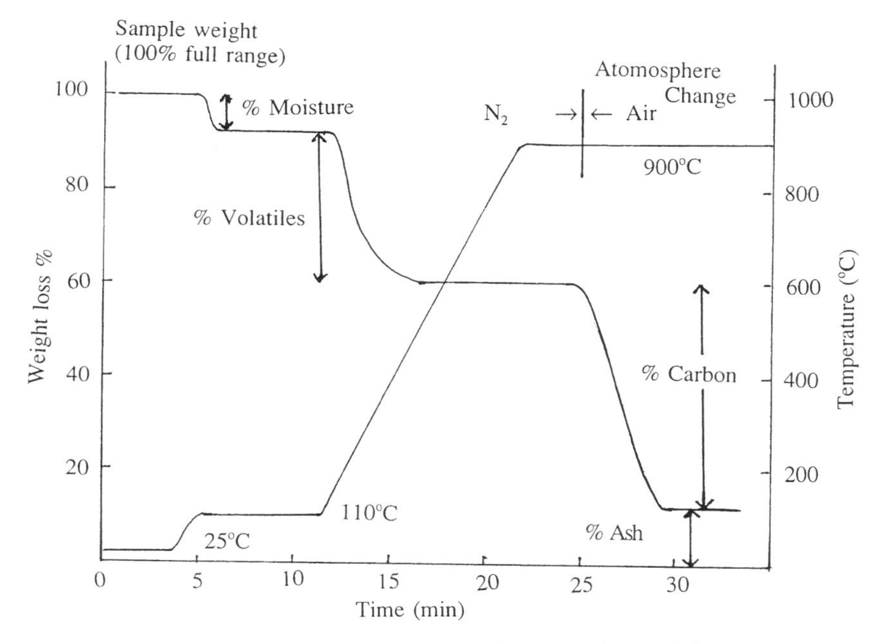

FIGURE 2B. Proximate analysis of a coal by TGA.

would be the composition of the ash if a blend of the coals was burned A:B of 2:1?

11. The proximate analysis of a coal can be determined automatically by Thermal Gravimetric Analysis (TGA). This method involves the automatic recording of the mass of a coal sample while the temperature of the sample is programmed and the atmosphere of the sample controlled. This is illustrated in Fig. 2.B where the mass of the sample is recorded at 25, 110, and 900°C under the inert atmosphere of nitrogen. Air or oxygen is then introduced while the sample is still at 900°C. (a) Describe the changes which occur at each temperature stage. (b) Determine the proximate analysis of the sample from the results in Fig. 2.B.
12. Explain why the elements listed in Table 2.4 are classified into the five specific groups.
13. Explain why the calorific value of the volatile fraction of a coal (per unit mass) is normally greater than that of the fixed carbon of the coal.
14. What is coal gas?
15. Explain why coal is often used as a source of active carbon.
16. Why do some coals have a higher calorific value than pure carbon?
17. Explain why using peat for fuel has an adverse effect on the CO_2 balance of our atmosphere.
18. The equilibrium constant, Kp for the reaction $N_2 + O_2 = 2NO$ as a function of temperature was determined experimentally to be: $\log \mathrm{Kp} = a - b/T$ where $a = 1.63$, and $b = 9.452$ (K). Calculate the equilibrium pressure of NO when air is heated to 800, 1000, and 1200 K.

FURTHER READING

J. A. Pajares and J. M. D. Tascon, *Coal Science*, 2 Vol., Elsevier, Amsterdam (1995).
S. Rogers et al., *Coal Liquefaction and Gas Conversion: Contractor's Review Conference: Proceedings*, 2 Vol., DIANE Pub. Co., Upland, Pennsylvania (1995).
National Research Council Staff, *Coal: Energy for the Future*, Natl., Acad. Press, Washington DC (1995).
J. Tomeczek, *Coal Combustion*, Krieger, Philadelphia, Pennsylvania, (1994).
J. G. Speight, *The Chemistry and Technology of Coal*, 2nd Ed., Marcel Dekker Inc., New York (1994).
Clean and Efficient Use of Coal: The New Era for Low-Rank Coal, OECD, Washington DC (1994).
K. L. Smith et al., *The Structure and Reaction Processes of Coal*, Plenum Press, New York (1994).
N. Berkowitz, *An Introduction to Coal Technology*, 2nd Ed., Academic Press, New York, (1993).
R. F. Keefer and K. S. Sajwan, Editors, *Trace Elements in Coal and Coal Combustion Residues*, Lewis Publ., El Cajon, California (1993).
B. E. Law and D. D. Rice, Editors, *Hydrocarbons from Coal*, American Assoc. Petroleum Geologists, Tulsa, Oklahoma (1993).
Y. Yurum, Editor, *Clean Utilization of Coal*, Kluwer Academic, Norwall, Massachusetts (1992).
H. H. Schobert et al., Editors, *Coal Science II*, ACS Symposium Ser. No. 461; ACS, Washington DC (1991).
H. H. Schobert, The Geochemistry of Coal: (1) The Classification and Origin of Coal. (II) The Components of Coal, *J. Chem. Educ.* **66** 242, 290 (1989).
D. W. Van Krevelen et al., Editors, *Coal Science Technology*, 16 Vol., Elsevier, NY (1981–1989).
C. R. Ward, Editor, *Coal Geology and Coal Technology*, Blackwell Sci. Publ., Malden Massachusetts (1984).
M. L. Gorbaty, J. W. Larson, and I. Wender, Editors, *Coal Science*. 2 Vol., Academic Press, New York (1983).
V. Valkovic, *Trace Elements in Coal*, 2 Vol., C.R.C. Press, Boca Raton, Florida (1983).
R. A. Meyers, Editor, *Coal Structure*. Academic Press, New York (1982).
R. A. Meyers, Editor, *Coal Handbook*. Marcel Dekker Inc., New York (1981).
R. P. Greene and J. M. Gallagher Editors, *World Coal Study-Future Coal Prospects*, Ballinger Publ. Co., Cambridge, Massachusetts (1980).
Derek Ezra, *Coal and Energy*, Ernest Benn Ltd., London (1978).
D. A. Tillman, *Wood as an Energy Resource*, Academic Press, New York (1976).
Wood Heat Organization, http://www.woodheat.org
International Peat Society, http://www.peatsociety.fi/
International Energy Agency Clean Coal Center, http://www.iea-coal.org.uk/
World Coal Institute, http://www.wci-coal.com/
American Coal and Coke Chemical Institute, http://www.accci.org/
Coal Association of Canada, http://www.coal.ca
Retexo-RISP, http://www.flyashrecycling.com
ISG Resorces, http://www.flyash.com
International Ash Utilization Symp. Ser., http://www.flyash.org/
ASTM-Coal and Coke Term. , http://www.ASTM.org?DATABASE.CART/PAGES/0121.htm

3

Crude Oil

3.1. INTRODUCTION

Oil is a major liquid fuel and it is also the basis of most other liquid fuels. It is formed by refining petroleum or crude oil which is a very complex mixture of components composed of many different types of hydrocarbons of various molecular weights. Crude oils are usually classified by the major type of hydrocarbons in the oil. The three major classes are: (1) paraffinic (alkanes); (2) aromatic with naphthenic and asphaltic components; and (3) mixed oils containing significant quantities of both aliphatic and aromatic compounds. Typical gas chromatograms of two crude oils are shown in Fig. 3.1. Such chromatograms are often used as a "fingerprint" of an oil in identifying the origin of oil spills. The origin of petroleum is not known with certainty. It is believed to have formed from the accumulation of various marine organic deposits which by partial bacterial decay, heat, and pressure, eventually formed the crude oil which migrated through the pores, cracks and fissures in the rocks, forming oilfields in underground structures illustrated in Fig. 3.2.

Crude oil varies in its properties and composite from one country to another and from field to field. The differences are due to viscosity, volatility, and composition (e.g., sulfur content). A list of 14 different crude oils and their primary distillation products is given in Table 3.1.

3.2. EARLY HISTORY

Petroleum and its products, notably asphalt, has been used to waterproof boats since 6000 B.C.. The Bible refers to many applications from Noah's Ark to Moses' basket. Job refers to "the rock poured me out rivers of oil" (Job 29:2) and even today one can find asphalt floating up to the surface of the Dead Sea. The Egyptians used asphalt in the construction of the pyramids. It is believed that the Chinese were drilling for oil in the third century A.D. In North America the native Cree Indians used bitumen from the Alberta oil sands to seal their canoes as well as for medicinal purposes. Oil seepage on the surface were the earliest sources of crude oil. The first to use petroleum products commercially was Dr. Abraham Gesner of Nova Scotia. He distilled a kerosene fraction (about 1840) to use as an illuminating fuel. This coal oil replaced the smoky smelly fuel from whale oil and vegetable oils used for lighting. Seventy plants were producing coal oil for lighting by 1861. However by 1855 a Canadian company

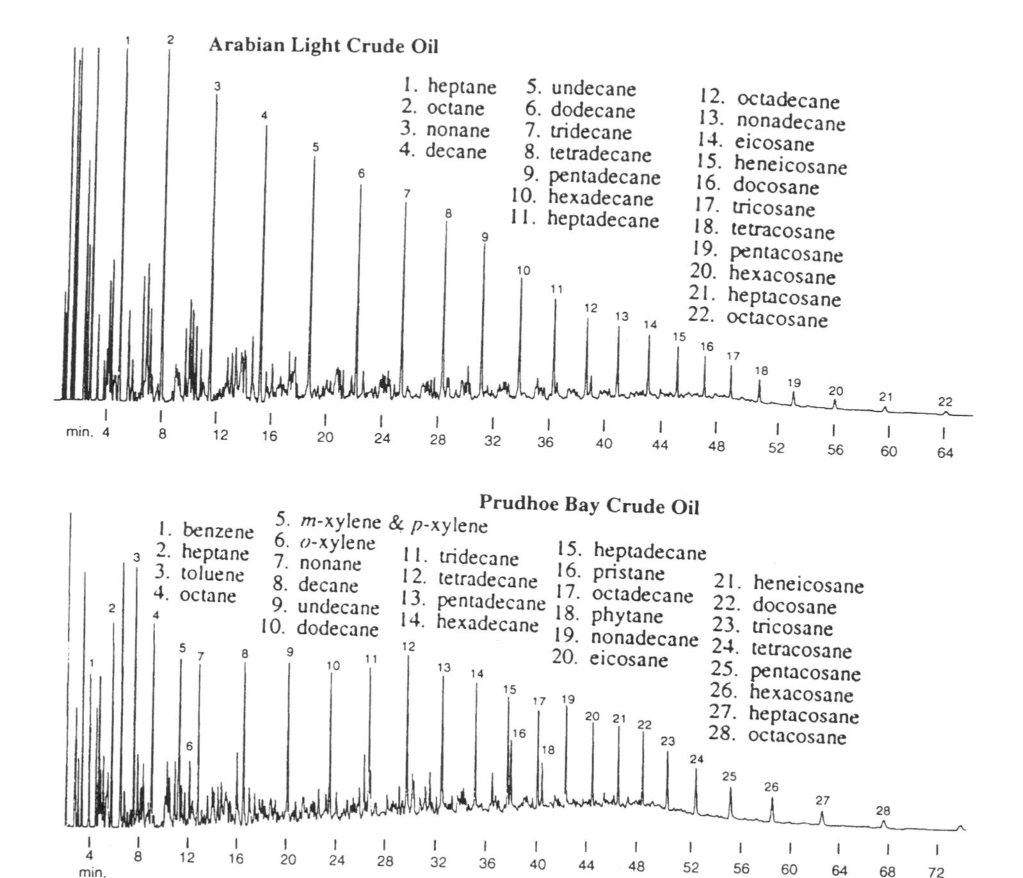

FIGURE 3.1. Typical gas chromatograms of two crude oils where each peak represents one (or more) individual compounds. The height (or area) of the peak is proportional to its concentration in the injected sample. The time taken for the substance to elute from the column, which is temperature programmed to speed up the analysis, is used to identify the substance. The presence and the ratio of the two branched hydrocarbons, pristane (2,6,10,14-tetramethylpentadecane) and phytane (2,6,10,14-tetramethylhexadecane) can be used to identify the source of the oil.

was producing a better grade of kerosene from oil seepages near Sarnia, Ontario where the first oil producing well was drilled (15 m) in 1858. By 1863 there were 30 refineries in Ontario producing oil for $4/bbl which rose to $11/bbl in 1865. However, the oil found in Pennsylvania proved to contain less sulfur and thus produced a better quality kerosene. The Canadian oil price then dropped to $0.50/bbl.

3.3. WORLD PRODUCTION OF CRUDE OIL

In 1950, world oil production was 519 Mtonnes or 10 million bbl/day. This doubled by 1960 (1.05 Gtonne) and doubled again by 1970 (2.3 Gtonne). The

ANTICLINE

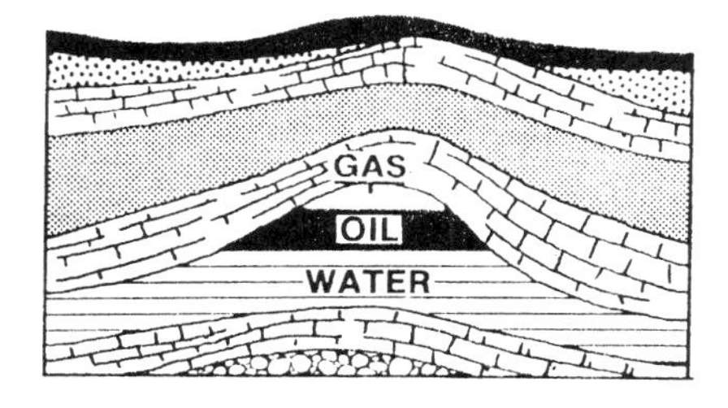

FAULT

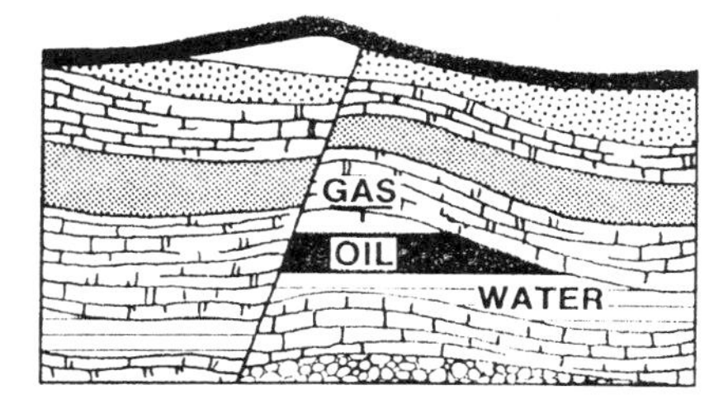

STRATIGRAPHIC TRAP

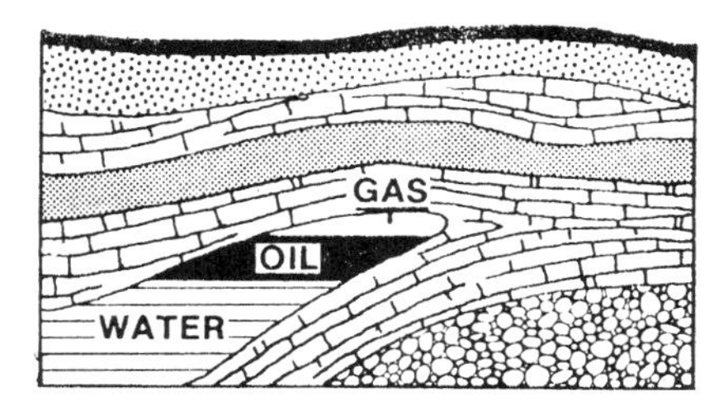

FIGURE 3.2. Oil formation in sedimentary basins where natural gas is usually trapped with the oil.

production continued to increase (3.0 Gtonnes or about 22 billion bbl in 1979) and seemed to level off at about 20 billion barrels during the 1980s reflecting the changing attitude towards energy. However, the world's annual production of crude oil has gone to over 22 billion bbl in 1992 in spite of an economic slump. The percentage contributed by different geographical areas in 1998 is shown in Fig. 3.3. The 1992 production and consumption levels for the major producers and consumers is given in Table 3.2.

The United States used to be the world's leading producer of crude oil (7.75 M bbl/day in 1965, approx. 8.7 M bbl/day in 1980, and 7.2 M bbl/day in 1992). The decrease in production was more than balanced by increased imports of 7.0 M bbl/day in 1992. Thus, the percentage contributed by the United States to the world supply, decreased from 45% in 1956 to 13.7% in 1979 to less than 12% in 1992. Even though

TABLE 3.1
Various Crude Oils and Yields of Main Products from Primary Distillation (% vol.)

Country (crude oil name)	Gases	Gasoline	Kerosene	Gasoil/diesel	Residue (fuel oil component)	Sulfur content
North America						
USA (Alaska)	0.8	13.4	11.6	21.5	53.3	Low 1.0% on crude, 1.5% on residue
Western Europe						
UK (Forties)	4.3	22.5	12.2	21.9	39.5	Very low 0.3% on crude, 0.6% on residue
Norway (Ekofisk)	3.3	31.2	13.6	21.6	30.8	Very low 0.1% on crude, 0.3% on residue
Middle East						
Saudi Arabia (Arabian Light)	1.7	20.5	12.0	21.1	45.1	Very high 3% on crude, 4.5% on residue
Qatar (Qatar)	4.4	29.1	15.9	20.6	30.7	Medium 1.1% on crude, 2.6% on residue
South and Central America						
Venezuela (Tia Juana Pesado)	—	1.4	3.6	14.7	80.8	High 2.7% on crude, 3.0% on residue
Mexico (Isthmus)	1.8	22.9	13.1	22.0	40.4	High 1.6% on crude, 3.0% on residue
Africa						
Nigeria (Nigerian Light)	2.9	25.8	14.4	27.7	29.4	Very low 0.09% on crude, 0.02% on residue
Libya (Libyan Light)	2.8	21.6	12.9	22.1	40.9	Low 0.5% on crude, $<1.0\%$ on residue
Far East and Australasia						
Indonesia (Sumatran Light)	0.5	11.5	9.5	20.6	58.3	Very low 0.08% on crude, 0.1% on residue
Malaysia (Miri Light)	1.9	28.1	16.7	32.1	21.3	Very low 0.04% on crude, 0.1% on residue
Australia (Gippsland)	2.3	36.0	13.8	24.7	23.5	Very low 0.08% on crude, 0.2% on residue
CIS China						
(Ural) Russia	2.2	20.9	14.7	19.5	43.1	High 1.5% on crude, 2.6% on residue
China (Daqing)	0.5	9.8	6.9	16.5	66.4	Very low 0.1% on crude, 0.15% on residue

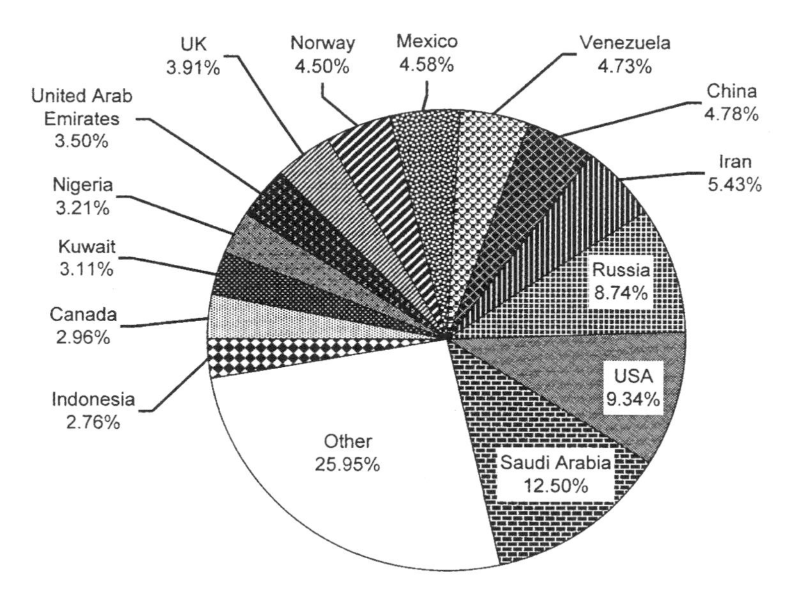

FIGURE 3.3. Percentage of world crude oil production for various countries in 1998 (World total = 75 Mbbl/day).

Canada increased its production during this same time period, 3% in 1956 went to 2.6% of the total world production in 1992. In the same interval new producing areas were developed, notably in the Sahara, the Arctic, Australia, and the North Sea. The Middle East also showed a marked increase in oil production from 11.6 M bbl/day in 1983 to 17.6 M bbl/day in 1992. Russian production of crude oil increased from 9.3% of the world's oil in 1956 to about 20% in 1992.

The Russian oil reserves are as vast as the country. The major producing areas are West Siberia and the Volga-Urals region which produce about 400 Mtonnes and 200 Mtonnes per year, respectively. Many of the oil fields of northern Siberia are not connected by pipelines and thus await further development. This is similar to Canada where arctic gas and oil fields are not as yet being tapped.

World oil production and consumption is projected to reach 77.4, 82.4, and 86.5 M bbl/day in the years 2000, 2005, and 2010, respectively. Total world reserves have been estimated to be about 10^{12} bbl. Based on projected uses and available reserves at present prices, it would appear that oil will be available for only 30–50 years or until the year 2040. It must be pointed out that all these predictions are based on present technology and in the USA alone there is an additional 3×10^{11} bbl of oil which can only be economically recovered when the price of oil has risen to about $30/bbl.

Another source of oil is oil shales and tar sands which require special processing to extract the oil. Tar sands are found throughout the world and deposits are known to exist in 49 countries. Major deposits are found in Venezuela and Canada. Alberta's tar sands contain an estimated 9×10^{11} bbl of crude bitumen from which about 3×10^{11} bbl of synthetic crude oil can be obtained. The tar sands consist of particles of sand held together by a water and oil coating as shown in Fig. 3.4. The process by which the oil is separated from the water and sand involves subjecting the mixture to

TABLE 3.2
World Crude Oil Reserves and Marketing (1997)

	Reserves 10^9 bbl	Production capacity 10^6 bbl/day	Consumption 10^6 bbl/day
Algeria	11.2[b]	1.2	0.23
Australia	2.0[b]	0.6	0.82
Brazil	6.0[a]	—	1.8
Canada	5.1[c]	1.9	1.8
China	29[e]	3.2	3.8
Egypt	3.7[a]	0.8	0.5
India	3.9[d]	0.6	1.8
Indonesia	7[b]	1.5	0.9
Iran	91[b]	3.7	1.1
Iraq	106[f]	1.1	0.5
Japan	—	—	5.7
Kuwait	94[b]	2.1	0.1
Libya	28[a]	1.4	0.2
Mexico	40[a]	3.1	1.9
Nigeria	19[b]	2.3	0.3
Norway*	11[a]	3.0	0.2
Russia	51[c]	5.9	2.8
Saudia Arabia	262[b]	8.6	1.2
South Korea	—	—	2.4
UK	5[a]	2.5	1.8
United Arab Emirates	85[g]	2.3	0.3
USA	27[a]	6.4	18.6
Venezuela	60[g]	3.5	0.5
Other	51		
Total	1020	66.1	73.0

*North Sea
Uncertainty. $\pm$: $a = 1$, $b = 2$, $c = 3$, $d = 4$, $e = 5$, $f = 6$, $g = 10$.

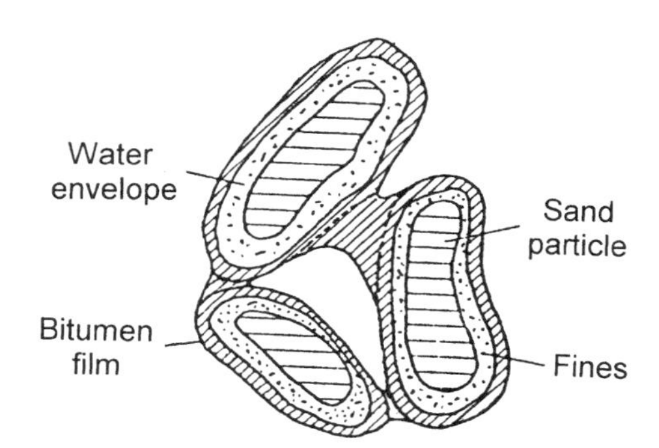

FIGURE 3.4. Typical structure of tar–sand consisting of an oil–sand–water mixture.

hot water and steam and skimming off the oil. Since over 90% of the oil sands cannot be surface mined economically, more elaborate techniques are being developed. For *in situ* mining, the steam can be injected into the drill holes to heat up the sands and fluidize the oil. Other processes involve controlled ignition of the tar sands and air pressure to drive the freed oil towards the production holes.

3.4. CRUDE OIL PROCESSING

Crude oil is treated by physical and chemical processes to produce the various petroleum products. The early use of oil was in the preparation of kerosene. This was accomplished by batch distillation which separated the mixture of hydrocarbons by boiling points (vapor pressure). The modern distillation process (see Fig. 3.5) is designed to operate continuously. The temperature gradient of the column separates the crude oil into fractions according to specific boiling point ranges. These are shown in Table 3.3.

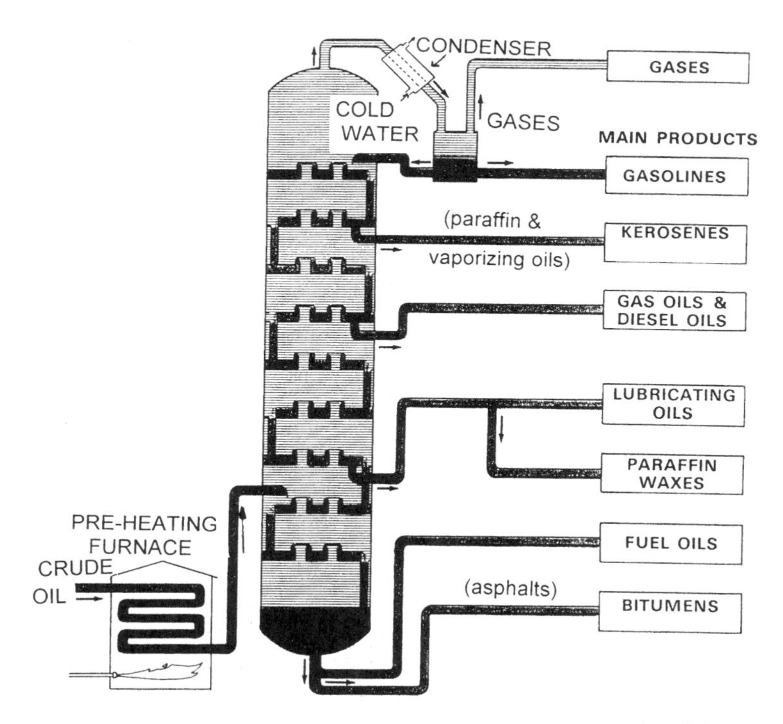

FIGURE 3.5. Distillation, the first step in oil refining, separates crude oil into a number of products. First, the crude oil is heated by being pumped through pipes in a furnace. The resulting mixture of vapors and liquid goes to a tower where the vapors rise, condense on trays, and are drawn off through pipes as products. The part of the crude that does not boil in the first distillation step is reheated and distilled in a vacuum. Again, vapors rise, condense, and are drawn off as products. Some of the very heavy oil, which does not boil even under the reduced pressure, is used in factories and ships, or made into asphalt.

TABLE 3.3
Primary Distillation Fractions from Crude Petroleum

Fraction	Boiling point range (°C)	Composition	Uses
Gas	up to 25	C_1–C_4	Petrochemicals fuel
Petroleum ether	25–60	C_5–C_7	Solvent
Gasoline	0–17	C_6–C_{11}	Fuel for ICE
Kerosene	17–220	C_{11}–C_{13}	Jet fuel
Heating oil	220–250	C_{13}–C_{16}	Fuel oil
Lubricating oils	350–400	C_{16}–C_{20}	Lubricants and grease
Paraffins	400–500	C_{20+}	Candles
Asphalts	>500	C_{30+}	Road surfaces

The demand for the various fractions in a crude oil seldom coincide with the distillation yields. Hence it is necessary to chemically alter the various proportions of the natural oil fractions. This is called *reforming* and is accomplished by the use of various catalysts. The main reactions are:

(a) the dehydrogenation of cyclic paraffins to aromatic hydrocarbons and H_2
(b) isomerization of normal paraffins to branched paraffins (alkylation)
(c) hydrocracking of paraffins and cycloparaffins to short chain alkanes
(d) visbreaking reduces the viscosity of the oil by cracking the heavy components
(e) polymerization is the formation of large molecules from the catalytic recombination of two or more smaller ones.

These reaction are designed to provide the fuels required and the chemicals needed for the petrochemical industry. Other reactions which are important in reformulation is the removal of sulfur from the oil as H_2S by catalytic reactions of Co and Mo on Al_2O_3. Nitrogen forms NH_3 on Ni/Mo catalysts.

3.5. PETROLEUM PRODUCTS

The principal consumer of petroleum products is the transportation industry with the Internal Combustion Engine (ICE) the major application. This is shown in Table 3.4. The ICE includes the spark ignition and diesel engines, common for the automotive vehicle, and the gas turbine used for aviation and industrial applications. Each class of engine requires a special type of fuel. The heats of combustion of the fuels may not be significantly different but the composition of the components and their rates of reactions with oxygen could determine its application. The characteristics of gasoline and diesel fuels are discussed in Chapter 4 and lubrication oils and greases are discussed in Chapter 8. Aviation fuel has not changed much since its early use in the jet engine. The stability of the fuel has improved, and specific corrosion inhibitors, antioxidants, and anti-static additives have been included in formulations. Combustion in a gas turbine is continuous in contrast to the intermittent combustion of the ICE and thus the "anti-knock" quality of the fuel is not important. The thrust of the engine is generated by the energy released during the combustion process which takes place at

TABLE 3.4
Consumption of Petroleum Products in Canada (in thousands m^3/day)

	1983	1990	1994
Motor gasoline	91.0	103.5	109.1
Aviation turbo fuel	10.7	15.7	13.0
Diesel fuel	38.4	46.2	49.8
Light fuel oil	23.5	17.6	15.5
Heavy fuel oil	25.2	29.1	19.8
Lubricating oil	2.6	2.5	2.4
Lubricating grease	0.10	0.083	0.062
Total domestic demand	236.1	267.3	271.3

constant pressure. The heat generated thus causes an increase in volume and gas flow to the turbine blades. An important restriction is the material of construction which must withstand the high temperatures developed.

The jet fuel must have a low tendency to deposit carbon which can lead to high thermal radiation causing damage to the engine components. Likewise, sulfur in fuel must be very low since it enhances carbon deposition and corrosion. These factors and the physical properties require the fuel to have a low freezing point and be free of wax which at low temperatures can precipitate out and clog fuel lines. At high altitudes temperatures as low as $-50°C$ are normal and $-60°C$ is the specified maximum freezing point for aviation fuel. Similarly water must be absent. Anti-icing additives are present in the fuel to ensure that at low temperatures any water which does come out of the fuel will not form large ice crystals which can clog the fuel lines.

Aviation fuel (avogas), which is designed for use in piston engines, still contains lead [0.53 mL of tetraethyl lead (TEL) per liter of fuel]. One hazard, tolerable on the ground but deadly in the air, is the vapor lock, and special care is taken to ensure that this does not happen. Hence automotive fuel, though often equivalent to avogas but half the cost, cannot be used as a substitute.

Oil seepages on the surface were the early sources of crude oil. Kerosene was initially used primarily for lamps. Today it is generally used for heating as space heaters, where a catalytic surface maintains a hot radiating source of heat. In countries where electricity is expensive, e.g., Japan and Israel, it is not uncommon to see such appliance heaters in homes and offices. Such open flames are sufficiently hot to produce NO_x as well as some hydrocarbon products which are readily detected by their odor. The main characteristics of a domestic grade of kerosene, which burns on a catalytic surface, is low sulfur, and controlled smoke point and volatility.

Coal for heating was replaced by oil during the 1940s and oil was replaced by natural gas soon after. The natural gas pipeline however cannot reach every home and, as a result, fuel oil as well as bottled gas (propane) is still used in many locations.The fuel oil differs only slightly from diesel fuel and because its ignition property is not important, it usually contains more aromatics and olefins than diesel fuel. The most important property of fuel oil is its storage stability, and additives prevent sludge formation. Viscosity is another controlled characteristic of the fuel because the size of the droplets in the burning spray is critical for efficient combustion.

World production of petroleum wax is about 2 Mtonnes per year. The major product is candles (about 25%), replacing beeswax and tallow used a century ago. The present applications include coating on cartons for juice, milk and other food products, and for water-proofing and preserving freshness. The major concern with the use of wax or any petroleum product for food and edibles is the need to have it completely free of the Polynuclear Aromatic Hydrocarbons (PAH). Trace amounts have been detected in liquid petroleum used for oral consumption. However, our daily diet usually contains much more PAH than what is in a spoonful of liquid petroleum taken occasionally. Bitumen or asphalt is the black or dark brown solid residues from the distillation of crude oil. Naturally occurring bitumen is found in USA, Europe and other parts of the world. Commercial bitumen comes in various degrees of hardness and softens at a temperature of 25–135°C. When air is blown into molten bitumen, oxidation or dehydrogenation occurs and the material becomes rubbery and more penetrating than the regular grades. Bitumens can be applied as a melt, emulsion or from a solution. The application of bitumen to roads usually involves the inclusion of sand and gravel and in North America this is normally referred to as asphalt.

The inclusion of other aggregates which have been tested and added for various road systems include sulfur, shredded discarded tires, limestone (<200 mesh), and crushed concrete. Other applications of bitumen are in roofing, flooring, and as an anticorrosion coating on surfaces exposed to corrosive atmosphere, aggressive soils, and chemicals. The main advantage of bitumen, in many of its applications, is that it is cheaper than alternative materials. The world production of bitumen in 1993 was about 17 Mtonnes.

The reserves of oil are limited and its uses as a fuel for its heat value may eventually become a luxury which few will be able to afford. The petrochemical industry supplies the plastics and resins we use daily, the synthetic fibre for our clothes, and the detergents for our soaps and washings as well as the chemicals and solvents for industrial use. World petrochemicals amount to about 1/7 of total steel production and about 7 times the aluminum produced by weight.

3.6. SYNTHETIC OIL

It is possible to produce oil from coal either by direct hydrogenation at high temperatures and high pressures or by the syngas route followed by the Fischer-Tropsch process. Though these processes can make an oil which is more expensive than well-head crude, in exceptional strategic circumstances (e.g., war, oil embargo, etc.) it has been produced and used as a viable substitute.

The first process was studied by Berthelot in 1867 and was further developed in Germany by Bergius in 1910. The early Bergius process involved the reaction of H_2 under atmospheric pressure with pulverized coal suspended in an oil heated to about 450°C in the presence of a catalyst such as stannous formate or Mo. The liquid oil product is separated from the solid residue and processed as ordinary crude oil. Modern developments in this coal liquefaction approach include: (1) Exxon Donner Solvent (EDS) process, (2) the HRI H-Coal process, and (3) the Gulf Solvent Refined Coal SRC-II process. The major improvement of these processes over the Bergius process is in the catalyst used, allowing for milder reaction conditions.

TABLE 3.5
Some Potential Donor Components in Solvent

Compound	Structure	Compound	Structure
Tetralin	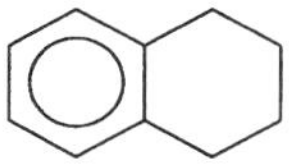	Tetrahydrophenanthrene	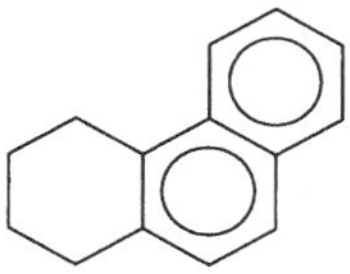
Tetrahydroaccnaphthene	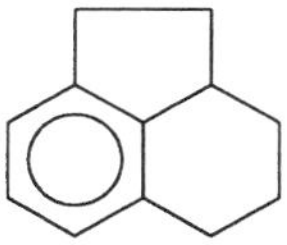	Octahydrophenanthrene	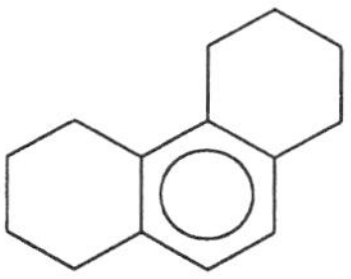
Hexahydrofluorene	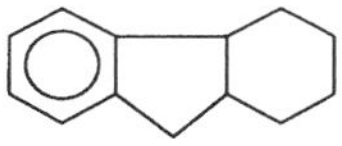	Hexahydrophyrene	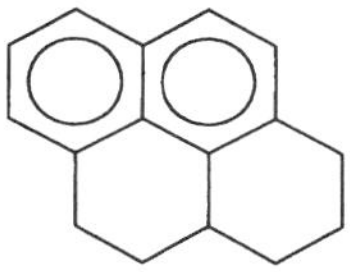

The EDS system is a direct liquefaction procedure in which coal is chemically reacted at about 540°C for 30–100 min with a recycle solvent that is rehydrogenated between passes to the liquefaction reactor. Each tonne of dry coal can give up to 3 bbl of synthetic crude oil. Some examples of donor solvents are given in Table 3.5. A schematic block diagram of the overall process is shown in Fig. 3.6. A pilot plant processing 225 tonnes/day has been in operation since June 1980 and the product yields obtained are given in Table 3.6. The advantage of the EDS process is that it makes maximum use of petroleum refining technology, uses a wide range of coals, and has no catalyst to be poisoned or recycled.

TABLE 3.6
EDS Products and Their Disposition

Stream	Disposition	Yield (% of liquids)
C_4–177°C naphtha	Motor gasoline	40–80
	Chemicals	
177–204°C distillate	Jet fuel	
	Turbine fuel	10–30
204–399°C solvent	Home heating oil	
	Fuel oil blendstock	
399–538°C vacuum gas oil	Fuel oil blendstock	
Coker gas oil		10–30
Scrubber liquids	Fuel oil blendstock	

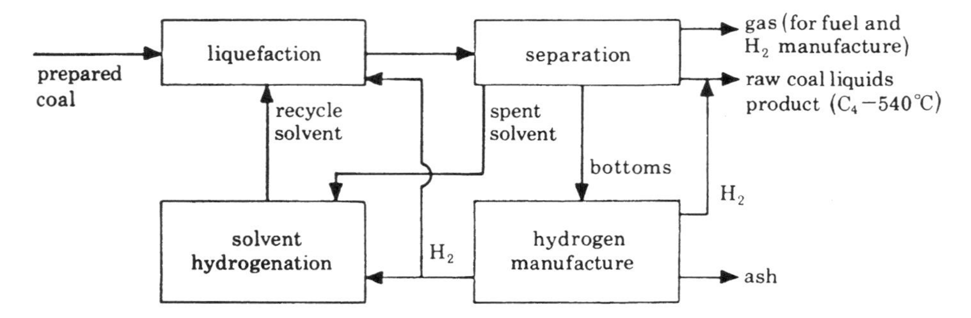

FIGURE 3.6. Simplified block diagram of the EDS process.

The H-coal process was developed by Hydrocarbon Research Inc. and uses a catalytic ebullated reactor. The pulverized coal-oil slurry is mixed with hydrogen (150 atm) fed into the reactor (450°C) where the Co/Mo catalyst converts the coal and hydrogen to oil.

The SRC-II process, like the EDS process, is claimed to be free of catalysts and the problems associated with their use. A schematic diagram of the process is shown in Fig. 3.7. The coal is dissolved in a process-derived solvent at elevated temperatures and reacts with hydrogen under pressure to form the oil. The solvent is recovered by vacuum distillation and the residue (ash plus insoluble carbon) is separated.

The second method by which synthetic oil can be made is by the Fischer-Tropsch process. This can use either coal or natural gas as the starting material since the process uses synthesis gas (a mixture of CO and H_2) which can be made from coal by the water gas reaction.

$$C + H_2O \rightarrow CO + H_2 \qquad \Delta H^0 = 131.28 \text{ kJ/mol} \tag{3.1}$$

or natural gas or other hydrocarbon blends by steam reforming

$$CH_4 + H_2O \rightarrow CO + 3H_2 \qquad \Delta H^0 = 206.1 \text{ kJ/mol} \tag{3.2}$$

The syngas is then balanced for a preferred H_2/CO ratio of 2 since the reaction is

$$2H_2 + CO \rightarrow (CH_2)_x + H_2O \tag{3.3}$$

The reaction occurs on various catalysts such as iron oxide which is the most commonly used material. Some oxygenated compounds such as alcohols and acids (1%) are also formed.

The Fischer-Tropsch process involves the following steps:

1 Manufacture of syngas.
2 Gas purification-removal of S, H_2O, etc., and H/C balance.
3 Synthesis of oil on fresh catalyst.
4 Condensation of liquids and removal of light fractions (gasoline).
5 Distillation of remaining products: 35% gas–oil and 30% paraffins.

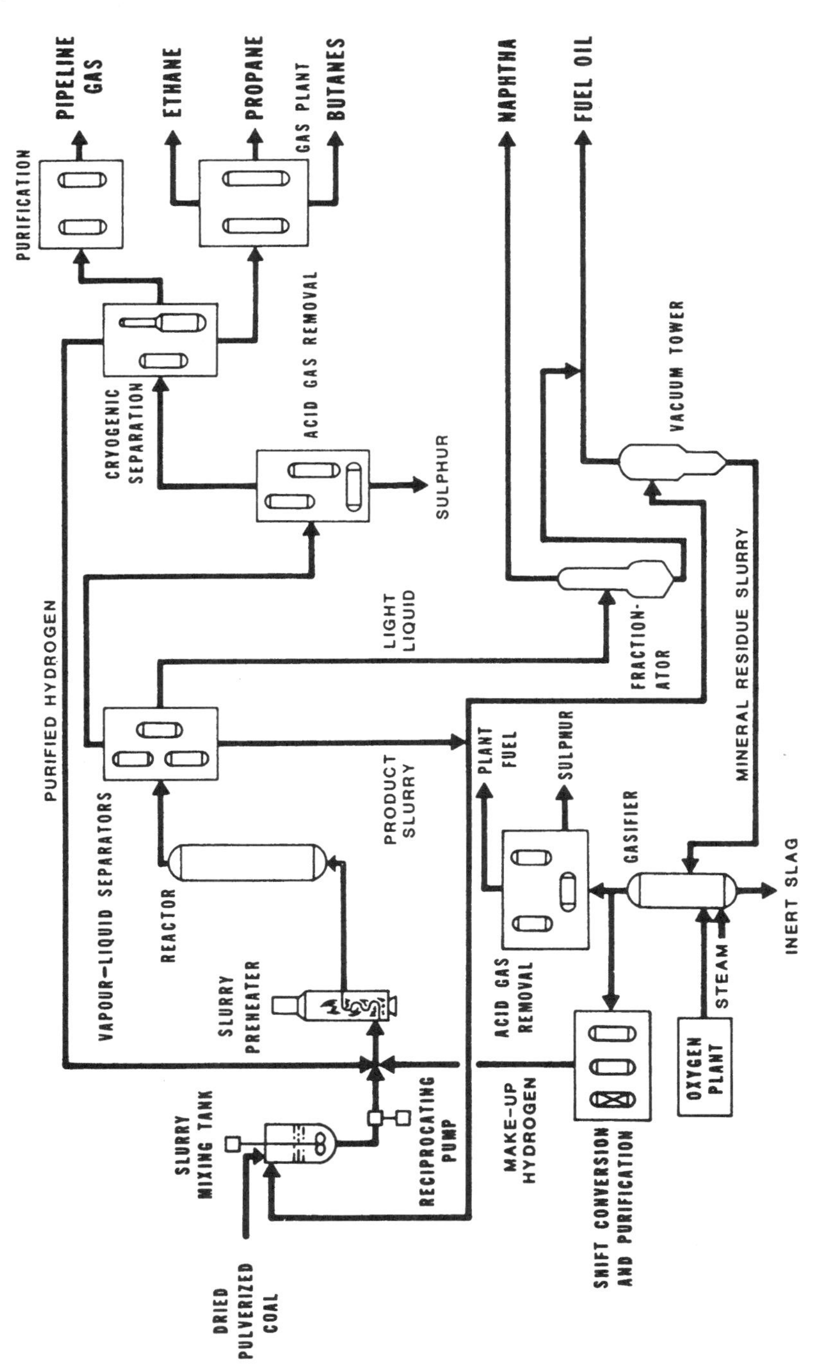

FIGURE 3.7. The SRC-II process.

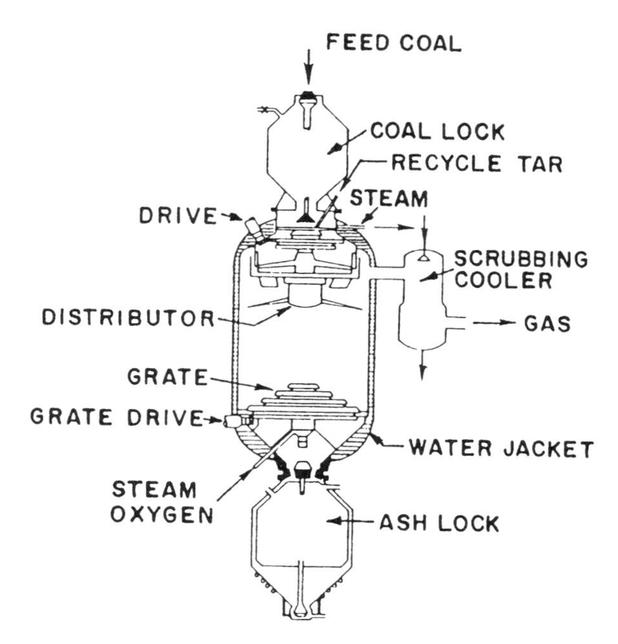

FIGURE 3.8. A Lurgi dry ash gasifier.

The gasoline is primarily straight chain hydrocarbons of low octane rating and must therefore be reformulated for use.

The only commercial plant converting coal into oil is in South Africa and known as the Sasol I and II. Sasol I was built in 1956 and produced 10,000 bbl/day of synthetic crude oil using fixed bed (with precipitated iron catalyst) and fluidized bed (with powdered iron catalyst) reactors. A typical gasifier is shown in Fig. 3.8.

The construction of Sasol II plant was started in 1976 and completed in 1980, it cost $\$3 \times 10^9$ and produced 40,000 bbl/day oil from 36 gasifiers each of which processed 1200 tonnes/day of coal. The third plant, Sasol III was designed to be the most advanced conversion plant in the world, it cost $\$5.5 \times 10^9$ and is set to produce slightly more than Sasol II. Total synthetic crude oil production was expected to be 120,000 bbl/day and represented over half of the country's needs. A simplified flow sheet of Sasol II and III is shown in Fig. 3.9. A photograph of Sasol II is shown in Fig. 3.10.

It should be noted that the production cost of the synthetic crude oil is higher than the world price for well-head crude but the value of independence cannot be assigned a simple price.

EXERCISES

1. The populations of USA, Canada, and Japan in 1992 were approx. 250 M, 28 M, and 124 M, respectively. Calculate the annual per capita oil consumption for each country (see Table 3.2) and comment on the differences.

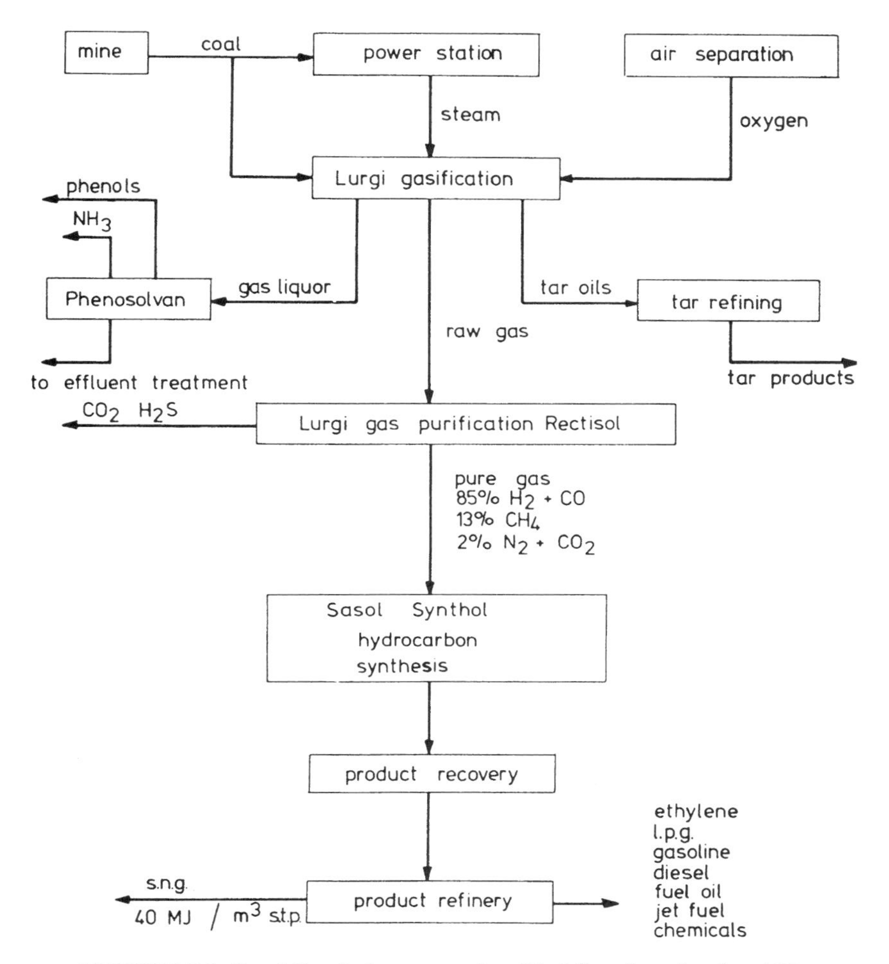

FIGURE 3.9. Sasol Synthol process: simplified flowsheet for Sasol II.

2. Calculate the time required to exhaust the world's oil reserves if the 6 billion people on earth all consume oil at the rate of 28 bbl each per year.
3. Explain the shape of the theoretical depletion curve of world oil reserves. (Fig. 3A).
4. Discuss the problems associated with the use of waste glass as an aggregate in asphalt for roads.
5. Write the reaction for the syngas production from ethane.
6. The reaction $CO + H_2O \rightarrow H_2 + CO_2$ is called the water-shift reaction and is often used in the Fischer-Tropsch process. Explain why this reaction is used in this process.
7. Why is the direct liquefaction of coal with hydrogen to form oil a more sensible process than the production of crude oil via the Fisher-Tropsch process for the production of synthetic fuel?
8. Estimate the standard heat of reaction for reaction 3.3.

FIGURE 3.10. General view of Sasol II at Secunda in the Eastern Transvaal.

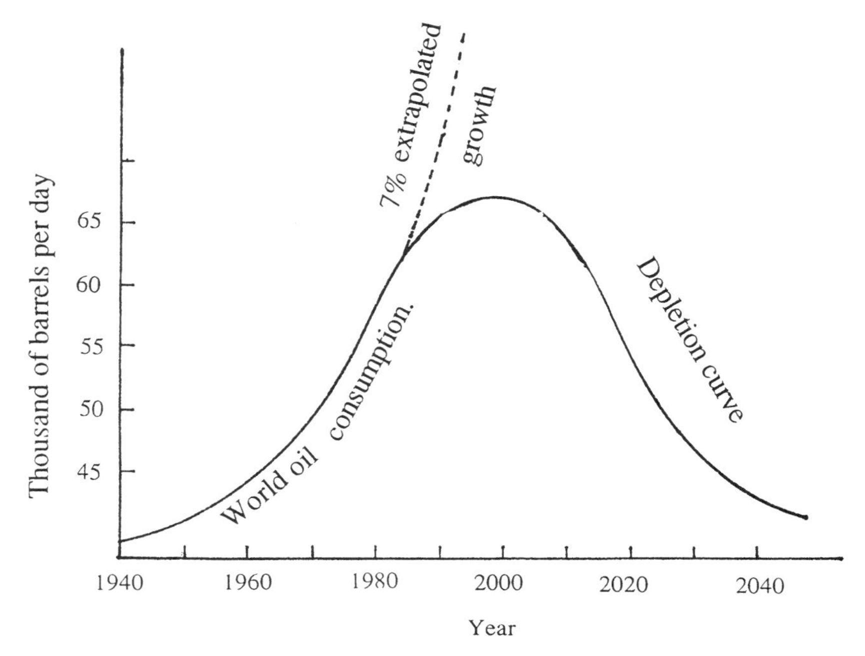

FIGURE 3.A. World consumption of oil and theoretical depletion curve for world oil reserves.

FURTHER READING

N. Berkowitz, *Fossil Hydrocarbons*, Academic Press, San Diego, California (1997).
H. Kopsch, *Thermal Methods in Petroleum Analysis*, VCH Pub., New York (1996).
P. H. Ogden, Editor, *Recent Advances in Oilfield Chemistry*, CRC Press., Boca Raton, Florida (1995).
S. Matar and L. F. Hatch, *Chemistry of Petrochemical Processes*, Gulf Pulb. Co. (1994).
D. Liebsen, *Petroleum Pipeline Encyclopedia*, OPRI, Boulder, Colorado (1993).
J. L. Kennedy, *Oil and Gas Pipeline Fundamentals*, 2nd Ed., Penn Well Bks, Tulsa, Oklahoma (1993).
J. G. Speight, *The Chemistry and Technology of Petroleum*, 2nd Ed., M. Dekker Inc., New York (1991).
C. J. Campbell, *The Golden Century of Oil, 1950–2050: The Depletion of a Resource*, Kluwer Academic, Norwall, Massachusetts (1991).
Petroleum Handbook. (Shell), 6th Ed., Elsevier, New York (1983).
M. B. Hocking, The Chemistry of Oil Recovery from Bituminous Sands, *J. Chem. Educ.* **54**, 725 (1977).
N. Berkowitz and J. G. Speight, The Oil Sands of Alberta, *Fuel* **54**, 138 (1975).
Oil and Gas News and Prices, http://www.petroleumplace.com/
Canadian Association of Petroleum Producers, http://www.capp.ca/
Alberta Syncrude, http://www.syncrude.ca
American Petroleum Institute, http://www.api.org/
Institute of Petroleum, http://www.petroleum.co.uk/
World Petroleum Congress, http://www.world-petroleum.org/
Organization of Petroleum Exporting Countries (OPEC), http://www.opec.org/
National Petroleum Council, http://www.npc.org/

4

Liquid Fuels

4.1. INTRODUCTION

Liquid fuels are a major energy factor which determines the course of transportation and the automobile is one of the most important consumers of such fuel. These liquid fuels include diesel oil, gasoline, liquid propane, alcohol (both methyl and ethyl) as well as the less common liquids such as ammonia and hydrazine. Though the boiling point of propane (−42°C) and ammonia (−33°C) are below ambient temperature (in most places) these substances are still classed as liquids because they can be stored as liquids at room temperature (25°C) at the modest pressure of about 10 atm.

Whale oil was used for lighting and heating long before the extensive use of petroleum oil. Similarly, vegetable oils (or biomass liquids and saps) are also potential fuels (see Chapter 1) but their modest production at present precludes their wide spread use. Perhaps when the energy farm has developed it will be possible to consider biomass fuels as an alternative to fossil-based fuels.

Of the various liquid fuels, diesel oil requires the minimum of preparation from petroleum crude oil and therefore is usually the cheapest of the fuels.

4.2. DIESEL ENGINE

The diesel engine was described in a patent in 1892 by Dr. Rudolf Diesel who originally designed it to operate using coal dust as the fuel. However, the characteristics of such fuel were not very reproducible and it was quickly replaced by oil.

In the diesel engine, air is pulled into the cylinder and compressed to approx. 35 atm. This compression is effectively adiabatic and causes the temperature of the air to increase to about 550°C. At the end of the compression stroke, when the piston is at the top of the cylinder [Top Dead Center (TDC)] an oil spray is injected into the hot air where it ignites on vaporization. The heat of combustion raises the temperature of the gas mixture which now expands at constant pressure as the piston moves down, increasing the volume of the gases. The heat generated and the larger volume of product gases thus further increases the volume as the pressure drops.

An ideal PV diagram for a diesel engine is shown in Fig. 4.1 where the 4 strokes are:

1 Compression $1 \rightarrow 2$
2 Expansion/combustion $2 \rightarrow 3 \rightarrow 4$
3 Exhaust $4 \rightarrow 1 \rightarrow 5$
4 Air intake $5 \rightarrow 1$

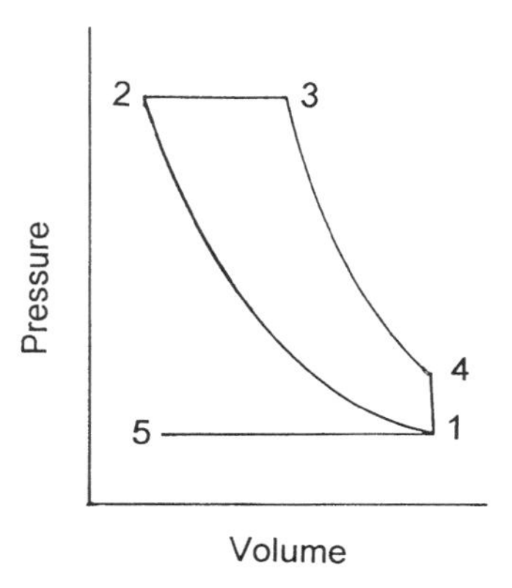

FIGURE 4.1. Ideal PV diagram of a 4-stroke diesel engine. (1) Compression $1 \rightarrow 2$. (2) Expansion/combustion $2 \rightarrow 3 \rightarrow 4$. (3) Exhaust $4 \rightarrow 1 \rightarrow 5$. (4) Air intake $5 \rightarrow 1$.

In the compression stage air is compressed to about 1/20 of its initial volume, i.e., the engine has a compression ratio (CR) of at least 20 and this high ratio accounts for the high efficiency of the engine.

Diesel engines are classified into indirect and direct injection engines. The former have a precombustion chamber where the fuel is initially injected and where combustion starts after an induction delay. The fuel rich flame then expands into the main chamber. Such engines are common in high speed diesel passenger cars.

Direct injection engines are normally used in larger engines with lower speed and with cylinder bores greater than 12 cm in diameter. They are more efficient and easier to start than indirect injection engines of comparable size.

Diesel engines are also classified according to their speed (rpm). Low-speed engines run at 100–500 rpm and are usually large stationary installations or marine engines. Medium-speed engines operate from 500–1200 rpm and are usually installed in power generators, power shovels, tractors, and locomotives. The diesel–electric locomotive runs at about 20–28% efficiency, compared to the coal fired steam engine which was only 5–8% efficient, or the electric powered engine which is about 23% efficient. The high-speed diesel engines run at 1200–2000 rpm and are found in the automobile, light trucks and buses, and aircraft. The type of fuel used in a diesel engine is determined, to a great extent, by the speed of the engine.

4.3. DIESEL FUEL

The diesel fuel injected into the hot compressed air ignites only after a short delay. This ignition delay depends on the composition of the fuel. It is longer for aromatic hydrocarbons and cycloparaffins than for olefinic and paraffinic fuels. Thus the best grade of diesel fuel consists of long straight chain hydrocarbons which can spontaneously ignite in the hot compressed air. The absence of an induction period is important since it results in a loss of efficiency. Long ignition delay times result in rapid combustion and sharp pressure rise, causing the engine to run roughly. The grade of diesel fuel is determined by an empirical scale and is based on the ignition characteristics of the fuel, which is compared with a blend of hexadecane (cetane) rated as 100 and 1-methyl naphthalene (α-methyl naphthalene) which is rated at 0. The recent use of 2,2,4,4,6,8,8-heptamethyl nonane (HMN), CN = 15, has been introduced as a low cetane standard because it can be obtained easily in high purity. Low grade fuels have a cetane number (CN) equal to about 20 which is suitable for low speed engines whereas a high grade fuel will have CN = 70. High speed engines require a fuel with CN = 50 or more. Table 4.1 shows some of the performance characteristics of diesel fuel.

A crude indication of the CN of a fuel can be obtained from its density. A plot of CN against density for various grades of fuel is shown in Fig. 4.2. In general the CN is reduced by the presence of aromatics in the fuel.

Corrosion inhibitors and gum reducers are other additives designed to improve the quality of a fuel. In cold weather it is customary to add ethanol to the fuel (not more than 1.24 mL/L) to prevent ice from clogging fuel lines.

The addition of barium compounds to diesel fuel has been shown to reduce the emission of black smoke. However, its use is limited, and proper engine maintenance can do much to reduce smoke in the exhaust.

Additives can be incorporated to improve the quality of diesel fuel. For example, ethyl nitrite, ethyl nitrate, and isoamylnitrite when added to an oil will increase its CN value. The addition of 2.5% by volume of amylnitrite to a diesel oil (CN = 26) increased the CN to 44. Additives such as amylnitrate ($C_5H_{11}ONO_2$) when added at about 0.1% by volume will increase the CN by 4 and 0.25% will add 7 to the CN of the fuel. Other nitrates such as heptyl and octyl nitrates are also ignition improvers. A similar effect is obtained when ammonium nitrate is added to oil. A 2% by weight addition of a solution of 5 M NH_4NO_3 in water can increase the CN of a diesel fuel from 39 to 42. Other additives will prevent gum formation, decrease surface tension permitting a finer spray, or reduce the change in fuel properties due to changes in temperature.

Diesel fuel is a complex mixture of hydrocarbons which includes paraffins, naphthenes, and aromatics, and at low temperatures phase separation can occur causing engine failure. The temperature, Tps, at which phase separation occurs is called the *cloud point* and is determined under standard conditions of cooling, e.g., 1°C/min. Long chain polymer additives at as low as 0.1% can lower the cloud point, Tps, by several degrees.

The recommended cloud point of a fuel is 6°C above the pour point, which is the temperature at which the fuel ceases to flow readily. A fuel is normally blended so as to make the pour point at least 6°C below normal driving temperatures. The ASTM (D-975-78) specification of diesel fuel oils is given in Table 4.2.

Vegetable oils are composed primarily of glycol esters of fatty acids with the general formula

$$
\begin{array}{l}
\overset{\displaystyle O}{\overset{\|}{C}}-R' \\[-4pt]
H_2C-O-\\
\;\;|\\
\;HC-O-C-R''\\
\;\;|\qquad\quad\;\; \| \\
\;\;|\qquad\quad\;\; O\\
H_2C-O-C-R\\
\qquad\qquad\;\;\|\\
\qquad\qquad\;\;O
\end{array}
$$

TABLE 4.1
Comparison of Components in Diesel Fuel and Performance Characteristics

	Conventional diesel fuel (Wt.%)	Cracked gas oil (Wt.%)	Synthetic diesel fuel (Wt.%)	Ignition quality	Cold flow character	ΔH comb by volume	Density	Smoking tendency
n-Paraffins	39	19	17	Good	Poor	Low	Low	Low
							Low	Low
iso-Paraffins				Low	Good	Low		
Naphthenes	34	16	37	Moderate	Good	Moderate	Moderate	Moderate
Olefins				Low	Good	Low	Low	Moderate
Alkyl benzenes	18	34	36	Poor	Moderate	High	High	High
2-Ring aromatics	8	28	8	Poor	Moderate	High	High	High
3-Ring aromatics	1	3	2	Poor	Moderate	High	High	High

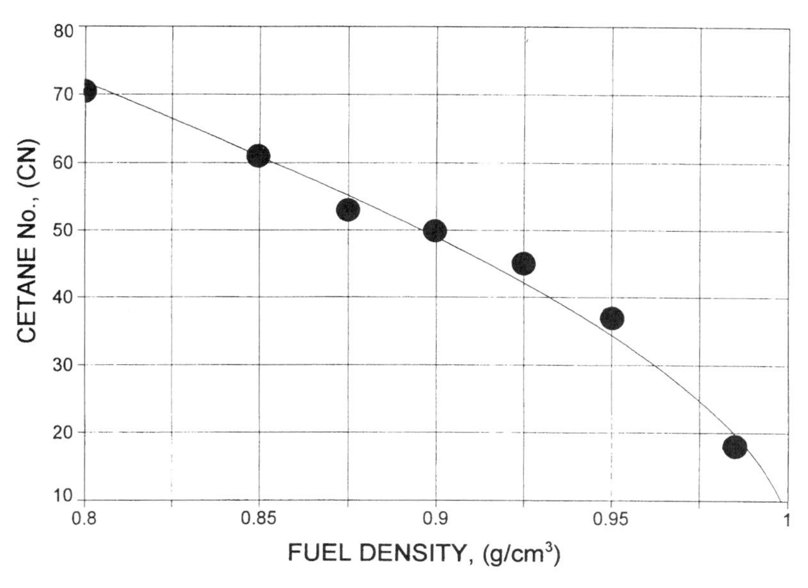

FIGURE 4.2. General relationship between cetane number (CN) and density of the fuel.

where R′, R″, and R are the same or different alkane or alkene radicals with carbon chains from 16 to about 20. The triglycerides are too viscous and of too high a molecular mass and too low a vapor pressure to be a useful diesel fuel. However, it is possible to convert the triglycerides to monomethyl esters by a transesterification process which can be represented by the equation

$$\begin{array}{l} H_2C{-}O{-}\overset{\displaystyle O}{\overset{\|}{C}}{-}R' \\ \;| \\ HC{-}O{-}\underset{\displaystyle O}{\underset{\|}{C}}{-}R'' \\ \;| \\ H_2C{-}O{-}\underset{\displaystyle O}{\underset{\|}{C}}{-}R \end{array} + 3CH_3OH \xrightarrow{H^+} \begin{array}{l} H_2C{-}OH \\ \;| \\ HC{-}OH \\ \;| \\ H_2C{-}OH \end{array} + 3R{-}\overset{\displaystyle O}{\overset{\|}{C}}{-}O{-}CH_3 \qquad (4.1)$$

The methyl esters are usually straight chain hydrocarbons and have relatively high cetane values shown in Table 4.3. The ethyl esters are between 2 and 5 units higher than the corresponding methyl esters. Such diesel fuel, though effective, is still too expensive and not competitive with petroleum based fuels.

4.4. IGNITION TEMPERATURE, FLASH POINT, FIRE POINT, AND SMOKE POINT

Four important tests which are used to characterize an engine fuel are the spontaneous ignition temperature (SIT), flash point, fire point, and smoke point. These

TABLE 4.2
North American Specification for Diesel Fuel (average of US and Canadian)

	Type AA	Type A	Type B	ASTM 1-D	ASTM 2-D	ASTM 4-D
Flash point °C min.	40	40	40	38	52	54
Cloud point °C max.	−48	−34	—	—	—	—
Pour point °C max.	−51	−39	—	—	—	—
Kinematic viscosity 40°C cSt	min 1.2 to	1.3	1.4	1.3	1.9	5.8
	max—	4.1	4.1	2.4	4.1	26.4
Distallation						
90% recovered °C max.	290	315	360	288	282[a] min to 338 max	
Water and sediment, % vol. max.	0.05	0.05	0.05	0.05	0.05	0.5
Total acid number, max.	0.10	0.10	0.10	—	—	—
Sulfur, % mass max.	0.2	0.5	0.7	0.5[a]	0.5[a]	2
Corrosion, 3hr @ 100°C max.	No. 1	No. 1	No. 1	No. 3	No. 3	—
Carbon residue (Ramsbottom) on 10%[a] bottoms, % mass max.	0.15	0.15	0.20	0.15	0.35	—
Ash, %Wt., max.	0.01	0.01	0.01	0.01	0.01	0.1
Ignition quality, CN, min.	40	40	40	40[a]	40[a]	30

[a]Specifications provide for modification of these requirements appropriate for individual situations.

tests are standardized and specalized fuels have specific requirements as defined by these tests. The SIT is dependent on the composition of the fuel and the conditions of the walls of the cylinder. Diesel fuels require low SIT with short delay times of the order of 1–2 msec. The SIT of heptane (CN = 60) is 330°C whereas benzene with CN = −10 as a SIT of 420°C.

The flash point of a fuel is obtained by slowly increasing the temperature (5.6°C/min) of the liquid fuel in a standard container (flash cup) until sufficient vapor is given off to produce a flash as a flame is passed over the mouth of the cup every 30 sec. The temperature of the oil at which this occurs is the flash point. This is an index of the volatility of the oil or liquid. It is used as an indication of the fire hazard of combustible liquids. For example, the Canadian specification for heating fuel oil

TABLE 4.3
Experimental CNs for Selected Transesterified Vegetable Oils (Methyl esters)

Oil	CN
Babasu	63
Palm	62
Peanut	54
Soybean	45
Sunflower	49

TABLE 4.4
Boiling Point (BP), Flash Point (FP), Spontaneous Ignition Temperature (SIT), and Smoke Point (SP) of Selected Substances

Substance	BP (°C)	FP (°C)	SIT[a] (°C)	SP (mm)
Acetone	56.5	−18		
Benzene	80	−11	560	8
n-Pentane	36	−49	308	150
n-Hexane	69	−23	247	149
n-Heptane	98	−4	447	147
iso-Pentane	28	−51	420	~135
Neopentane	9.5	—	440	~140
1-Pentane	30	−28	298	84
Toluene	110	4	536	6
p-Xylene	138	27	464	5
PCB	380 ± 10	222	—	—
Kerosene	260 ± 10	54	—	27
Diesel 2-D	—	52	—	—
10-W-30	—	226	—	—

[a]In air.

stipulates a minimum flash point of 43°C whereas the flash point of diesel fuel varies from 38 to 52°C. The flash point and boiling point of various substances are compared in Table 4.4.

The fire point is the temperature to which the oil must be heated so that the vapor pressure is sufficient to maintain the flame after the flame source is removed.

The smoke point of a fuel is an arbitrary scale related to the height of a flame of the fuel burning in a standard lamp without smoking. Some values are given in Table 4.4 and show that aromatic hydrocarbons have low smoke point values whereas saturated normal alkanes have the highest smoke points. Additives such as ferrocene $[Fe(C_5H_5)_2]$ increase the smoke point. Fuels that have low smoke points tend to deposit carbon during the combustion process.

Water emulsified in diesel fuel can reduce smoke in the exhaust and carbon deposits in the engine. It was shown that, while 5% or less water increased the fuel consumption, more water showed an increase in the thermal efficiency, the maximum increase being 9.7% when the fuel contained 22.5% water.

The internal combustion engine is a notorious polluter but the diesel engine has the additional emission of particulate matter (10–25 $\mu g/m^3$) which is rich in polynuclear aromatic hydrocarbons such as phenanthrene, fluoranthene, benzo(a)pyrene, and benzoperylene—all of which are carcinogenic.

4.5. THE SPARK IGNITION INTERNAL COMBUSTION ENGINE

The four-stroke cycle spark ignition (SI) internal combustion engine (ICE) was initially proposed by Beau de Rockas in 1862 and first built by N. A. Otto in 1876. This engine has become the major piston engine in use today. The PV cycle of the

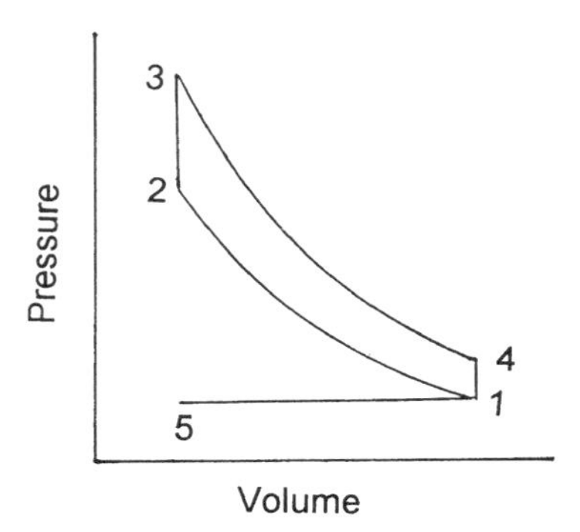

FIGURE 4.3. Ideal PV diagram of the 4-stroke spark ignition ICE. (1a) Compression 1 → 2. (1b) Spark ignition, constant volume combustion 2 → 3. (2a) Expansion of product gases 3 → 4. (2b) Exhaust at constant value 4 → 1. (3) Exhaust of cylinder 1 → 5. (4) Induction stroke air/fuel intake 5 → 1.

engine is shown in Fig. 4.3 where the 4 strokes are indicated.

1a	Compression of the air/fuel mixture 1 → 2
1b	Spark ignitions, constant volume combustion 2 → 3
2a	Expansion of product gases 3 → 4
2b	Exhaust at constant volume 4 → 1
3	Exhaust of cylinder 1 → 5
4	Induction stroke air/fuel intake 5 → 1

The efficiency of the engine is a maximum of about 25% at a CR of about 15. Fuel injection, as in the diesel engine, means better control of the air/fuel ratio under variable temperature conditions.

The use of fuel injection with the 2-stroke SI engine has added a new dimension to small and efficient engines. The major advantages over the 4-stroke engine are lower cost, and low weight/power ratio. These engines have been primarily used in motorcycles, outboard motors, lawn mowers, chain saws, and similar lightweight engines. Similarly the rotary 2-stroke Wankel engine has reached the production stage in the Mazda car but further extension of its use has not materialized. Two-cycle diesel engines have also been produced.

4.6. GASOLINE FUEL

The fuel for the SI–ICE was in the early years centered around alcohol. In 1895 Nikolaus Otto recommended that alcohol be used in his engine. One of Henry Ford's first models, the quadaicycle, was meant to run on alcohol. His later model "T" was designed to run on either alcohol or gasoline requiring only a simple adjustment of the carburetor. Efforts in 1906 to extend the use of alcohol in the USA as a fuel for the automobile by rescinding the 40 cent/gal. liquor tax failed. A 50/50 mixture of alcohol gasoline was used in France after World War I due to the oil shortage. The Great Depression of the 1930s made farm-produced alcohol a cheap fuel which was available in 2000 midwestern service stations.

However, the cheaper oil eventually predominated and gasoline has become the common fuel for the automobile. The quest for greater power and efficiency has resulted in an increase in the CR from about 5 in 1928 to over 13 in 1995. This has resulted in the need to reformulate the fuel to meet the stringent demands of the newer engines.

The problem in the pre-1920 years was the tendency of the engine to "knock"—a violent explosion in the cylinder which, at times, cracked the cylinder head and pistons. This knock was shown to occur after ignition by the spark and it was believed that the delayed vaporization of the fuel droplets caused the post-ignition explosion. Thomas Midgley Jr., a mechanical engineer working for Dayton Engineering Laboratory Co. (Delco), decided to add a colored component to the fuel and selected iodine. This had a beneficial effect and eliminated the engine knock. When several red dyes were tested they showed no improvement whereas ethyl iodide reduced the knock, proving that the iodine was the effective element. However, iodine and its compounds were too expensive and corroded the engine, and so a search was on for a simple alternative. By 1919 the research team found that 2 mL/L of aniline was better than 1 g of iodine. On December 9, 1921 after testing about 33,000 compounds they found that 0.025% tetraethyl lead was a superior antiknock substance when compared to the 1.3% aniline which was the comparison standard. Thus, a smooth running fuel became a reality.

One difficulty caused by the added lead was the buildup of the yellow lead oxide (PbO) in the engine which coated the spark plugs and valves. This problem was solved by adding ethylene dibromide to the fuel, which converted the lead to the more volatile lead bromide ($PbBr_2$). The first sale of "ethyl" gas was on February 1, 1923. A list of antiknock additives is shown in Table 4.5. The present interpretation of knock has recently been reached by photographing the combustion process through a glass-top cylinder and observing the smooth propagation of the flame front from the spark to the farthest part of the combustion chamber. The gas to be burned last is called the end gas and is usually located in the combustion chamber furthest from the spark plug. This end gas is heated by compression, by the combustion taking place and by the approaching flame front. The spontaneous ignition of this end gas results in an explosion and knock unless the normal flame reacts and consumes the end gas before it ignites. This is shown in Fig. 4.4.

Thus, antiknock additives are inhibitors of autoignition. The lead oxide aerosol which forms in the combustion is a free radical trap and prevents the chain reaction from branching and progressing to explosive rates.

4.7. GRADING GASOLINE

As the automobile engine developed it became necessary to grade the gasoline on a realistic scale. This was established by determining the behavior of various organic substances as a fuel in a standard engine and observing the onset of knock as the CR is increased. Figure 4.5 shows the effect of structure on the critical CR for the various isomers of heptane. The best isomer—trimethyl butane (called *Triptane*)—proved to

TABLE 4.5
Relative Effectiveness of Antiknock Compounds and Some Antiknock Fuels (Based on Aniline = 1)

Benzene	0.085
Isooctane (2,2,4-trimethylpentane)	0.085
Triphenylamine	0.090
Ethyl alcohol	0.101
Xylene	0.142
Dimethyl aniline	0.21
Diethylamine	0.495
Aniline	1.00
Ethyl iodide	1.09
Toluidine	1.22
Cadmium dimethyl	1.24
m-Xylidine	1.40
Triphenylarsine	1.60
Titanium tetrachloride	3.2
Tin tetraethyl	4.0
Stannic chloride	4.1
Diethyl selenide	6.9
Bismuth triethyl	23.8
Diethyl telluride	26.6
Nickel carbonyl	35
Iron carbonyl	50
Lead tetraethyl	118
MMT	2000

be too difficult to make in the large quantities required for testing though it is superior to isooctane. The octane was selected as the rating of 100 on the octane scale. The compound *n*-heptane was selected as the 0 rating.

Other octane enhancers such as methyltertiarybutyl ether (MTBE) can act in a different manner by suppressing cool-flame reactions by consuming OH free radicals.

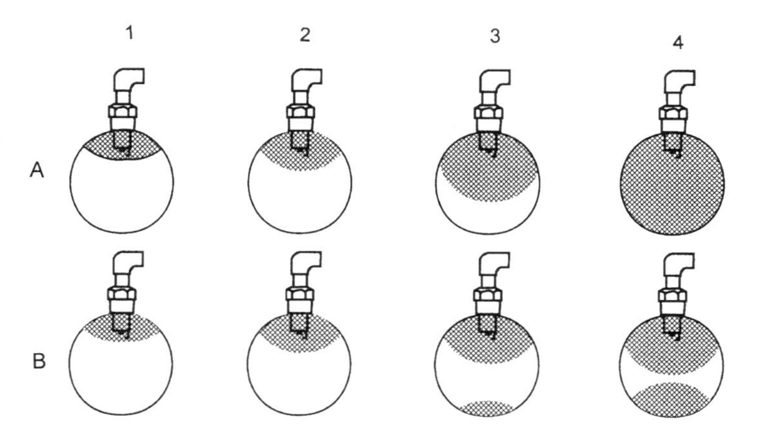

FIGURE 4.4. Flame propagation from a spark in a SI–ICE. (Time increases from left to right.) A: Normal process—the flame move uniformly from the spark to the end gas. B: Knock condition where the end gas ignites at B3 before the flame reaches the end gas.

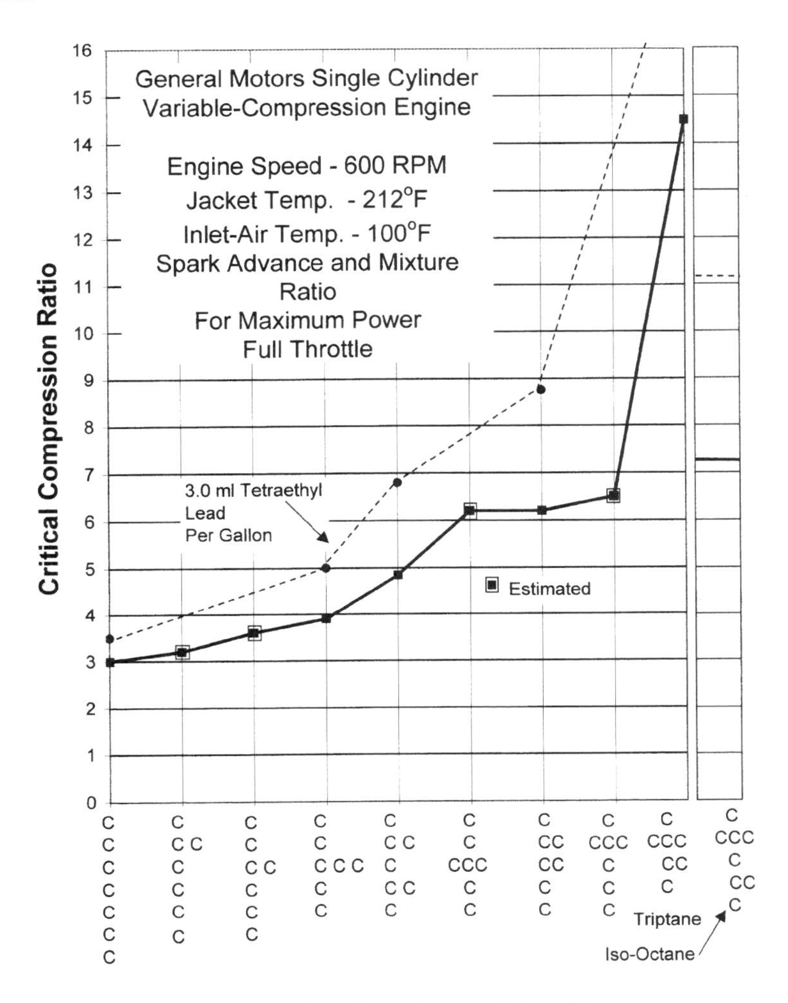

FIGURE 4.5. Knocking characteristics of heptanes.

MTBE is sufficiently soluble in water to contaminate the aquifer when MTBE/fuels have been spilled. As a result efforts are under way to ban MTBE as a fuel additive. Ethanol is considered to replace it.

A good illustration of the global contamination by lead is a comparison of the lead and mercury content of the Greenland ice field at various depths. This is shown in Table 4.6 where the level of Hg in the ice has not changed significantly in 3000 years whereas the level of Pb has increased 200 fold, showing the direct effect of the automobile. This implies that the contamination by mercury is of regional interest and not of global importance.

Because of the poisonous properties of lead and its compounds, some countries have replaced lead by other octane enhancers. One such additive is methyl cyclopen-

TABLE 4.6
Mercury and Lead Concentrations in the Glacial Samples[a]

Time of deposition	Mercury content[b]	Lead content[b]
800 B.C.	62	1
1724	75	10
1815	75	30
1881	30	—
1892	66	—
1946	53	—
1952	153	200
1960	89	—
1964 (fall)	87	—
1964 (winter)	125	—
1965 (winter)	94	—
1964 (spring)	230 ± 18	
1965 (summer)	98	—

[a]The sample deposited in 1724 was recovered from Antarctica, the others from Greenland.
[b]Nanograms (Hg/Pb) content (per kilogram of water).

tadiene manganese (III) tricarbonyl (MMT).* This substance is probably as toxic as the lead compounds but since much smaller quantities are used (1/20 of lead) it is argued that it is safer than lead. Actually manganese will be contaminating the environment just as lead has done if its use is continued. Aromatic hydrocarbons such as benzene and toluene also act as octane enhancers even though they are known carcinogens.

The addition of water to gasoline is reported to improve the quality of the fuel. The water is sucked up into the carburetor and vaporized as it passes around the hot exhaust manifold and enters the intake manifold. As with diesel fuel the mechanism by which the water acts is not fully understood.

The mechanism of oxidation of hydrocarbon (which can be represented by RH and $RCH_2CH_2CH_3$) involves the formation of peroxides (ROO) and hydroperoxides (ROOH) by the following series of reactions:

1 Initiation by spark

$$RCH_2{-}CH_2{-}CH_3 \rightarrow RCH_2{-}CH_2^{\cdot} + {}^{\cdot}CH_3$$

$$O_2 \rightarrow 2O$$

2 Chain branching

$$RCH_2{-}CH_2{-}CH_3 + O \rightarrow RCH_2{-}\dot{C}H{-}CH_3 + {}^{\cdot}OH$$

3 Peroxide formation

$$RCH_2{-}CH_2^{\cdot} + O_2 \rightarrow RCH_2CH_2O_2^{\cdot}$$

$$RCH_2\dot{C}H{-}CH_3 + O_2 \rightarrow RCH_2\underset{\displaystyle O_2^{\cdot}}{\underset{|}{C}}HCH_3$$

$$CH_3^{\cdot} + O_2 \rightarrow CH_3O_2^{\cdot}$$

*MMT, as well as lead, is no longer used as fuel additives in Canada.

4 Hydroperoxide formation

$$RCH_2{-}CH_2O_2^{\bullet} + RH \rightarrow RCH_2CH_2O_2H + R^{\bullet}$$

$$RCH_2{-}\underset{\displaystyle O{-}O^{\bullet}}{\underset{|}{CH}}{-}CH_3 + RH \rightarrow RCH_2{-}\underset{\displaystyle O{-}O{-}H}{\underset{|}{CH}}{-}CH_3 + R^{\bullet}$$

$$CH_3O_2^{\bullet} + RH \rightarrow CH_3OOH + R^{\bullet}$$

5 Hydroperoxide decomposition

$$RCH_2CH_2O_2H \rightarrow RCH_2CH_2O^{\bullet} + OH$$

$$RCH_2{-}\underset{\displaystyle O_2H}{\underset{|}{CH}}{-}CH_3 \rightarrow RCH_2{-}\underset{\displaystyle O^{\bullet}}{\underset{|}{CH}}{-}CH_3 + OH$$

$$CH_2OOH \rightarrow CH_3O^{\bullet} + OH$$

$$RCH_2CH_2O^{\bullet} \rightarrow RCH_2^{\bullet} + CH_2O$$

$$RO^{\bullet} + RH \rightarrow ROH + R^{\bullet}$$

Recent kinetic modeling of the combustion mechanism involving about 2000 reactions has made it possible to calculate RON values. A list of comparable values is given in Table 4.7.

TABLE 4.7
Comparison of Calculated and Measured Research Octane Number for Various Hydrocarbons

Fuel	Formula	Measured RON	Predicted RON	Ignition time, ms after TDC
n-Heptane (PRF 0)	C_7H_{16}	0	0	55.0
n-Hexane	C_6H_{14}	25	20	55.2
PRF 25	25% C_8H_{18}+75% C_7H_{16}	25	25	55.3
2-Methyl hexame	C_7H_{16}	42	40	55.5
PRF 50	50% C_8H_{18} + 50% C_7H_{16}	50	50	55.8
n-Pentane	C_5H_{12}	62	55	55.9
3-Ethyl pentane	C_7H_{16}	65	80	57.4
2-Methyl pentane	C_6H_{14}	73	80	57.3
3-Methyl pentane	C_6H_{14}	74	80	57.2
PRF 75	75% C_8H_{18} + 25% C_7H_{16}	75	75	56.7
3,3-Dimethyl pentane	C_7H_{16}	81	50	55.7
2,4-Dimethyl pentane	C_5H_{12}	86	75	56.4
2,2-Dimethyl propane	C_7H_{16}	83	70	56.2
PRF 90	90% C_8H_{18} + 10% C_7H_{16}	90	90	58.0
2-Methyl butane	C_5H_{12}	92	100	59.1
2,2-Dimethyl butane	C_6H_{14}	92	100	59.9
2,2-Dimethyl pentane	C_7H_{16}	93	>90	No ignition
n-Butane	C_4H_{10}	94	85	57.6
2,3-Dimethyl butane	C_6H_{14}	100	90	58.0
iso-Octane (PRF 100)	C_8H_{18}	100	100	59.5
iso-Butane	C_4H_{10}	102	>90	No ignition
Propane	C_3H_8	112	>90	No ignition
2,2,3-Trimethyl butane	C_7H_{16}	112	>90	No ignition
Ethane	C_2H_6	115	>90	No ignition

TABLE 4.8
Gasoline Prices and Tax Component in the OECD,[a] 1992 and 2000[†]

Country	Gasoline prices ($US/L)	Tax component (% of total)
Australia	0.499	46.2
Austria	0.970	64.8
Belgium	0.987	70.0
Canada	0.455 (0.48)	46.2
Denmark	0.961	67.2
Finland	1.013	68.0
France	0.992 (1.04)	77.2
Germany	0.981 (0.97)	72.4
Greece	0.820	69.1
Ireland	1.001	66.6
Italy	1.236 (1.00)	75.8
Japan	0.977 (0.94)	46.1
Luxembourg	0.746	62.0
Netherlands	1.141	72.4
New Zealand	0.5411	46.6
Norway	1.284	71.4
Portugal	1.083	75.4
Spain	0.943 (0.76)	69.8
Sweden	1.137	69.2
Switzerland	0.759	62.5
Turkey	0.745	63.7
United Kingdom	0.882 (1.00)	69.5
United States	0.298 (0.36)	33.9

[a]OECD = Organization for Economic Cooperation and Development.
†Values in ( ) refer to year 2000.

The price of gasoline in various parts of the world is given in Table 4.8. The major single component of the cost to the motorist is a government tax which varies from about 33% to 75%. The cost of gasoline should be compared to other liquid products such as bottled spring water which can sell for $0.50–$0.75 per liter or distilled water at about $2.00 per gallon (4.5 L) in Canada. Considering the processing involved in marketing gasoline compared to water, one must conclude that gasoline is still a bargain in most countries.

The problem of environmental contamination by vehicle exhaust is further illustrated by the presence of lead compounds in French wines from a vineyard at the intersection of two major autoroutes. It was recently shown that triethyl lead and trimethyl lead were present in older wines and a graph of their concentration for different vintage wines is shown in Fig. 4.6. The trimethyl lead and triethyl lead originate from the tetramethyl lead (TML) and tetraethyl lead (TEL) which were added to gasoline as antiknock agents. The fall-off in the lead is due to the discontinued use of lead in gasoline which was phased out in the beginning of the 1980s. The sharp rise in TML in 1962 is associated with the introduction of TML as an octane enhancer in 1960. The levels of lead in the wines are still at least 10–100 times less than the 0.5 μg/L limit of lead in drinking water.

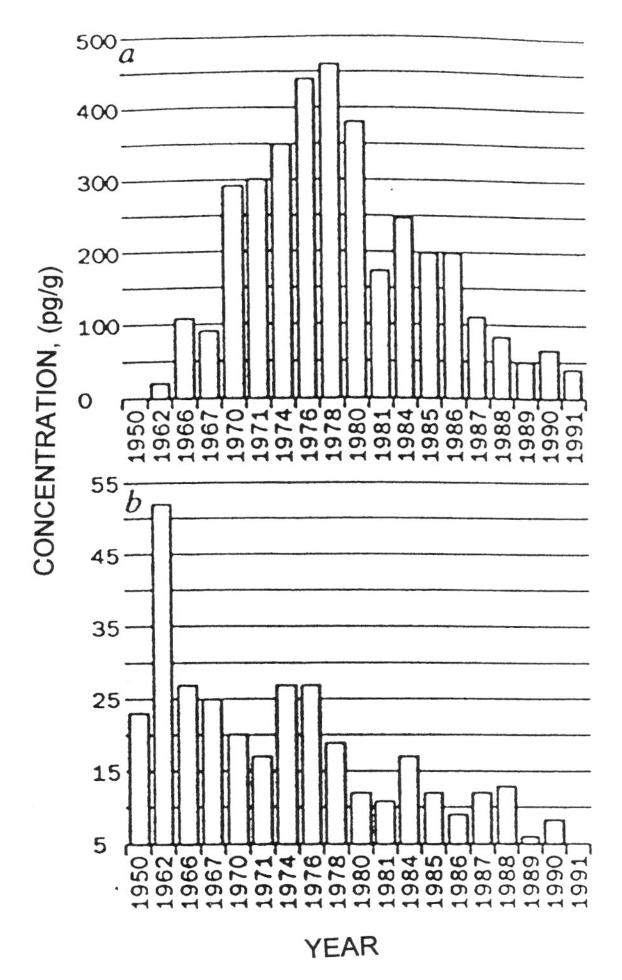

FIGURE 4.6. Changes in the concentration of (a) trimethyl lead and (b) triethyl lead as a function of vintage.

Some general properties of gasoline and diesel fuel compared to some alternate fuels are given in Table 4.9. A major parameter which is missing from the table is the unit cost, which to a great extent, determines the choice to be made.

EXERCISES

1. Smoke emission from a diesel engine is environmentally undesirable. Explain why, and discuss methods which can eliminate this aspect of diesel fuel use.
2. How can the ignition temperature (SIT) of a fuel be altered?
3. Calculate the air/fuel ratio for gasoline. (Note: Assume gasoline can be represented by nonane [C_9H_{20}].)

TABLE 4.9
Fuel Comparison Chart

	Gasoline $(CH_2)_x$	Methanol CH_3OH	Ethanol C_2H_5OH	Methane CH_4	Propane C_3H_8	Hydrogen H_2	Ammonia NH_3	Gasahol 10% C_2H_5OH	Diesel $(CH_2)_x$
RON	90–96	106	132		101			128	CN
MON	82–87	92	103	105	95			99	45
Heat of combustion MJ/kg	43.9	21.0	26.8	50.1	46.5	120.0	18.6	43.1	42.5
Density (gm/mL) temp. (°C)	0.73 @ 25°	0.791 @ 25°	0.789 @ 25°	0.466 @ −161°	0.50 @ 20°	0.071 @ −252°	0.771 −33°		0.85 25°
Heat of combustion MJ/L	32.3	16.6	21.1	23.3 liq.	23.3 liq.	8.5 liq.	14.3		35.3
Latent heat of vaporization MJ/kg	0.35	1.18	0.92	0.51	0.34	0.45	1.37		
VP @ 20°C (Torr) (atm)	450–680	99	38	45 atm @ −82°C[a]	8 (atm)	12.8 atm @[a] −239°C[a]	8 atm		1.1 @ 100°C
F.P. (°C)	−80	−94	−117	−182	−187	−259	−77	−80	−50
B.P. (°C)	30–150	65	78	−161	−42	−253	−33	30–150	150–300
Ignition temp. (°C)	371	446	422	540	432	500	650	300	250
Flash point (°C)	−43	11	13					−40	>40
Explosion limits %vol. in air	1.4–7.5	6.0–36	3.3–19	5.3–14	2.1–9.5	4.0–74	16–25	1.4–7.5	
Stoichiometric air/fuel ratio (kg air/kg fuel)	14.7	6.4	9.0						

[a]Critical temperature and critical pressure.

4. The TLV (Threshold Limit Value) of mercury in Canada and USA is 0.1 mg/m^3. If a droplet of mercury (density = 13.6g/mL) 3 mm in diameter is spilled and allowed to evaporate completely in a room 3 m × 3 m × 2.5 m. What will its concentration be, relative to the TLV?
5. Show how dimethyl ether can be prepared from methanol.
6. Compare the energy density of dimethyl ether with diesel fuel.
7. What advantages are there in using dimethyl ether in a diesel engine?
8. Discuss the special characteristics and requirements of aviation fuel in comparison to automotive fuels.

FURTHER READING

E. Ralbovsky, *Introduction to Compact and Automotive Diesels*, Delmar, Albany, New York (1997).
F. J. Thiessen and D. N. Dales, *Diesel Fundamentals*, 3rd Ed., Prentice Hall, New York (1996).
H. Halberstadt, *Modern Diesel Locomotive*, Motorbooks Intl., Osceola, Wisconsin (1996).
Motor Gasoline Industry—Past, Present, and Future, Gordon Pr., New York (1995).
J. F. Dagel, *Diesel Engines*, 3rd Ed., Prentice Hall, New York (1993).
Diesel Fuels for the Nineties, Soc. Auto. Engineers, Warrendale, Pennsylvania (1993).
C. F. Taylor, *The Internal-Combustion Engine in Theory and Practice*, 1 and 2 Vol. 2nd Ed., Rev. MIT Press, Cambridge, Massachusetts (1985).
C. G. Moseley, Chemistry and the First Great Gasoline Shortage, *J. Chem. Educ.* **57**, 288 (1980).
J. R. Pierre, The Fuel Consumption of Automobiles, *Scientific American*, January **34** (1975).
Universal Diesel Liquifier (fuel atomizer), http://www.universaldiesel.com/
Caterpillar, http://www.cat.com/
Detroit Diesel, http://www.detroitdiesel.com/
Fuel Technologies Inc., http://www.fueltechnic.com/
Biodiesel Fuel, http://www.veggievan.org/
Canadian Renewable Fuels Association, http://www.greenfuels.org/
Fuel prices, http://www.cheap-fuel.com/
Gasoline prices, http://www.chartoftheday.com/20000913b.htm

5

Alternate Fuels

5.1. INTRODUCTION

The search for alternate fuels has been stimulated by environmental concerns about the emissions in exhaust from gasoline and diesel fuel engines. Urban air pollution due to the automobile is due to excessive production of CO, NO_x, and hydrocarbons which in sunlight can give rise to ozone and smog. NO_x is formed during all high temperature combustions using air (see Exercise 2.18). Thus low temperature combustion processes result in lower NO_x emission because of the equilibrium

$$N_2 + O_2 \rightleftharpoons 2NO \tag{5.1}$$

Similarly the presence of oxygen in the fuel, e.g., methanol or ethanol, reduces the emission of smog-forming hydrocarbons.

The relative emission characteristics of fuels are shown in Table 5.1 and indicate that the electric vehicle is the least polluting energy source. However, this implies that the origin of the electrical energy used in recharging the batteries is environmentally benign.

The alternate fuels in use today are propane, methanol and ethanol.

5.2. PROPANE

The fuel properties of propane, also referred to as bottled gas, are shown in Table 4.9 and show that the boiling point of the liquid is $-42°C$. At 20°C the vapor pressure is 8 atm and thus the fuel must be stored in a pressurized cylinder meant to withstand at least 21 atm, 62°C if mounted in an enclosed space.

About 80% of the propane is obtained from natural gas where it occurs in concentrations which vary from 1 to about 5%. The other 20% is formed during the catalytic cracking of petroleum oil. The world production of propane in tabulated in Table 5.2 and shows an average annual growth rate (AAGR) of about 1.5%. Canada's consumption of about 3×10^9 L/year is less than half its production. In 1990 Canada exported 4.1×10^9 L/year to USA. For comparison the annual Canadian consumption of gasoline and diesel fuel is 3.5×10^{10} L and 1.8×10^{10} L, respectively. Hence all the propane available can only satisfy about 1/5 of the total gasoline usage in Canada. The

TABLE 5.1
Qualitative Summary of Emissions from Alternative-Fuel Vehicles vs. Gasoline-Fuel Vehicles

Environmental issues	Gasoline[a]	Reformulated gasoline	Diesel	Methanol[b]	M85[b]	Ethanol	Liquid petroleum gas	Compressed natural gas	Hydrogen[b]	Electric[c]
Tailpipe, evaporative emission levels[d]										
NMOG*	Base	Lower	Lower	Lower	Lower	Lower	Lower	Lower	None	None
CO	Base	Lower	Lower	Equl	Equal	Equal	Lower	Lower	None	None
No_x	Base	Equal	Higher	Equal	Equal	Equal	Higher	Higher	Higher	Lower
Impact on urban ozone relative to gasoline										
Ozone-forming potential	1.00	0.80 ± 0.2	0.80 ± 0.2	0.65 ± 0.2	0.70 ± 0.2	0.80 ± 0.2	0.65 ± 0.2	0.40 ± 0.2	0.00 ± 0.2	0.00 ± 0.2
Reduction in beak ozone[e] in Los Angeles	0%	1.4%	1.4%	2.4%	2.0%	1.4%	2.4%	4.1%	6.8%	6.8%
Levels of concern over airborne toxic emissions										
Particles	Low	Low	High	Low	Low	Low	Low	Low	—	—
1,3-Butadiene	Medium	Medium	Unknown	Low	Low	Low	Low	Low	—	—
Benzene	High	Medium	Low-medium	Low	Medium	Low	Low	Low	—	—
Formaldehyde	Medium	Medium	Medium	High	High	Medium	Low	Low	—	—
Acetaldehyde	Low	Low	Low	Low	Low	High	Low	Low	—	—
Global warming factors										
gCO_2/BTU	0.094	0.094	0.096	0.093	0.094	0.083	0.081	0.071/0.09[f]	0.15	0.18
Fuel economy (miles/gal)	34	34	39/48[g]	22	22	28	31.2	8.0	11.3	—
Energy economy (BTU/mile)	3353	3353	3257/2614	2923	2923	2923	2829	3353	2591	1705
gCO_2/mile	315	315	313/252[g]	272	275	243	229	238/301[f]	389	307

*NMOG = Nonmethane organic gases.
[a]Current industry average.
[b]Natural gas as the feedstock source of methanol and hydrogen.
[c]No direct emissions from vehicles, but associated power plant No_x emissions may be substantial.
[d]Relative to gasoline vehicles with Tier 1 (Clean Air Act Amendments) standards, using 9 lbs/in.^{2} Reid vapor pressure gasoline.
[e]Year 2015 with the pre-1987 emissions control regulations, except LDV. These are semiquantitative estimates.
[f]There is a debate over global warming impact of methane as compared to carbon dioxide. If the atmospheric lifetime of methane is fully accounted, as we believe it should, then CNG vehicles have a substantial global warming benefit. However, if the lifetime is discounted, CNG vehicles have approximately the same greenhouse impact as gasoline vehicles.
[g]The fuel economy improvement of diesel compared to gasoline vehicles is uncertain; estimates range from 15% to 40% better.

TABLE 5.2
World Annual Production of Propane (GL)

	1990	1995	AAGR[a] (%)
United States	70.0	66.0	(1.6)
Western Europe	34.8	34.9	0.2
Japan	5.0	4.9	—
Canada	7.8	8.0	0.9
Latin America	26.9	29.5	2.9
Africa	11.2	14.5	8.7
Asia/Australia	18.7	19.8	2.9
Middle East	33.1	41.0	7.2
Eastern Block	0.8	0.8	—
World total	208.3	219.4	1.4

[a]Average annual growth rate (%).

amounts produced are small in comparison to the gasoline used and therefore as an alternate fuel it has limited use. This applies to the other alternate fuels where production is still at a fraction of the full requirements. Propane is, like natural gas, a clean burning automotive fuel with a low ozone-forming potential, good fuel economy, and low emission of toxic organic hydrocarbons such as benzene and formaldehyde. However, acetaldehyde and olefins are some of the adverse components found in the exhaust.

The cost of propane is usually less than gasoline on a volume basis and major users are the fleet operators such as taxis, delivery trucks, and vans. Liquid propane is readily available in most locations. Its use is however restricted to ICE-SI engines and not for the diesel engines.

5.3. METHANOL

Methanol was produced by the destructive distillation of wood until about 1923 when BASF in Germany showed that methanol can be formed by the catalytic reduction of CO with H_2.

$$CO + 2H_2 \xrightarrow{\text{catalyst}} CH_3OH \qquad \Delta H^0 = -90.77\,\text{kJ/mol} \tag{5.2}$$

This is the method used commercially today where the syngas is usually prepared from natural gas by the steam reforming reaction

$$CH_4 + H_2O_{(g)} \xrightarrow{\text{catalyst}} CO + 3H_2 \qquad \Delta H^0 = +206.2\,\text{kJ/mol} \tag{5.3}$$

The energy intensive step is the formation of syngas and as a result the economics of the overall process is favorable only for large scale plants producing more than 1500 tonnes/day methanol. It seems wasteful to break 4 C–H bonds to form $CO + H_2$ and then reform 3 of the C–H bonds. This can be avoided by the direct partial oxidation of CH_4 to CH_3OH at high pressures. It was shown by Bone in 1930 that at high pressures (50–100 atm) methanol is a major intermediate product in the oxidation of natural gas. This has been extensively studied at the University of Manitoba since 1982 by the author. A small pilot plant has operated in Kharkow, Ukraine from 1985 to 1990 producing 100 tonnes/year methanol. A larger plant designed to produce 10,000 tonnes/year is in the planning process. The bench top optimum efficiency is achieved at about 50 atm, 10% O_2, and 400°C where 80% selectively and 12% conversion results in a 10% yield of methanol. The process is represented by the reaction

$$3CH_4 + 2\tfrac{1}{2}O_2 \rightarrow 2CH_3OH + CO + 2H_2O \tag{5.4}$$

The present uses of methanol are shown in Fig. 5.1 and given in Table 5.3 which shows that the formation of formaldehyde is still the major application of methanol. The direct use of methanol as a fuel is expected to surpass this in the near future. The octane enhancer methyltert-butylether (MTBE) is capable of satisfying the oxygen requirements in gasoline and to replace lead (TEL) as an additive. It is made by reacting methanol with isobutene

$$CH_3OH + CH_2{=}C(CH_3)_2 \rightarrow CH_3OCH_2{-}CH(CH_3)_2 \tag{5.5}$$

The ethanol equivalent, ethyltertbutylether (ETBE) is less soluble in water and thus less likely to contaminate water supplies when the fuel is spilled. Its higher cost, however, limits its use in fuel.

The present optimistic predictions of methanol production and demand are shown in Fig. 5.2 where it can be seen that the demand may soon exceed production. The construction of a full sized conventional methanol plant (2000 tonnes/day) takes at least 3 years from start to full time on-stream and requires a suitable source of natural gas. Small gas fields which usually accompany crude oil production are normally vented or flared. The extent of such wasted energy is shown in Fig. 5.3 and it is this lost gas which can be readily converted to methanol by the direct partial oxidation process.

The main apparent disadvantage of methanol is its low energy density by volume, 16.6 kJ/L, about half that of gasoline. This does not imply that the fuel tank must be twice as large because if the engine was designed for methanol consumption, the methanol fuel economy would be close to that of gasoline.

The use of gasohol or methanol (or ethanol) with gasoline (M-85) has become the intermediate stage in the extensive use of alcohol fuels. Because of the possibility of phase separation when water is present in 10% methanol in gasoline, a cosolvent such as tertbutyl alcohol (TBA) must be added. M-85 does not require a cosolvent though the emissions are worse than M-100.

Cold starting difficulties with methanol fuel are due to its high heat of vaporization and low volatility. Proper engine design can make up for such deficiencies. Another

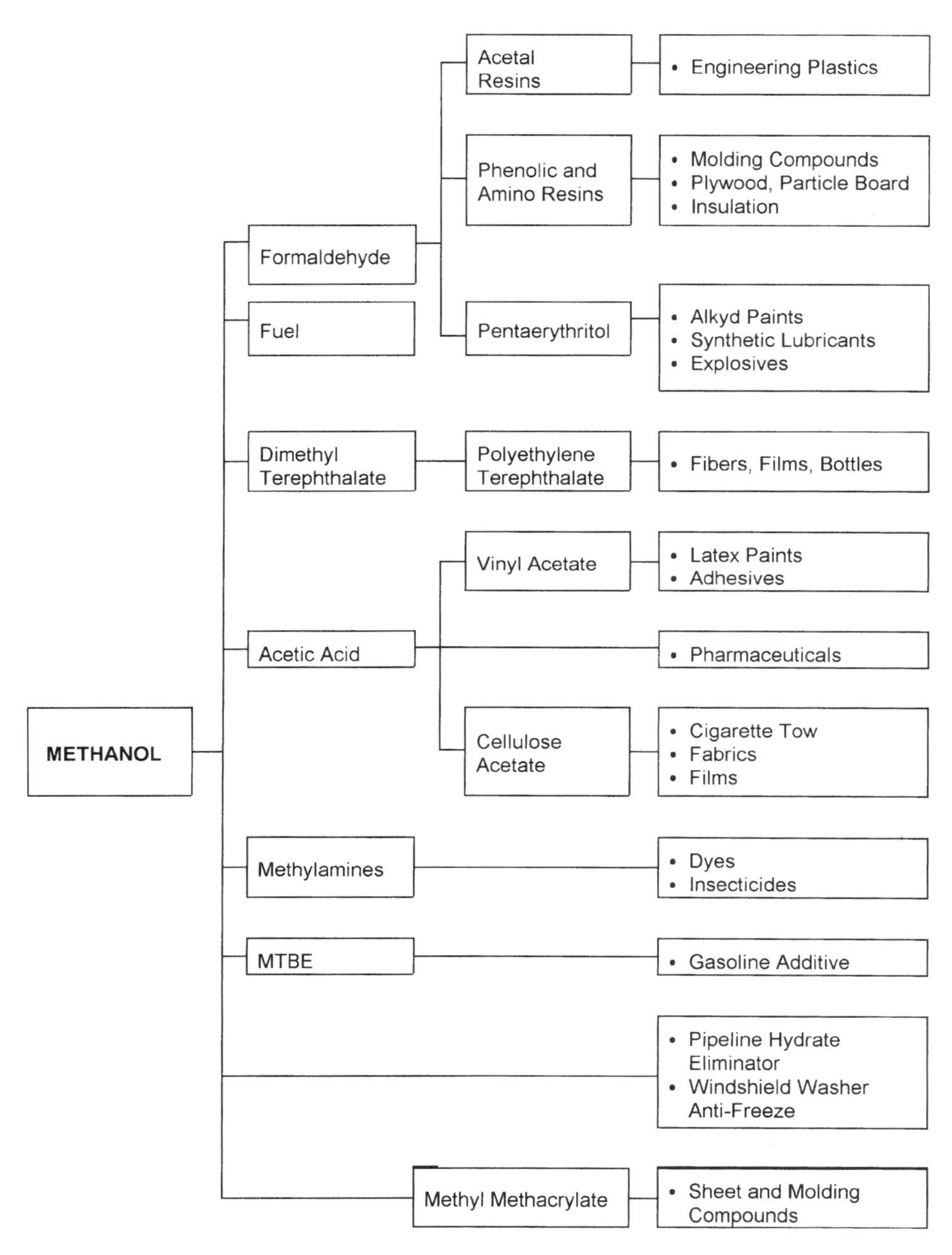

FIGURE 5.1. Uses of Methanol.

problem is corrosion due to the acid (formic acid) formed by the oxidation of the methanol and all fuel and engine components must be carefully selected. The methanol flame is not readily visible whereas M-85 (15% gasoline) has the advantage of appearing more like a yellow gasoline flame.

TABLE 5.3
Conversion of Methanol into Various Products (%)

	1982	1994
Formaldehyde	31	39
Methylamine	4	—
Chloromethanes	9	
Acetic acid	12	7
Methyl ester	8	6
Solvent	11	7
Fuel–antifreeze	25	2
MTBE	—	13
Miscellaneous	—	26

The aldehyde emission from alcohol fuel must be destroyed by a catalytic afterburner if environmental contamination is to be avoided. Methanol has been used in diesel engines though it has a CN between 0 and 5. However, because it is not readily self-ignited special engine design must make up for such deficiencies. A glow-plug or fuel additive can aid self-ignition. Partial conversion of the methanol to dimethyl ether is another approach to the problem which has recently been studied.

A major problem in the use of gasohol is in the materials of construction of the fuel system. Alcohol tends to dissolve the oxidation products and gums from gasoline and special precautions must be taken when first switching to gasohol in an old vehicle. This is even more important when pure methanol is used as the fuel. Special gaskets

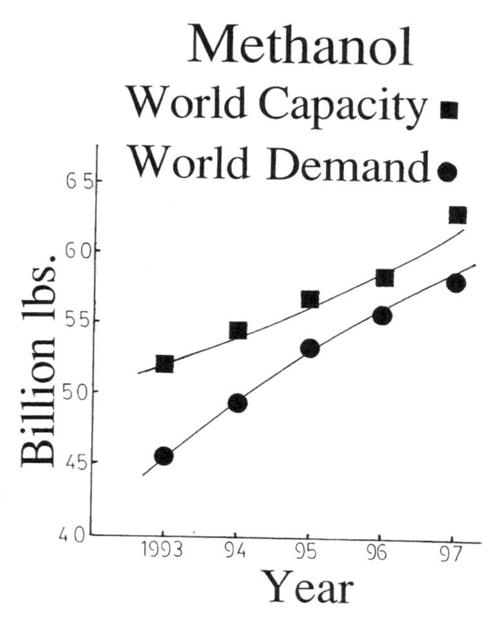

FIGURE 5.2. The world total capacity for methanol production (■) and the world demand (●).

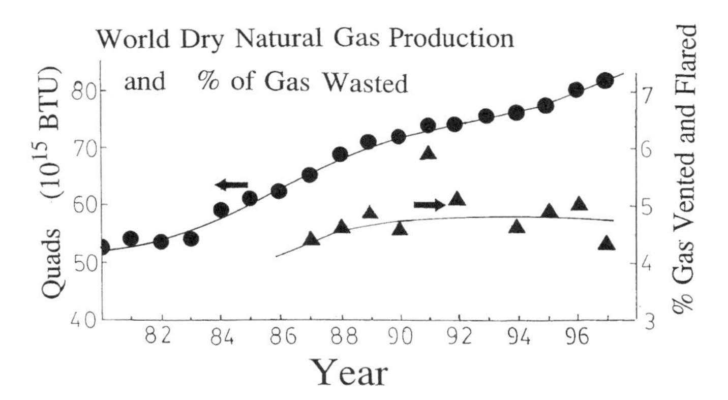

FIGURE 5.3. The total world production of dry natural gas (●) at various years and the percentage of gas vented and flared (■).

and O-rings are required. However, these present no problem if design and construction is started with such fuel in mind. Besides, being independent of petroleum, alcohol offers other inherent advantages, such as cooler cleaner combustion, improved power, reduced carcinogens and NO_x emission, reduced hydrocarbon emission which can be photochemically activated and smog forming, and smaller and lighter engine designs are possible.

Ethanol and methanol have cetane values of from 0 to 5 making them poor compression ignition fuels for a diesel engine. However, alcohol–diesel fuel blends have been used successfully and emulsifiers have been added to help blend the two components which have limited miscibility.

When methanol is catalytically converted to CO and H_2

$$CH_3OH_{(g)} \rightarrow CO_{(g)} + 2H_{2(g)} \qquad \Delta H^0 = 90.7 \text{ kJ/mol} \tag{5.6}$$

the resulting synthesis gas has a higher heat of combustion (−766 kJ) than the methanol (−676 kJ). If the exhaust heat is used to decompose the methanol then a higher energy fuel is obtained, i.e., approx. 20% increase in fuel economy has been proposed. This is valid if only thermal energy is considered. In an internal combustion spark engine the air and fuel is compressed (which requires work) and the combustion then expands the gas and work is done. Hence the ratio, R = moles of gaseous products/moles of gaseous reactant, must be used to normalize differences in fuels. Assuming $N_2/O_2 = 4$ and no excess O_2:

$$\begin{gathered} CH_3OH + 1.5O_2 + 6N_2 \rightarrow CO_2 + 2H_2O + 6N_2 \\ 8.5 \text{ moles} \rightarrow 9 \text{ moles} \qquad R_1 = 1.058 \end{gathered} \tag{5.7}$$

$$\begin{gathered} CO + 2H_2 + 1.5O_2 + 6N_2 \rightarrow CO_2 + 2H_2O + 6N_2 \\ 10.5 \text{ moles} \rightarrow 9 \text{ moles} \qquad R_2 = 0.857, R_1/R_2 = 1.235 \end{gathered} \tag{5.8}$$

The product gases have the same volume in each reaction. Hence the additional work of compression (W_c) is essentially due to the $P\Delta V$ or ΔnRT where $n = 2$ moles.

The remaining question is, what is the value of T? We can select 400°C as an estimated value of the temperature. This value can be calculated from an adiabatic compression of the gas (mostly air) from $T_1 = 350$ K and assuming that CR = 10.

It is worth considering an extension to this argument by considering the increased fuel economy if we remove the nitrogen from the combustion process. First, a higher temperature is obtained and second, less work is done in compressing the gas. If we assume that gasoline is represented by octane, C_8H_{18}, then in air, the reaction is:

$$C_8H_{18} + 12.5O_2 + 50N_2 \rightarrow 8CO_2 + 9H_2O + 50N_2 \qquad \text{63.5 moles} \rightarrow \text{67 moles} \quad R_3 = 1.055 \tag{5.9}$$

$$C_8H_{18} + 12.5O_2 \rightarrow 8CO_2 + 9H_2O \qquad \text{13.5 moles} \rightarrow \text{17 moles} \quad R_4 = 1.259,\ R_4/R_3 = 1.19 \tag{5.10}$$

The difference in moles of compression, Δn, of reactions (5.9) and (5.10) is 63.5–13.5 or $\Delta n = 50$.

Hence $\Delta nRT = 50 \times 8.314 \times 673 = 280$ kJ. The heat of combustion of octane is 5470 kJ/mol. Thus the extra work of compression is only about 5% of the combustion energy. However, the calculated adiabatic combustion flame temperature for octane in air is about 1400 K whereas in oxygen the flame temperature is over 9000 K. This represents a substantial increase in energy of combustion due to the removal of nitrogen. Besides obtaining more heat from the reaction (since the heat capacity of nitrogen absorbs some of the energy to reach the high temperature) the removal of nitrogen gives rise to about 20% higher fuel economy. This is now being done for stationary furnaces where only thermal improvement is obtained. It would be interesting to run an automobile on enriched oxygen which can be obtained by using suitable membranes. More will be said about this later.

One aspect which is important to note is that methanol is toxic and its threshold limit value (TLV) is 200 ppm requiring its dispensing in well ventilated areas.

The ultimately efficient fuel system would be an electric vehicle running on a methanol fuel cell. Such a system could meet the most stringent environmental emission requirements as well as high energy efficiency. This will be discussed in more detail in Chapter 9.

5.4. ETHANOL

Ethanol is one of the few possible nonfossil fuels which can be made from a variety of renewable sources such as grapes, corn, straw, and sugar cane. The fermentation of sugar and starch from various agricultural sources has been known for centuries. The industrial production of ethanol from the acid catalyzed hydration of ethylene was developed by Union Carbide Corp. in 1930

$$\begin{array}{c} CH_2{=}CH_2 + H_2SO_4 \rightarrow CH_3{-}CH_2{-}OSO_3H \\ \downarrow H_2O \\ H_2SO_4 + CH_3{-}CH_2OH \end{array} \tag{5.11}$$

The sulfuric acid, initially 96–98%, is diluted to about 50% during the reaction, and has to be reconcentrated. This is one of the more costly steps in the process. Other solid phase acidic catalysts have been used to replace the H_2SO_4 in the hydration process.

The ethylene is formed by the thermal cracking of ethane by the reaction

$$C_2H_{6(g)} \rightarrow C_2H_{4(g)} + H_{2(g)} \quad (5.12)$$

Ethanol from ethane is a fossil-based fuel and must be distinguished from grain alcohol. Ethanol from syngas has been extensively studied but catalysts have not been found, as yet, to be efficient enough to make the process cost effective. Similarly the conversion of syngas to ethanol by alternate routes have been considered.

$$\begin{aligned} CH_3OH_{(g)} + CO_{(g)} + 2H_{2(g)} &\rightarrow CH_3CH_2OH_{(g)} + H_2O_{(g)} \\ 2CO_{(g)} + 4H_{2(g)} &\rightarrow C_2H_5OH_{(g)} + H_2O_{(g)} \end{aligned} \quad (5.13)$$

or

$$2CH_3OH_{(g)} \rightarrow CH_3OCH_{3(g)} + H_2O_{(g)} \quad (5.14)$$

$$CH_3OCH_{3(g)} \rightarrow C_2H_5OH_{(g)} \quad (5.15)$$

These catalytic processes are attractive because the relative value of ethanol/methanol is about 4 whereas the C_2H_6/CH_4 value is about 2, though in the year ending in June 1995 the price of methanol doubled. Hence the direct conversion of ethane to ethanol by the reaction analogous to reaction 5.4

$$3C_2H_6 + 3\tfrac{1}{2}O_2 \rightarrow 2C_2H_5OH + 2CO + 3H_2O \quad (5.16)$$

is a potentially attractive approach to the utilization of surplus ethane.

The total annual world ethanol production in 1998 was over 31×10^9 L (31 GL) with the distribution into potable, industrial and fuel grades of about 9%, 30%, and 61%, respectively. The USA produced about 7.6 GL or about 25% of the world's production from over 15 plants. Another 18 plants, either being expanded, built, or in the planning stages can almost double the production levels. The estimated annual US market is 100 GL. The 1998 production of ethanol in the European Union was 22 GL. Russia produced 2.5 GL of which 60% was potable though imports allowed 2.2 GL to be consumed as beverage grade alcohol.

The present use of ethanol as a fuel additive is made possible by the subsidy which varies from one location to another, but is approx. 54¢/US gal. of ethanol or approx. 14¢/L in the US. This subsidy applies only to ethanol formed from grain, i.e., nonfossil fuel sources. The present processes of production includes fermentation using yeasts as the enzyme catalyst which converts sucrose to CO_2 and ethanol

$$C_6H_{12}O_{6(aq)} \xrightarrow{\text{catalyst}} 2C_2H_5OH_{(aq)} + 2CO_{2(g)} \quad (5.17)$$

This is then followed by distillation of the alcohol which normally reaches about 12%

TABLE 5.4
Some Properties of Ethanol–Water (E/W)

[E] (wt%)	Solutions density (g/mL)	[E] (molarity)	F.P. (°C)
2.5	0.9953	0.539	−1.02
5.0	0.9911	1.974	−2.09
10.0	0.9838	2.131	−4.47
15.0	0.9769	3.175	−7.36
20.0	0.9704	4.205	−10.92
25.0	0.9634	5.21	−15.4
30	0.9556	6.211	−20.5
35	0.9466	7.145	−25.1
40	0.9369	8.120	−29.3
44	0.9286	8.853	−32.7
50	0.9155	9.919	−37.7
54	0.9065	10.607	−40.6
60	0.8927	11.606	−44.9
64	0.8786	12.565	−48.64
68	0.8739	12.876	−49.52

in the fermentation process. The distillation produces 95% alcohol. The residual 5% water is removed either by the addition of a third component to form a 2-phase azeotrope followed by separation and distillation or by passing the wet alcohol through a 3A molecular sieve which removes the water leaving 99.9+% alcohol.

There are, however, enzymes which can still function at up to almost 20% ethanol. The energy balance in ethanol production which takes into account the energy used to grow the corn, the energy to ferment and distill the alcohol, and the energy in the remaining mash, results in a ratio for the input energy/output energy of 1.3, which varies somewhat depending on the efficiency of the various steps. It would appear that it makes little sense to spend 1.3 kJ of fossil fuels to obtain 1 kJ of grain alcohol fuel and to consume grain which is an important food supply for an expanding world population.

Improved efficiency may be achieved by using pervaporation to separate out the alcohol instead of distillation. This consists of a membrane through which water and ethanol permeate with large different rates. When fermentation is continuously conducted in a membrane reactor it is possible to continuously remove the alcohol as it is formed, enriching it by a factor of 6–10 fold. This can reduce the cost of distillation, which is as much as half the production cost of the ethanol.

The separation of ethanol from water can also be effected by freezing. The effect of ethanol concentration on the freezing point is given in Table 5.4. Thus, 1 L of fermented brew with 12.5% ethanol by volume was completely frozen and then allowed to thaw. The first 500 mL of solution was 17% ethanol. When this 500 mL solution was frozen and allowed to thaw again the first 250 mL was 23% ethanol. Various freeze-thaw cycles can thus concentrate ethanol. Another process which has been studied extensively for more than 70 years is the conversion of cellulose from wood and straw to glucose by enzymes or by acid hydrolysis.

The reaction is

$$\underset{\alpha\text{-cellulose}}{(C_6H_{10}O_5)_x} + xH_2O \rightarrow x\underset{\text{glucose}}{(C_6H_{12}O_6)} \qquad (5.18)$$

Cellulose is the most abundant natural occurring biomass material on earth and its utilization as a fuel has not as yet been fully exploited. Paper, though presently recycled to some degree, can be included with straw and wood for conversion to ethanol fuel.

Ethanol has been used as a gasoline additive in Brazil since the 1920s. The conversion to straight alcohol during World War II has continued to the present. Ethanol is used in its pure form in the Otto cycle engine for cars and light trucks, and in ethanol–gasoline blends (22% ethanol) where it replaces lead as antiknock additive and octane enhancer. Ethanol is also used in diesel engines. This is accomplished by the addition of 4.5% of a diesel ignition improver such as isoamylnitrate, hexylnitrate, and the ethylene glycol dinitrate to the fuel which consumes 65% more fuel volume than regular diesel fuel and 2% less than an Otto cycle engine running on pure ethanol. In some cases 1% castor oil is added to the ethanol to help lubricate engine parts.

The ethanol fuel program in Brazil is made possible by the large sugarcane crop as well as other sugar crops such as cassava and sorghum. Though liquid fuels are used for automotive fuels throughout the world, the precise choice is determined by local conditions which vary considerably.

EXERCISES

1. The standard heat of formation $\Delta H_f^0(\text{NO}) = 90.25$ kJ/mol, $\Delta G_f^0 = 86.57$ kJ/mol.
   (a) Calculate the value of the equilibrium constant Kp for the reaction

   $$N_2 + O_2 \rightleftharpoons 2NO$$

   at 800 K, 1000 K, and 1200 K and determine the %NO in air at these temperatures (see Exercise 2.18).
   (b) Determine the thermodynamic values of the constants **a** and **b** in the equation of Exercise 2.18.
2. Using the data in Table 5.A calculate the equilibrium constant for the reaction

   $$CO + 2H_2 \rightleftharpoons CH_3OH_{(g)}$$

   at 298 K, 400 K, and 600 K.
3. Using the data in Table 5.A calculate the equilibrium constant at 298 K, 400 K, and 600 K for reaction (5.3)

   $$CH_4 + H_2O_{(g)} \rightarrow CO + 3H_2$$

4. The air fuel ratio $(\text{AFR}) = \dfrac{\text{weight of air available for combustion}}{\text{weight of fuel available for combustion}}$

   Show that for complete combustion (stoichiometric) (a) the AFR = 15 for gasoline or diesel fuel; (b) calculate the AFR for pure methanol fuel.

TABLE 5.A
Selected Thermodynamic Values

	ΔH_f^0 (kJ/mol)	ΔG_f^0 (kJ/mol)	S^0 (JK^{-1} mol^{-1})
CO	−110.5	−137.2	197.6
H_2	0	0	130.6
O_2	0	0	205.0
$CH_3OH_{(g)}$	−201.6	−162.4	239.7
$CH_3OH_{(l)}$	−239.1	−166.4	126.8
$C_2H_5OH_{(g)}$	−234.4	−167.9	282.6
$C_2H_5OH_{(l)}$	−277.1	−174.9	160.7
$H_2O_{(g)}$	−241.8	−228.6	188.7
$H_2O_{(l)}$	−285.8	−237.2	70.0
$CH_{4(g)}$	−74.7	−50.8	186.2
$CH_3OCH_{3(g)}$	−184.4	−114.2	266.5

5. What are the reasons for pursuing the development of alternate fuels for the automobile?
6. How can NO_x formation be minimized when using air (with the N_2 present) in a combustion process?
7. Explain why propane is a "clean" fuel.
8. Explain why the heat of combustion by weight of propane is greater than that of gasoline (see Table 4.9).
9. Why would butane, if available in large quantities, be a better fuel than propane?
10. Why is it ill-advised to use methanol as a fuel in a vehicle designed to run on gasoline?
11. Can you suggest a nonfossil fuel source for methanol?
12. Calculate the temperature reached in the adiabatic compression (CR = 10) of an air/fuel mixture. Note: The heat capacity of air C_v(air) = 21.5 J/K mol.
13. Would it be an advantage to convert methanol to dimethyl ether for an SI-ICE?
14. Calculate the energy required (if any) to convert natural gas (CH_4) to ethanol by the reaction

$$2CH_{4(g)} + O_2 \rightarrow C_2H_5OH_{(l)} + H_2O_{(l)}$$

15. Discuss the energetics and feasibility of reactions (5.14) and (5.15) to produce ethanol from methanol.
16. Using Fig. 5.3, determine the rate of increase in natural gas production between 1985 and 1989 and between 1989 and 1995, and give an explanation for the change.

FURTHER READING

Alternate Fuels for IC Engines, Soc. Auto Engineers (1999).
The Alcohol textbook, 2nd Ed., Murtagh and Associates, Winchester, WV (1999).
S. Lee, *Alternate Fuels*, Taylor and Francis, Bristol, Pennsylvania (1996).

G. Boyle, Editor, *Renewable Energy: Power for a Sustainable Future*, Oxford Univ. Press, New York (1996).
Jana, Auto. *Alternate Fuels*, Delman Pubs., Albany, New York (1996).
McGlinchy, *Technical Guide to Alternate Fuels*, Van Nostrand-Reinhold, New York (1996).
D. Sperling and M. DeLuchi, *Choosing an Alternative Transportation Fuel: Air Pollution, and Greenhouse Gas Effects*, OECD, Washington, DC (1993).
Liquid Fuels from Renewable Resources: Proceedings Alternative Energy Conference, *Am. Soc. Ag. Eng.*, St. Joseph, Missouri (1992).
E. V. Anderson, Ethanol from Corn, *Chem. and Engin. News*, November 2, p. 7 (1992).
T. Y. Chang et al., Alternative Transportation Fuels and Air Quality, *Environ. Sci. Technol.* **25**(7), 1190 (1991).
Focus on Propane, Superior Propane Inc., Concord, Ontario (1991).
L. Morton, N. R. Hunter, and H. D. Gesser, Methanol: A Fuel for Today and Tomorrow, *Chem. and Indust.* July 16, p. 457 (1990).
Methanol Fuel Formulation and In-Use Experiences, Soc. Auto. Engineers, Warrendale, Pennsylvania (1990).
W. L. Kohl, Editor, Methanol as an Alternative Fuel Choice: An Assessment, Johns Hopkins Univ., Baltimore, Maryland (1990).
E. E. Ecklund and G. A. Mills, *Alternate Fuels: Progress and Prospects.* Part 1 and 2, Chemtech. September, p. 549, October, p. 626 (1989).
D. Semanaitis, *Alternate Fuels, Road and Track*, November, p. 72 (1989).
C. L. Gray and J. A. Alson, The Case for Methanol, *Sci. Amer.*, November, p. 108 (1989).
T. E. Van Koevering et al., The Energy Relationships of Corn Production and Alcohol Fermentation, *J. Chem. Educ.* **64**, 11 (1987).
S. S. Marsden Jr., Methanol as a Viable Energy Source in Today's World, *Annu. Rev. Energy*, **8**, 333 (1983).
Propane Carburetion, EMR Ottawa, Canada (1983).
Canadian Alternate Fuels, Biomass Energy Institute, Winnipeg MB (1983).
E. M. Goodger, *Alternate Fuels for Transport*, Alternate Fuels Technology Series, 1 Vol., Cranfield Press, UK (1981).
Switching to Propane, MTC Energy, Ontario (1981).
Alcohol Production from Biomass in the Developing Countries, World Bank, Washington, DC (1980).
J. W. Lincoln, *Methanol and Other Ways Around the Gas Pump*, Garden Way Publ., Charlotte, Vermont (1976).
H. F. Willkie and P. J. Kalachov, *Food for Thought—A treatise on the utilization of farm products for producing farm motor fuel as a means of solving the Agricultural Problem*, Indiana Farm Bureau Inc., Indianapolis, Indiana (1942).
Renewable Fuels Association, http://www.ethanolRFA.org/
Murtagh Associates: Fuel ethanol, http://www.distill.com/
Mitsubishi Motors: Methanol Vehicle, http://www.mitsubishi-motors.co.jp
Methanex: Methanol producer/dealer, http://www.methanex.com
Industrial Agriculture, http://www.ia-usa.org/
Information Center, http://www.fuelcells.org/
Propane Vehicle Council, .http://www.propanegas.com/vehicle/
American Methanol Institute, http://www.methanol.org/
National Corn Growers Association, http://www.ncga.com/
NRG Tech., (Clean energy), http://www.nrgtech.com/

6

Gaseous Fuels

6.1. INTRODUCTION

The combustion process is usually an oxidation reaction involving oxygen and the oxidant and is in many cases in the gaseous state. Thus, the burning candle or the alcohol burner are examples of solid and liquid fuels burning in the gaseous state. Coal however does not volatilize and the combustion of many solids, including some plastics, occurs at the surface. By gaseous fuels, consideration is given to the storage state at ambient temperatures even though the fuel in usage can be in the liquid state.

The gaseous fuels which will be considered are methane (or natural gas), synthesis gas (which is a mixture of carbon monoxide and hydrogen) and pure hydrogen.

6.2. NATURAL GAS

Natural gas is a clean burning fuel which is easily transported by pipeline or as liquefied natural gas (LNG). Figure 6.1 shows the structure of a ship for transporting LNG. The production, consumption, and reserves of natural gas for 1999 for various countries is given in Table 6.1. It can be noticed that some countries such as Norway are big producers with a large reserve in off-shore gas, but use little or no gas because in Norway hydroelectricity is the preferred type of energy used. Likewise, Japan has very little natural gas and no significant reserves, but is a major consumer bringing in the LNG from Australia, Saudi Arabia, Brunel, Qatar, Malaysia, and Indonesia. In the year ending April 1, 2000, Japan imported 52 Mtonnes of LNG. It is stored underground, as shown in Fig. 6.2. At the present rate of consumption the World resources of natural gas will last about 60 years or until about 2060.

The composition of natural gas varies considerably from one country to another and from one well to another in the same locale. The representative composition of natural gas for various countries is given in Table 6.2. Some extreme values are not shown and include 70% CO_2 in some gas in Indonesia and 40% C_2H_6 for some Siberian gas. Dry natural gas is gas freed of liquids of butanes and heavier hydrocarbons which are called natural gas liquids.

Natural gas is associated with petroleum deposits and usually accompanies the oil as it is drawn from the well. Often such gas cannot be conveniently stored or utilized, and as a result it is often wasted by venting or flaring. The proportion of this wasted

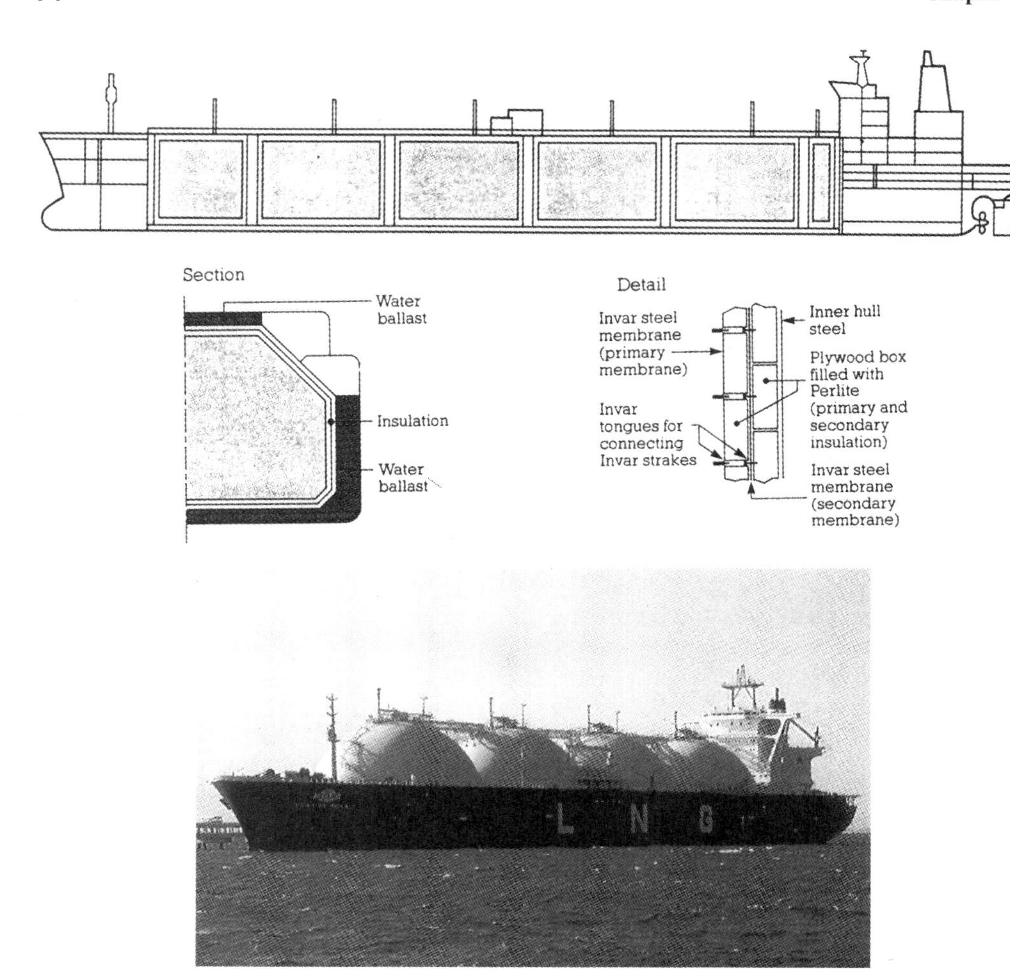

FIGURE 6.1. (A) Cross section of a LNG tanker and its detailed cell type structure. (B) LNG transport ship.

gas is increasing every year as shown in Fig. 5.3. It is this gas which can often be converted directly into methanol. The gas in large wells is cleaned, stripped of liquids and pumped into the pipeline for distribution.

The major contaminant in the gas is H_2S which is removed by the Claus process. Part of the H_2S is oxidized to SO_2 and then reacted by

$$2H_2S + SO_2 \rightarrow 2S + 2H_2O \tag{6.1}$$

The sulfur is stockpiled for sale, primarily to producers of sulfuric acid.

Methane mixed with carbon dioxide, usually in a 1:1 ratio, is formed by the anaerobic digestion of garbage and manure by microorganisms. Thus, landfill gas can be used as a source of heat when collected. Similarly, hog manure has been digested, forming biogas.

TABLE 6.1
World Annual Natural Gas Production, Consumption, and Reserves 1999 and 1992

	Production ($m^3 \times 10^{-9}$)	Consumption ($m^3 \times 10^{-9}$)	Reserves ($m^3 \times 10^{-12}$)
Canada	171 (27)	84 (73)	1.8
Mexico	36 (26)	36 (29)	1.8
USA	534 (503)	604 (560)	4.69
France	2 (3)	38 (33)	—
Germany	22 (19)	93 (76)	0.3
Italy	19 (18)	62 (49)	0.2
Netherlands	80 (87)	50 (48)	1.8
Norway	46 (28)	3.6 (3.7)	1.2
Russia	591 (641)	395 (457)	48.0
Ukraine	18 (21)	74 (111)	1.1
Turkmenistan	13 (60)	44 (8)	2.9
Uzbekistan	55 (40)	40 (38)	1.9
Indonesia	63 (54)	27 (18)	2.1
Japan	2 (2)	69 (57)	–[a]
Algeria	74 (56)	21 (22)	3.8
Saudi Arabia	47 (34)	47 (34)	5.8
UK	90 (56)	88 (62)	0.7
Venezuela	28 (22)	28 (22)	4.1
Iran	50 (25)	52 (24)	23.0
Iraq	3 (3)	3 (1)	3.1
Argentina	30 (20)	30 (22)	0.68
Qatar	20 (11)	15 (11)	9.9
United Arab Emirates	37 (29)	30 (24)	5.9
Other	318 (348)	434 (334)	17.6
Total	2349 (2124)	2327 (2116)	146 (136)

[a]Implies negligible amounts, blank refers to unknown amounts.

The anaerobic digestion is a two-stage biological process involving "acid forming" bacteria which convert the fats, carbohydrates and proteins into simple acids such as acetic and propionic acid. At the same time the "methane forming" bacteria convert the acids by decarboxylation into CO_2 and CH_4. The optimum temperature range is 30–45°C and economically between 25 and 35°C to reduce heating costs.

The biogas is composed of 60–70% CH_4 and 30–40% CO_2. Small amounts of H_2, H_2S and NH_3 are also formed. The gas has an energy content of 22–26 MJ/m^3. The H_2S at about 10 g/m^3 can be reduced to acceptable levels (<1.5 g/m^3) by passing the gas through iron oxide sponge which can be regenerated by exposing it to air.

Landfill gas and biogas can be readily converted to syngas either catalytically or by heated wires

$$CH_4 + CO_2 \rightarrow 2H_2 + 2CO \tag{6.2}$$

The syngas can then be used to form methanol or gasoline by catalytic reaction after readjusting the H_2/CO ratio to 2.

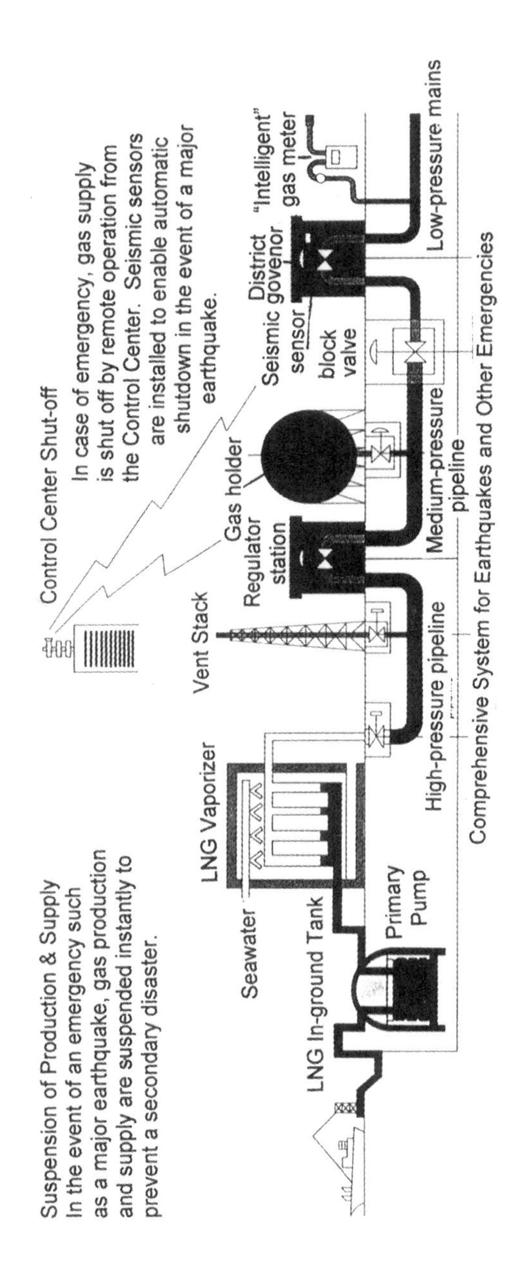

FIGURE 6.2. Distribution scheme for natural gas from stored LNG.

TABLE 6.2
Composition of Various Natural Gas Fields (% Vol.)

	Methane	Ethane	Propane	Butane	Pentane	N_2He	H_2S	CO_2
Europe								
France Lacq	69.0	3.0	0.9	0.5	0.5	1.5	15.3	9.3
Germany S. Oldenburg	89.6	1.7	—[b]	—	—	8.2	—	0.5
Netherlands Groningen	81.3	2.8	0.4	0.2	—	14.4	—	0.9
Italy Ravenna	99.5	0.1	—	—	—	0.4	—	—
UK Hewett	92.6	3.6	0.9	0.4	0.3	2.2	—	—
Norway EKOFISH	90.9	5.9		1.1[a]		0.6	—	1.5
U.R.S.S. Urengoy	85.3	5.8	5.3	2.1	0.2	0.9	—	0.4
Africa								
Algeria Hassi R'Mel	83.7	6.8	2.1	0.8	0.4	6.0	—	0.2
Nigeria Umuechem	79.6	7.6	5.2	2.7	3.3	0.5	—	1.1
Libya Amal	62.0	14.4	11.0	5.5	2.6	3.4	—	1.1
Middle East								
Saudi Arabia Ghawar	59.3	17.0	7.9	2.6	1.1	0.4	1.6	10.1
Iraq Kirkouk	55.7	21.9	6.5	3.9	1.7	—	7.3	3.0
Iran Agha Jari	76.1	11.1	6.1	2.2	1.1	—	0.3	3.1
AbuDhabi Murban OffS	76.4	8.1	4.7	2.6	1.8	0.1	1.7	4.5
Asia								
Pakistan Sui	88.5	0.9	0.3	0.4	—	2.5		
India								
Assam	83.3	10.9	1.5	0.3	0.9	0.4	7.4	
Bombay High (Off Shore)	87.0	6.4	4.5	0.7	0.4	0.01	—	2.6
Oceania								
Australia North Bankin	88.7	5.6	1.8	0.6	0.3	0.7		
Indonesia Arun	71.9	5.6	2.6	1.4	3.6	0.4	—	2.3
New Zealand Kapuni	44.2	11.6		11.6[a]		—	—	14.5
America								
US California	88.7	7.0	1.9	0.3	—	1.5		
Canada Alberta	91.9	2.0	0.9	0.3	—	4.9	—	0.6
Mexico Tampico	46.0	0.6		0.6[a]		2.4	—	—
Venezuela Maracaibo	82.0	10.0	3.7	1.9	0.7	1.5	2.2	48.8
Argentina Camp Duran	88.8	5.9	1.8	0.3	0.1	1.4	—	0.2
							—	1.7

[a]Includes $C_3 + C_5$ fractions.
[b]Insignificant amounts.

Another potential source of CH_4 is primordial gas (nonfossil origin) which Thomas Gold has proposed exists 10–20 km below the earth's surface—just below the earth's crust. This gas is believed to exist throughout the earth and in sufficient quantities to last for several centuries. Experimental drilling has been conducted in Sweden and could determine if the Gold hypothesis is correct.

Still another source of methane is the gas hydrate. At high pressure methane forms a clathrate (cage) complex with water which is stable at temperatures below 20°C and at pressures greater than 20 atm. This natural gas hydrate is present on the ocean floor, in the sediment below the sea floor as well as in the cold permafrost of the Arctic.

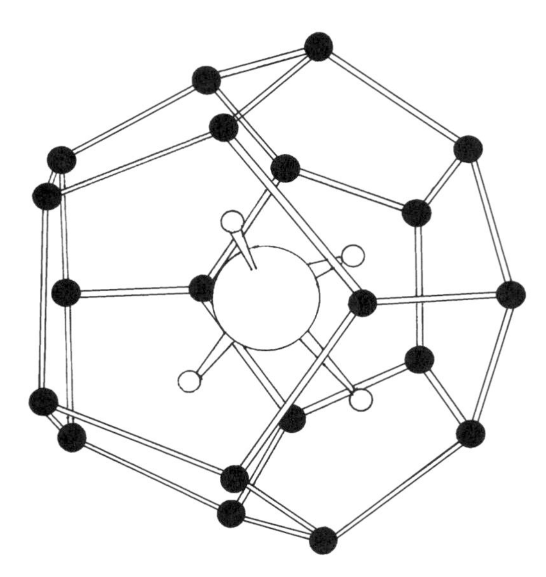

FIGURE 6.3. Model of part of a methane-hydrate cell showing the oxygen atoms (●) of the water molecules with the CH_4 in the center.

The structure of the methane-hydrate is illustrated in Fig. 6.3 where part of a unit cell is shown for 20 water molecules with one methane molecule. The unit cell contains two CH_4 molecules associated with 46 H_2O molecules to give about 3.7% CH_4 by weight. The estimated amounts of CH_4 bound as the hydrate in permafrost and deep ocean waters is 10 teratons or about 53% of the earth's total combustible carbon. Such a dilute system will require new technology and innovative engineering for the successful exploitation of this unique energy source.

Natural gas is usually saturated with water as it leaves the well and as a result the gas hydrates can form in the pipeline restricting the flow of gas. To destroy such plugs in the cold Siberian pipeline large amounts of methanol are introduced at the approximate rate of 1 kg/1000 m^3 of gas. This represents more methanol than can be produced by a standard methanol plant.

6.3. NATURAL GAS USES

Methane — the major constituent of natural gas — is not only an excellent fuel but an important chemical. This is illustrated by the various reactions in Fig. 6.4 where each of the products are important reactants themselves. The thermodynamics of methane pyrolysis reactions is shown in Fig. 6.5 for equilibrium conditions. If, however, the hydrogen produced is removed from the reaction system then the product yields can be substantially higher. This has been demonstrated using thin palladium membranes which allow only H_2 to pass through. Similarly when the pyrolysis occurs on a hot wire the hydrogen produced can diffuse out of the reaction zone faster than the other heavier

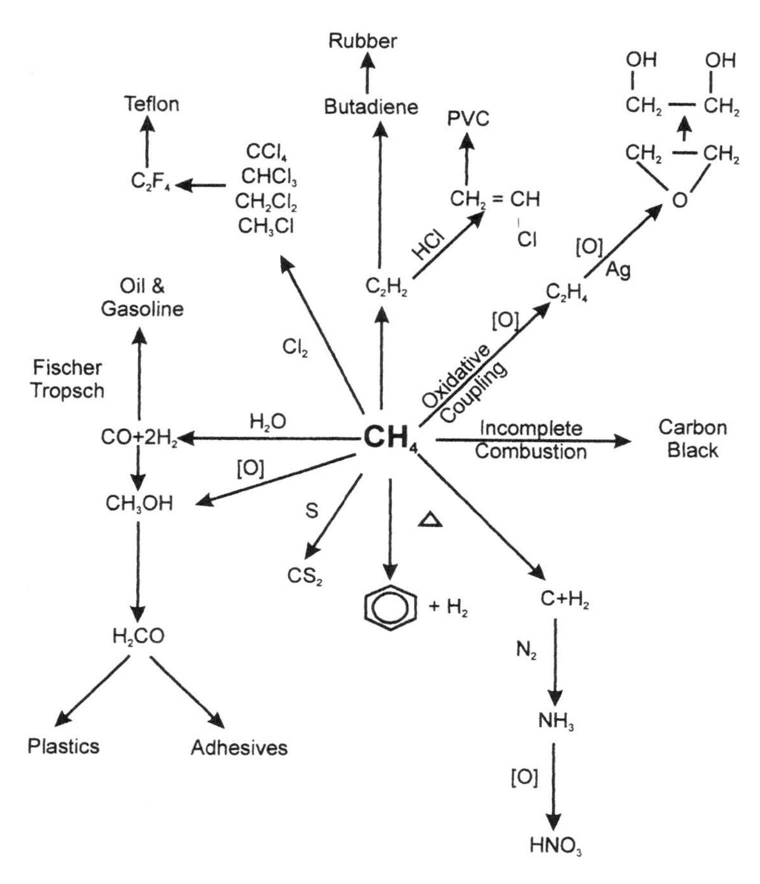

FIGURE 6.4. Chemical reactions and uses of methane.

product and therefore equilibrium conditions do not prevail. Thus, in Fig. 6.5, though reaction #1 is preferred thermodynamically, it is found that experimentally reaction #4 occurs on a hot wire forming an aromatic oil.

6.4. NATURAL GAS AS A FUEL

Natural gas is an excellent fuel which burns cleanly with little or no residue. The main problem with its use is its low energy density—it is not readily stored in sufficient quantities to power a vehicle for the normal distance of 300–400 km. The simplest storage method is as a compressed gas in cylinders. Buses in Vancouver have such storage cylinders on the roof. It has been shown that by filling a cylinder with active carbon it is possible to double the amount of CH_4 which can be stored in the cylinder. However, this still limits the use of compressed natural gas (CNG) to fleet vehicles which can routinely refill the cylinders with compressed gas. A more novel approach

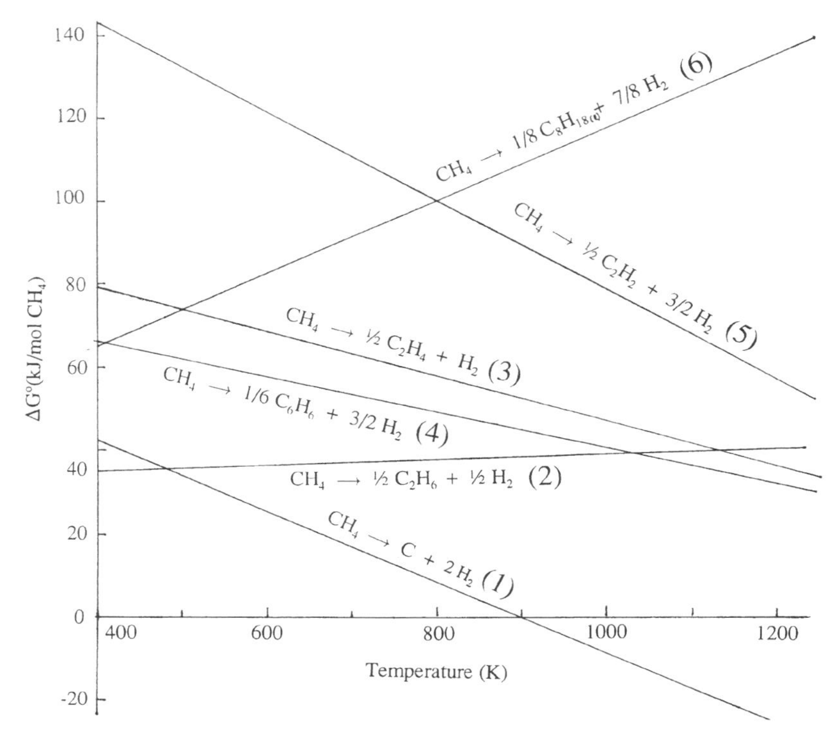

FIGURE 6.5. The thermodynamics of six CH_4 pyrolysis reactions.

FIGURE 6.6. Buses in Zigong, Sichuan, China, run on natural gas stored in inflatable bags on the roof. The bags can contain about 36 m^3 and permits the bus to travel about 80 km before refueling.

has been developed in China where an inflatable bag on the roof stores the natural gas. This is shown in Fig. 6.6.

Natural gas can be stored in insulated containers as a liquid at $-161^{\circ}C$, and as a liquid fuel the LNG has sufficient energy density. Its use has been tested in airplanes and locomotives. However, the use of natural gas as an automotive fuel is still limited unless higher densities can be achieved.

Recent designs of thin-walled polyethylene carbon-fiber over wrapped lightweight high-pressure cylinders encapsulated with energy-absorbing foam in a protective casing have enabled a Geo Prizm sedan to achieve a driving range of 315 miles (500 km) when fueled with natural gas. This is the most promising new approach to the clean fueled automobile.

6.5. OTHER CARBON BASED FUEL GASES

Natural gas is not the only carbon-based gas which can be used as a fuel. One of the simplest is derived from coal. The reaction of coal, coke, or charcoal with insufficient air for complete combustion forms Producer Gas which is a mixture of CO and N_2 in a 1:2 ratio

$$2C + O_2 + 4N_2 \rightarrow 2CO + 4N_2 \tag{6.3}$$

Any moisture in the air results in the formation of hydrogen by the water gas reaction

$$C + H_2O_{(g)} \rightarrow CO + H_2 \tag{6.4}$$

to form Water Gas. The water shift reaction can also occur to produce some CO_2

$$CO + H_2O_{(g)} \rightarrow CO_2 + H_2 \tag{6.5}$$

When coal is heated to about 500°C the CH_4 trapped in the pores is released. At higher temperature (1000°C) the organic matrix in the coal is decomposed, forming mostly H_2 with some CO_x. This Coal Gas is also formed during the coking process.

The addition of oil to the water gas process at approx. 30 L of oil per 100 m^3 of gas increases the calorific value of the gas produced 3-fold by cracking the oil to $C_2 + C_3$ hydrocarbons.

The gasification of coal with oxygen and steam under pressure is called the *Lurgi Process* and the gas is called *Lurgi Gas*. As the pressure is increased from 5 to 20 atm the CH_4 and CO_2 increases while the H_2 and CO decrease. This gas, once popular as a town gas, is seldom used today.

The approximate composition and heating values of the various fuel gases are given in Table 6.3. The flame speeds relative to that of H_2 are also shown and account for the need to change burner nozzles when fuel gas composition is significantly changed.

The TLV value of CO is 50 ppm and hence a level of 20 ppm is considered to be safe. However, it has been shown that such low levels of CO can inhibit the learning

TABLE 6.3
Approximate Composition and Calorific Values of Various Fuel Gases (%Vol.)

Gas comp.	Natural gas	Producer gas	Water gas	Coal gas	Lurgi[a] gas	Landfill gas	Bio gas	Sewage gas	Flame[b] speed	ΔH^0(MJ/m^3) combustion
CO	—	33	46	7	24	1	—	—	18	11.75
CO_2	1	1	2	2	26	58	63	60	—	—
H_2	—	4	48	53	45	—	—	1	100	11.88
CH_4	88	2	1	30	4	40	35	35	14	37.2
C_2+C_3	10	—	—	3	—	1	—	—	16	60
N_2	1	60	3	5	1	—	2	2		0
Calorific value, MJ/m^3	39	5	11	20	16	22				

[a] 2 atm pressure.
[b] Relative to H_2 + 2 atm pressure.

process in children and reduce the response time of adults. Thus, apparently safe levels may not be healthy if long term exposure is anticipated. A catalyst of gold supported on metal oxide surfaces has recently been shown to convert CO in air to CO_2 at room temperature.

6.6. EXPLOSION LIMITS

All combustible gases and vapors can be made to explode if mixed with the proper proportion of air or oxygen. There are two limits of gas concentration which are referred to as explosion limits. Below the lower limit (LEL) there is insufficient gas to carry the chain reactions which results in the rapid released of energy due to the exothermic heat of reaction. Above the upper limit (UEL) there is insufficient oxygen to permit the chain reaction to proceed. This is illustrated in Fig. 6.7 for methane and shows how the limits vary with pressure and temperature and how they differ for air and oxygen. The explosion limits and autoignition temperatures of various gases and vapors are listed in Table 6.4 which shows some wide ranges of limits for H_2, CO, and acetylene. Mixtures of gases such as water gas (CO + H_2) have two sets of limits. Gas which can explode over a wide range of concentrations presents a major hazard when these gases leak into air. However, light gases such as H_2 diffuse in air very rapidly and hence the danger of an explosion is reduced. Combustible solids in the form of dust can also explode. This is illustrated by coal dust, grain dust, and manufacturing dust which have all caused explosions. The very narrow explosion limits of methyl bromide are of interest since this substance can reduce the explosive characteristics of dust explosions.

6.7. HYDROGEN

Hydrogen is receiving a lot of attention lately as a fuel of the future. We are all aware of the difficulty the world will be facing as our nonrenewable energy resources become depleted. Oil is expected to last, by various estimates, into the early 22nd

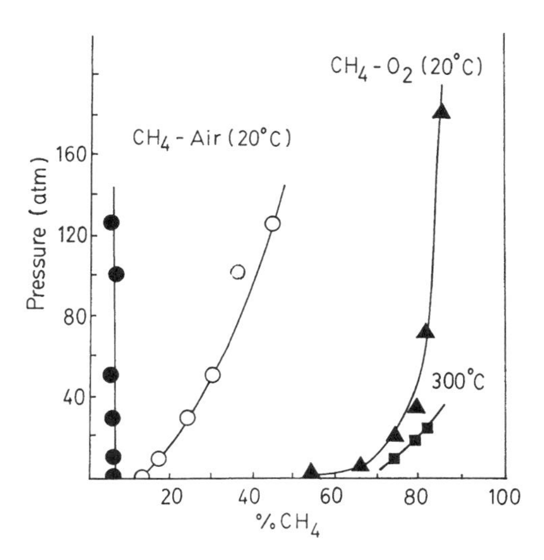

FIGURE 6.7. Explosion limits for CH_4 in air (circles, ● and ○) and in O_2, filled points, ● and ▲ at 20°C and for CH_4 in O_2 at 300°C (● and ■).

century and coal perhaps for another century or two depending on the alternate energy sources utilized. Natural gas will probably replace oil and its reserves will not last much longer than oil. Hence, hydrogen, which is readily made from water, is being considered as a replacement for natural gas and even as a substitute for gasoline. However, there are some problems and difficulties which should be considered, since hydrogen is not a primary fuel but must be made using another energy source such as hydroelectricity,

TABLE 6.4
Autoignition Temperatures and Explosive Limits of Some Gases and Vapors in Air as Vol.%

Substance	Autoignition Temp. °C	LEL/UEL	Substance	Autoignition Temp. °C	LEL/UEL
Acetone	465	2.6–12.8	Hexane	225	1.1–7.5
Acetylene	305	2.5–8.0	Hydrogen	400	4.1–74.0
Ammonia	651	16.0–26.0	Isopropanol	456	2.0–12.0
Benzene	562	1.3–7.1	Jet fuel		1.3–80.0
Butane	405	1.8–8.4	Kerosene	210	1.3–80.0
Carbon disulfide	125	1.3–50.0	Methane	650	5.3–14.0
Carbon monoxide	609	12.5–74.0	Methyl bromide	537	13.5–14.5
Cyclopropane	500	2.4–10.4	Naphtha	277	0.9–6.0
1,2-Dichloroethylene	460	9.7–12.8	Naphthalene	568	0.9–5.9
Dimethyl ether	350	3.4–27.0	*n*-Octane	220	0.8–3.2
Ethane	515	3.2–12.5	Propane	450	2.4–9.5
Ethanol	423	3.3–19.0	Toluene	480	1.2–7.1
Ethylene	490	3.0–36.0	Water gas		6.0–9.0, 55.0–70.0
Fuel oil		0.7–5.0	Xylene	530	1.0–6.0
Gasoline	280–456	1.4–7.6	Dust		~50 mg/L

coal, nuclear energy, or solar energy to name but a few. Hydrogen is thus classified as an energy currency like electricity which can be pumped from one location to another and stored. It can also be used as a fuel for the automobile and so can replace gasoline when our fossil fuels are depleted.

What do we know about hydrogen?

1 It is easily made by the electrolysis of water or appropriate thermochemical cycles.
2 It is lighter than air and was used in lighter-than-air dirigibles such as the Zeppelin.
3 It is highly explosive and with the destruction of the Hindenburg in 1937, commercial travel in H_2-filled dirigibles ended.

However, it must be pointed out that the Hindenburg disaster resulted in the death of only 36 people, most from flaming diesel fuel from the engines and from jumping to the ground. The Hindenburg had made 54 flights, 36 of which were across the Atlantic.

The physical properties of hydrogen are given in Table 6.5 and the general methods of preparation are outlined in Table 6.6.

6.8. METHODS OF PREPARATION OF H_2

6.8.1. Electrolysis

The most attractive method of producing hydrogen is by hydroelectricity at off-peak load. The minimum voltage required for water splitting at 25°C and 1 atm is 1.23 V (reversible voltage). Under these conditions hydrogen would be produced only if heat is added. If no heat exchange with the surroundings takes place, the cell cools down. The thermoneutral potential at 25°C is 1.48 V.

The electrolysis cells draw thousands of amperes (current densities of 1–4 kA/m^2 (100–400 A/ft^2) at high temperatures (up to 300°C) and pressures (above 200 atm). The electrolyte is usually potassium hydroxide (KOH) but acidic solutions (H_2SO_4) are also used occasionally.

At 25°C, $\Delta G^0 = -237.13$ kJ for water, but at 900°C, $\Delta G^0 = -182.88$ kJ for steam. Hence, the minimum voltage E_{th} (900°C) = 0.906 V at the higher temperature and the electrolysis of steam at high temperatures and pressures is also being studied as a means of producing hydrogen efficiently. At 200°C the voltage of 1.3 V is the minimum voltage at which electrolysis would occur. Above this voltage heat is produced due to the IR drop in the cell and due to the over-voltage on the electrodes. At 1 ¢/kwh the cost of 1 GJ of H_2 is 3.8¢ at 1.3 V and 4.9¢ at 1.8 V. Hence, great care is required in designing cells, electrodes, and electrode separators in order to reduce costs.

The production of oxygen which accompanies the hydrogen is a surplus product since oxygen is readily separated from nitrogen in liquid air or by the use of molecular sieves or membranes. Hence, by using a carbon anode, the following reactions would occur:

$$H_2O_{(l)} + C \rightarrow H_2 + CO \qquad \Delta G^0 = 100.0 \text{ kJ/mol} \tag{6.6}$$

$$H_2O_{(l)} + C \rightarrow H_2 + CO_2 \qquad \Delta G^0 = 40 \text{ kJ/mol} \tag{6.7}$$

TABLE 6.5
Selected Physical Properties of Hydrogen (H_2)

Property			Value
Molecular weight (g/mol)			2.01594
Natural isotopic abundance			
^1H	1.00794 g/mol	99.985%	
^2H (deuterium)	2.01355 g/mol	0.015%	
^3H (tritium)[a]	3.01550 g/mol	0	
Melting point (K)			14.2
Boiling point (K) para-H_2			20.27
Boiling point (normal-H_2; 25% p-H_2) (K)			20.39
Triple point (K, n-H_2)			13.96
Triple point liquid density (kg/m^3)			77.20
Triple point solid density (kg/m^3)			86.71
Triple point vapor density (kg/m^3)			0.131
Boiling point liquid density (kg/m^3)			71.0
Boiling point vapor density (kg/m^3)			1.33
Critical temperature (K, n-H_2)			33.19
Critical pressure (atm, n-H_2)			12.98
Critical volume (kg/m^3, n-H_2)			30.12
Latent heat of fusion at T.P.; p-H_2 (J/mol)			117.6
Latent heat vaporization at B.P. (J/mol, n-H_2)			897.3
Heat of combustion (gross) liquid H_2O (kJ/mol)			285.8
Heat of combustion (net) gaseous H_2O (kJ/mol)			241.8
Limits of flammability in air (vol.%)			4.0–75.0
Limits of detonability in air (vol.%)			18.0–59.0
Burning velocity in air (m/s)			up to 2.6
Burning velocity in oxygen			up to 8.9
Limits of flammability in oxygen (vol.%)			4.0–95.0
Limits of detonability on oxygen (vol.%)			15.0–90.0
Detonation velocity			
15% H_2 in O_2 (m/s)			1400
90% H_2 in O_2 (m/s)			3600
Nonflammable limits, air-H_2			$<8\%$ H_2
Nonflammable limits, O_2-H_2			$<5\%$ H_2
Maximum flame temperature @ 31% H_2 in air (K)			2400
Autoignition temperature in air (K)			847
Autoignition temperature in oxygen (K)			833

[a]Radioactive $\lambda_{1/2} = 12.26$ yr.; n-H_2 refers to normal H_2 (25% p-H_2 and 75% o-H_2).

This can be compared with $\Delta G_f^0(H_2O)_l = -237.2$ kJ/mol and represents a substantial energy saving. However, when hydrogen is prepared by this method it no longer is an environmentally friendly fuel since CO and CO_2 will be formed.

The main problem involved in hydrogen production by electrolysis is the materials of construction since reliability is essential if explosions are to be avoided.

6.8.2. Thermal Methods

Several closed-cycle thermal processes have been developed whereby hydrogen can be produced from water using, e.g., coal as the source of heat. Three common schemes are presented in Table 6.7. The corrosive nature of some of the products implies rather

TABLE 6.6
Preparation of H_2

1. Electrolysis of water	$H_2O \rightarrow H_2 + \frac{1}{2}O_2$
2. Thermal methods	$2HI \rightarrow H_2 + I_2$
3. Natural gas	$CH_4 \rightarrow C + 2H_2$; $CH_4 + H_2O \rightarrow CO + 3H_2$
	$CH_4 + \frac{1}{2}O_2 \rightarrow CO + 2H_2$
4. Thermal, nuclear, electrical	$CH_4 + H_2O \rightarrow CO + 3H_2$
5. Photoelectrolysis of water	$H_2O + h\nu + \text{catalyst} \rightarrow H_2 + \frac{1}{2}O_2$

elaborate precautions in construction and design. The total energy required for the reaction

$$H_2O_{(l)} \rightarrow H_2O_{(g)} \rightarrow H_2 + \tfrac{1}{2}O_2 \qquad \Delta H^0 = 285.8 \text{ kJ} \tag{6.8}$$

is 285.8 kJ if one starts with liquid H_2O and 241.8 kJ if steam is available. In either case this energy can be supplied in a series of steps as shown in Table 6.7. The efficiencies given represent the thermal efficiencies of the sum of the steps as well as the Carnot (reversible) efficiency.

6.8.3. Natural Gas

The major use of hydrogen today is in the production of ammonia via the Haber reaction

$$\tfrac{1}{2}N_2 + \tfrac{3}{2}H_2 \xrightarrow[500^\circ\text{C } 300\text{ Atm}]{\text{Catalyst}} NH_3 \qquad \Delta H^0 = -46.2 \text{ kJ} \tag{6.9}$$

TABLE 6.7
Three Schemes for the Thermal Generation of Hydrogen by Closed-Cycle Processes with Thermal Efficiency (TE) and Carnot Efficiency (CE)

Agnes		
TE = 41–58%	$3FeCl_2 + 4H_2O = Fe_3O_4 + 6HCl + H_2$	450–750°C
CE = 58%	$Fe_3O_4 + 8HCl = FeCl_2 + 2FeCl_3 + 4H_2O$	100–110°C
	$2FeCl_3 = 2FeCl_2 + Cl_2$	300°C
	$Cl_2 + Mg(OH)_2 = MgCl_2 + \frac{1}{2}O_2 + H_2O$	50–90°C
	$MgCl_2 + 2H_2O = Mg(OH)_2 + 2HCl$	350°C
Beluah		
TE = 53–63%	$2Cu + 2HCl = 2CuCl + H_2$	100°C
CE = 63%	$4CuCl = 2CuCl_2 + 2Cu$	30–100°C
	$2CuCl_2 = 2CuCl + Cl_2$	500–600°C
	$Cl_2 + Mg(OH)_2 = MgCl_2 + H_2O = \frac{1}{2}O_2$	80°C
	$MgCl_2 + 2H_2O = Mg(OH)_2 + 2HCl$	350°C
Catherine		
TE = 64–83%	$3I_2 + 6LiOH = 5LiI + LiIO_3 + 3H_2O$	100–190°C
CE = 83%	$LiIO_3 + KI = KIO_3 + LiI$	0°C
	$KIO_3 = KI + 1\frac{1}{2}O_2$	650°C
	$6LiI + 6H_2O = 6HI + 6LiOH$	450–600°C
	$6HI + 3Ni = 3NiI_2 + 3H_2$	150°C
	$3NiI_2 = 3Ni + 3I_2$	700°C

The nitrogen is prepared by the fractional distillation of liquid air (b.p. $N_2 = -196°C$, b.p. $O_2 = -183°C$) whereas the hydrogen is usually prepared by the thermal cracking of natural gas

$$CH_4 \rightarrow C + 2H_2 \qquad \Delta H^0 = 74.8 \text{ kJ} \tag{6.10}$$

Hydrogen, as a component of synthesis gas, is also made from carbon and methane by the following reactions:

$$C + H_2O \rightarrow CO + H_2 \qquad \Delta H^0 = 131.3 \text{ kJ} \tag{6.11}$$

$$CH_4 + H_2O \rightarrow CO + 3H_2 \qquad \Delta H^0 = 206 \text{ kJ} \tag{6.12}$$

$$CH_4 + \tfrac{1}{2}O_2 \rightarrow CO + 2H_2 \qquad \Delta H^0 = -36 \text{ kJ} \tag{6.13}$$

$$CH_4 + 2H_2O \rightarrow CO_2 + 4H_2 \qquad \Delta H^0 = 165 \text{ kJ} \tag{6.14}$$

The CO_2 can be separated from H_2 by its solubility in water at 25 atm.

6.8.4. Thermal–Nuclear–Electrical

Water is readily dissociated into its elements H_2 and O_2 by an electrical discharge or by thermal energy. The separation of H_2 from O_2 is then required. An interesting proposal presented several years ago was to detonate a nuclear explosive device in a land-fill (garbage dump). A calculated yield of the products, H_2, CO, CO_2, and CH_4 showed that as the organic content of the garbage reaches 80% the H_2 and CO each approach 50%. The obvious problem in such a scheme is that the gases cannot be used until the radioactivity of the gases have decayed to acceptable levels.

6.8.5. Photoelectrolysis

About 25 years ago it was first shown that ultraviolet light can dissociate water on a suitable semiconductor catalyst. Since then it has been possible to photocatalytically dissociate water into H_2 and O_2 at separate electrodes (hence not requiring a separation of the gases) or to generate electrical energy for direct use. This rapidly expanding field will eventually achieve what nature does in plants, i.e., using visible light in multiple steps to convert CO_2 and water into cellulose. Efficiencies are usually low and long term stability has not as yet been achieved but with continued efforts it will be possible to produce hydrogen from sunlight and water.

6.9. TRANSPORTATION AND STORAGE OF H_2

The cost of pipeline transportation of hydrogen is higher (20–30%) than that for natural gas for equivalent energy flow. Another major difference is due to the small molecular size of H_2 (diameter = 0.289 nm) compared to CH_4 (diameter = 0.38 nm). Hence H_2 pipelines are more susceptible to leaks.

Another problem anticipated with hydrogen is the hydrogen embrittlement of metals, usually at elevated pressures and at ambient temperatures. Hydrogen has a tendency to diffuse through metals along grain boundaries and, in some cases, forming metal hydrides. This makes the metal brittle, causing fractures and failure. This was first noted about 100 years ago and special alloys have been made specifically for H_2 service. Thus the use of natural gas pipelines to transport hydrogen may not be the safest thing to do.

The storage of hydrogen as a liquid (LH_2) is well known in the nuclear field where the liquid hydrogen bubble chamber (5–1500 L) is used to detect nuclear particles. Liquid hydrogen is also used as a fuel.

The hydrogen molecule exists in two forms: ortho-H_2 and para-H_2. In ortho-H_2, the nuclear (proton) spins of the two hydrogen atoms are parallel whereas in para-H_2, the nuclear spins are antiparallel. It is possible to catalytically convert hydrogen to the pure lower energy p-H_2 at low temperatures ($T < 20$ K) but at higher temperatures ($T > 200$ K) only an equilibrium mixture of 25% p-H_2 and 75% o-H_2 can be obtained. The conversion of o-H_2 to p-H_2 is an exothermic process which at 10 K, $\Delta H = -1.062$ kJ/mol and at 300 K, $\Delta H = 55.5$ J/mol.

The liquefaction of hydrogen produces 25% p-H_2 + 75% o-H_2 and the slow conversion of o-H_2 to p-H_2 adds an additional heat source to the storage system. Hence it is desirable to convert the o-H_2 to p-H_2 either completely in the liquid state (by adding charcoal) or preferably partially in the precooled gas phase (at 77°K, liquid N_2, the equilibrium mixture of H_2 vapor is 60% p-H_2) followed by complete conversion in the liquid state. This reduces the losses on storage from 25%/day for 25% p-H_2 to 0.02%/day for 98% p-H_2.

Liquid hydrogen, like liquid helium, is usually stored in double Dewar flasks, an outer Dewar containing liquid nitrogen (b.p. = 77 K) into which is placed the second Dewar containing the liquid hydrogen (b.p. 20.3 K). However, in transport by truck or rail, single horizontal cylindrical Dewars are used with multilayer insulation, resulting in typical boil-off losses of 0.25%/day.

Table 6.8 lists some thermal conductivities of evacuated powders used in large Dewars for LH_2. Since the dielectric powders are transparent to room temperature radiation the effectiveness of the powders can be increased significantly by the addition of metal powders (e.g., Al, Cu, Ag) which reflect the radiation. Since the metal powders conduct, their addition is an optimum at certain levels.

It is obvious that liquid hydrogen is practical only for large continuous users such as ships, trains, and even aircraft, but impracticable for intermittent users such as automobiles. Hydrogen is not very soluble in liquids but dissolves in some metals, forming hydrides which readily dissociate reversibly back to the metal and hydrogen at elevated temperatures. The formation and dissociation cycle of a typical system is shown in Fig. 6.8 for a $LaNi_5$ alloy which reacts with 3 molecules of H_2 to form $LaNi_5H_6$

$$LaNi_5 + 3H_2 \rightleftharpoons LaNi_5H_6 \tag{6.15}$$

though usually more hydrogen can be "dissolved" if the pressure is increased. Several metal systems can be used in this way to store hydrogen. A comparison is given in Table 6.9 where cryoadsorption of H_2 on active carbon at liquid nitrogen temperatures

TABLE 6.8
Thermal Conductivities (λ) of Some Evacuated Insulating Powders

Powder	Particle size	Density (g/mL)	Vacuum (Torr)	λ (μW/cm, K)
Perlite, expanded	+30 mesh	0.06	$<10^{-4}$	21
Perlite, expanded	−80 mesh	0.14	$<10^{-4}$	10
Silica aerogel	250 Å	0.10	$<10^{-4}$	21
Diatomaceous earth	1–100 μm	0.29	$<10^{-4}$	10
Cab-O-Sil	200–300 Å	0.04	$<10^{-6}$	30
Cab-O-Sil + 50% Metal powder $<44\ \mu$m	200–300 Å	—	$<10^{-6}$	2.0

(−196°C) is included along with encapsulated H_2 in zeolite. (Zeolites are natural and synthetic minerals of silicon and aluminum which have interconnected channels and cages.) By modifying the size of the opening of a type 3A molecular sieve (approx. 3 Å cage size) by replacing Na^+ ions by the larger Cs^+ ion it is possible to force H_2 into the cage at 250°C and 10,000 psi (500–4000 atm). When quenched to room temperature the gas remains in the cage even when the pressure is released. The loss is insignificant for short periods (approx. 1 week) and could be reduced by further modification of the zeolite. In the case of methane encapsulation it has been shown that up to about 7% by weight CH_4 can be stored in an unmodified type 3A molecular sieve for 5 months without noticeable loss.

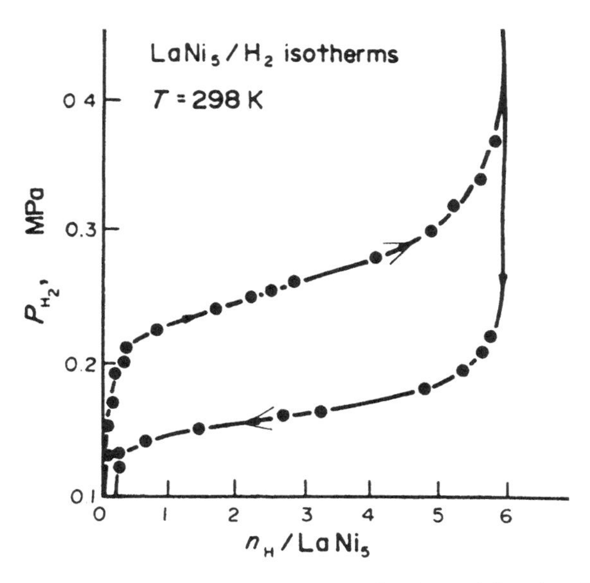

FIGURE 6.8. Equilibrium pressure-temperature curves (isotherms) for the alloy $LaNi_5H_n$ where n changes from 0 to 6 as the pressure of H_2 increases from 0 to over 2 atm. Note the hysteresis effect.

TABLE 6.9
A Comparison of Some H_2 Storage Systems

Storage medium	Storage capacity Wt.%	Storage capacity Vol. g/mol	Energy density Wt. kJ/g	Energy density Vol. kJ/mL
MgH_2	7	0.10	9.93	14.3
Mg_2NiH_4	3.16	0.08	4.48	11.5
VH_2	2.07	0.095	2.93	13.9
$FeTiH_{1.95}$	1.75	0.096	2.48	13.6
$LaNi_5H_7$	1.37	0.089	1.94	12.6
H_2 (liquid) $-253°C$	100	0.07	141.8	9.93
H_2 (gas) 100 atm	100	0.008	141.8	1.13
H_2 (carbon)[a] $-196°C$	6.8	0.024		
H_2 (zeolite)[b]	1.5	0.009		
H_2 (NiO; 2.5 SiO_2) $-196°C$	3.4	0.024		

[a]At 40 atm — mass and volume of liquid nitrogen is not included.
[b]Encapsulated by high pressures greater than 1000 atm but stored at ambient temperatures (25°C).

It should be pointed out that the density of hydrogen in the metal hydride is similar to that of liquid H_2, implying that hydrogen in the metal is in an atomic form. This has been verified by other methods. The relative cost of the various storage systems is estimated in Table 6.10 but this too will greatly depend on the quantities involved.

Hydrogen has been used in surface vehicles such as automobiles and buses, and plans are in progress to convert a Lockheed L-1011 to burn LH_2 and fly between Pittsburgh, Montreal, Birmingham, Frankfurt. and Riyadh.

6.10. SAFETY

It is constantly reported that hydrogen is a fuel which is so safe — it requires no chimney since its only product is water (H_2O). This is not always correct since it has been shown that when used as a fuel in an internal combustion spark engine hydrogen also forms hydrogen peroxide (H_2O_2) which was initially shown to be at about 225

TABLE 6.10
Approximate Comparative Costs for Alternative Hydrogen Storage Systems

Facility	Cost $/k Wh
Cryoadsorber (carbon)	0.025
Metal hydride (FeTl alloy)	0.028
Compressed gas (11 atm)	0.028
Liquid H_2	0.035

TABLE 6.11
Ignition Energy of H_2-Air Mixtures

$\%H_2$ in air	10	15	20	30	40	50	60
Ignition energy, mJ	0.2	0.06	0.025	0.02	0.03	0.07	0.3

ppm in the exhaust. The TLV value for H_2O_2 is 1 ppm and though it is easy to destroy H_2O_2, its removal to below 1 ppm is not a simple task. A more recent determination of the H_2O_2 formed in the exhaust of a hydrogen fueled engine showed that very little H_2O_2 is present and that the previously reported high levels were due to the method of analysis which was affected by NO_x in the exhaust.

Hydrogen is also one of the easiest explosives to ignite when suitably mixed with air or oxygen. For example, the ignition energy of hydrogen (H_2), methane (CH_4), and propane (C_3H_8) are 0.02, 0.45, and 0.25 mJ, respectively. This value depends on the composition of the mixture and for hydrogen the values are given in Table 6.11.

Though hydrogen is extremely dangerous because of its wide range of explosive limits and low ignition energy, it has a low mass and therefore diffuses very rapidly away from a source. Thus, a spilled tank car full of LH_2 caused no problem. With reasonable precautions we will be able to handle hydrogen with as much ease as we now manage gasoline.

6.11. HELIUM

The composition of natural gas (Table 6.1) indicates the presence of up to 0.5–1% helium. The concentration of helium in air is much less than 0.001% and it is impractical to extract helium from air. Argon, another noble gas, is present at up to 1% in air and it is obtained as a by-product in the liquefaction of air and the production of O_2 and N_2. Thus, since He is only available economically from natural gas it is obvious that when natural gas is exhausted, so will be the inexpensive supply of He. Hence it is imperative that wherever possible He should be extracted from natural gas before it is put into pipelines, burned, or used for chemical processes. The He is usually stored in abandoned mines (storage in cylinders as a compressed gas is too costly). It must be stressed that if natural gas with even 0.5% He is burned, the He is lost and will only be recoverable from air at a very great expense.

EXERCISES

1. How can the heat value of hydrocarbon gases be determined?
2. What volume of (a) air and (b) oxygen is required to burn 1L of octane (density = 0.7025 g/mL at 25°C?
3. Explain how a ratio of $H_2/CO = 1$ can be increased to 2 without the addition of hydrogen from an external source.

4. If the heats of combustion of carbon monoxide, hydrogen, and methane are 11.75, 11.88, and 37.28 MJ/m^3, respectively, calculate the heating value of the water gas and bio-gas given in Table 6.3.
5. What is the theoretical weight of steam necessary to convert 1 tonne of coke containing 2% ash into water gas at 900°C?
6. Comment on the statement "Natural gas is more explosive than manufactured gas."
7. Write equations for the removal of sulfur from natural gas.
8. What burner adjustments would have to be made in changing from a bottled gas to natural gas?
9. Hydrogen is claimed to be the ideal fuel since it is made from water and burns to form water. Comment on the validity of this statement.
10. When helium is exhausted, hydrogen would be ideal for lighter-than-air ships. Do you agree?
11. Comment on the prevalence of explosions caused by dust.
12. Some prepared gases such as syngas have two sets of explosion limits. Explain!
13. How can the calorific value of a producer gas be increased?
14. Vehicles in New Zealand are required to run on at least two types of fuels which include CNG, propane, methanol, gasoline, gasohol, and diesel. Comment on the two fuels you would select for your vehicle if living in New Zealand.
15. Some metal oxides can be reduced with hydrogen to form an activated metal $MO_x + H_2 = M + H_2O$. The activated metal can be reacted with water to produce hydrogen and the metal oxide, i.e., the reverse of the above reaction. This means that water is effectively the fuel that produces the combustible hydrogen. The metal oxide can be regenerated to metal similarly to the charging of a battery. Iron sponge has recently been proposed and tested as a typical metal for such a system. Calculate the amount of iron and water needed for a trip of 35 km using H_2 in a SI-ICE vehicle with a thermal efficiency equivalent to a gasoline engine which uses 10 L/100 km.
16. Consider a small vehicle fueled by hydrogen in a SI-ICE. The efficiency of the engine is assumed to be thermally equivalent to gasoline which uses 10 L/100 km. The hydrogen is prepared by the electrolysis of water during the evening (7 p.m.–7 a.m.) and stored for use to drive 35 km. What current is required to produce this hydrogen in a one cell (1.5 V) electrolysis unit?

FURTHER READING

G. D. Berry, *Hydrogen as a Transportation Fuel: Costs and Benefits.* Bus. Tech. Bks. (1997).

R. W. Willett, *The 1996 Natural Gas Handbook*, Wiley, New York (1996).

H. Pohl, *Hydrogen and other Alternate Sources of Energy for Air and Ground Transportation*, Wiley, New York (1995).

S. R. Bell and R. Sekar, Editors, Natural Gas and Alternate Fuels for Engines, Vol. 21, ASME, *Am. Soc. Mech. Eng.*, Fairfield, New Jersey (1994).

G. D. Brewer, Editor, *Hydrogen Aircraft Technology*, CRC Press, Boca Raton, Florida, (1991).

R. E. Billings, *The Hydrogen World View*, Int. Acad. Science, Independence, Missouri (1991).

L.Marinescu-Pasoi et al., Hydrogen Metal Hydride Storage with Integrated Catalytic Recombiner for Mobile Application, *Int. J. Hydr. Energy* **16**(6), 407–412 (1991).

M. R. Swain et al., Hydrogen Peroxide Emissions from a Hydrogen Fueled Engine, *Int. J. Hydr. Energy*. **15**(4), 263–266 (1990).
A. Melvin, Gas Taps its Natural Flare, *New Scientist*, October 7, p. 59 (1989).
Natural gas — An Alternative Transportation Fuel, EMR Ottawa, Ontario (1988).
D. L. Klass, Energy from Biomass and Wastes — 1983 Update, Inst. Gas Technol., Chicago, Illinois (1984).
D. L. Klass, Methane from Anaerobic Fermentation, *Science* **223**, 1021 (1984).
J. O'M. Bockris, *Energy Options. Real Economics and the Solar-Hydrogen System*, Taylor and Francis Ltd., London (1980).
C. Bell et al., *Methane: Fuel of the Future*, Prism Press, Dorset, United Kingdom (1980).
T. Gold and S. Soter, The Deep-Earth-Gas Hypothesis, *Sci. Amer.* **242**(6), June p. 154 (1980).
J. J. Reilly and G. D. Sandrock, Hydrogen Storage in Metal Hydrides, *Sci. Amer.* February, 118 (1980).
Hydrogen: Its Technology and Implications, 5 Vol., CRC Press, Cleveland, Ohio (1977–1979).
R. L. Woolley and D. L. Hendrickson, Water Induction in Hydrogen Powered I.C. Engines, *Int. J. Hydr. Energy* **1**, 401 (1977).
R. L. Woolley, Performance of a Hydrogen Powered Transit Vehicle, *Proc. 11th Intersociety Energy Conversion Conference*, Lake Tahoe (1976).
R. E. Billings et al., Ignition Parameters for Hydrogen Engines, *Proc. 9th Intersociety Energy Conversion Engineering Conference*, San Francisco, CA (1974).
J. G. Finegold et al., *The U.C.L.A. Hydrogen Car: Design, Construction, and Performance*, Paper SAE 730507, Soc. Auto. Eng, New York (1973).
R. E. Billings and F. E. Lynch, *Performance and Nitric Oxide Control Parameters of the Hydrogen Engine*. Publ. No. 73002, Energy Research, Provo, Utah (1973).
R. O. King et al. The Hydrogen Engine: Combustion Knock and Related Flame Velocity, *Transactions E.I.C.* **2**, 143 (1958).
R. O. King and M. Rand, The Oxidation, Decomposition, Ignition, and Detonation of Fuel Vapors and Gases, XXVII. The Hydrogen Engine, *Can. J. Technol.* **33**, 445 (1955).
Biomass to gases and liquid fuels, http://www.gpc.peachnet.edu/
National Fuel Gas Co., http://www.natfuel.com/
Pan American Enterprises Inc. (NG, H2), http://www.panent.com/
US Geological Survey, resources, http://www.usgs.gov/
HyWeb--H2 Fuel Cell, http://www.hydrogen.org/index-e.html
Gaseous explosions, http://www.directedtechnologies.com/energy/safety.html
Helium source and reserves, http://members.truepath.com/gr/helium.html

7

Nuclear Energy

7.1. INTRODUCTION

Nuclear energy represents a viable choice as an alternate source of energy to fossil fuels and as a means of supplying the energy needed to produce some of the alternate fuels and fuel currencies which are environmentally acceptable. Though the hazards and dangers of nuclear energy are known to be enormous, a world with a population growth 2–3% per annum,* will require more energy at a reasonable cost in order to raise the standard of living of the developing nations, to meet the needs of the growing population, and at the same time preserve our environment. Though solar energy has the potential of meeting the energy needs of the world the scattered nature of its power makes the present and foreseeable costs too great to be competitive with nuclear energy on a global basis.

Several countries have already embarked on a program to develop their nuclear energy and this is illustrated in Table 7.1 where we see that France (57 reactors) and Belgium (7 reactors) produce 77% and 60% of their electricity by nuclear energy. A map showing the location of nuclear power stations in Europe is given in Fig. 7.1. In 1993 there were 430 nuclear power reactors producing 337 GWe of electrical power. Canada has 22 power reactors and the USA 109 power reactor producing about 17% and 21%, respectively, of these nations' electricity.

In spite of a moratorium in some countries on the construction of new nuclear power plants, predictions indicate that the present world nuclear energy production of 22.7 EJ will reach about 30 EJ by 2010. This growth in the use of nuclear power is primarily due to its attractiveness as a "clean" nonfossil energy source.

7.2. BASIS THEORY OF NUCLEAR ENERGY

Nuclear energy is based on the Einstein Equation

$$E = mc^2 \tag{7.1}$$

During some nuclear reactions a small loss in mass, Δm, means that a corresponding liberation of energy, ΔE, occurs. During a chemical reaction some heat may be liberated

*The world population is 6.1×10^9 (July 2000) and was 5.64×10^9 in 1994, 4.48×10^9 in 1980 and 2.56×10^9 in 1950. It is estimated to be 7.0×10^9 in 2010 and 8.05×10^9 in 2020.

TABLE 7.1
Nuclear Power Status Around the World, January 2000 (December 1994)

	In operation			Under construction	
	No. of units	Total net GWe	Nuclear share (%) of electricity generation	No of units	Total net GWe
Argentina	2 (2)	0.94 (0.94)	9.0 (14.2)	1	0.692
Armenia	1	0.38	36.3		
Belgium	7 (7)	5.5 (5.5)	57.7 (58.9)		
Brazil	1 (1)	0.63 (0.63)	0.2 (0.2)	1	1.23
Bulgaria	6 (6)	3.54 (3.54)	47.1 (36.9)		
Canada	14 (22)	10.0 (15.8)	12.7 (17.3)		
China	3 (2)	2.17 (1.19)	1.2 (0.3)	7 (1)	5.4 (0.91)
Czech Republic	4 (4)	1.65 (1.65)	20.8 (29.2)	2 (2)	1.8 (1.8)
Finland	4 (4)	2.65 (2.31)	33.1 (32.4)		
France	59 (57)	63.1 (59.0)	75.0 (77.7)	(4)	(5.82)
Germany	20 (21)	22.3 (22.7)	31.2 (29.7)		
Hungary	4 (4)	1.73 (1.73)	38.3 (43.3)		
India	11 (9)	1.90 (1.59)	2.65 (1.9)	3 (5)	0.61 (1.01)
Iran				2 (2)	2.11 (2.39)
Japan	53 (48)	43.7 (38.0)	36.0 (30.9)	4 (6)	3.82 (5.65)
Korea, Rep. of	16 (9)	13.0 (7.22)	42.8 (42.8)	4 (7)	3.82 (5.77)
Lithuania	2 (2)	2.37 (2.37)	73.1 (87.2)		
Mexico	2 (1)	1.31 (0.65)	5.0 (3.0)	(1)	(0.65)
Netherlands	1 (2)	0.45 (0.50)	4.0 (5.1)		
Pakistan	1 (1)	0.13 (0.13)	1.2 (0.9)	1 (1)	0.30
Romania	1 (0)	0.65 (0)	10.7 (0)	1 (5)	(0.30)
Russia (CIS)	29 (29)	19.8 (19.8)	14.4 (12.5)	4 (4)	0.65 (3.16)
South Africa	2 (2)	1.84 (1.84)	7.4 (4.5)		3.38 (3.38)
Slovakia	6 (4)	2.41 (1.63)	36.2 (53.6)	2 (4)	
Slovenia	1 (1)	0.63 (0.63)	38.2 (43.3)		0.78 (1.55)
Spain	9 (9)	7.47 (7.11)	40.0 (36.0)		
Sweden	11 (12)	9.43 (10.0)	46.8 (42.0)		
Switzerland	5 (5)	3.08 (2.99)	36.0 (37.9)		
United Kingdom	35 (35)	13.0 (11.9)	28.9 (26.3)	(1)	(1.19)
Ukraine	16 (15)	13.8 (12.7)	43.8 (32.9)	4 (6)	3.8 (5.7)
USA	104 (109)	97.1 (98.8)	19.5 (21.2)	(2)	(2.33)
World total[a]	436 (430)	351.7 (337.8)		38 (55)	31.7 (44.4)

[a]The total includes Taiwan with 6 operating reactors totalling 4.89 GWe and 2 under construction.

or absorbed but there is no measurable change in mass. The law of conservation of mass is effectively valid. The simplified model of the atom is that of a nucleus surrounded by electrons. The nucleus is composed of protons, Z, and neutrons, $N = (A - Z)$, where A is the mass number of the atom and represents the sum of protons and neutrons (nucleons) in the nucleus. The mass and charges of these fundamental particles are given in Table 7.2. Isotopes of an element have the same number of protons (and electrons) but different numbers of neutrons. Thus, natural uranium consists of 3 isotopes whereas there are 5 isotopes of zirconium which are shown in Table 7.3.

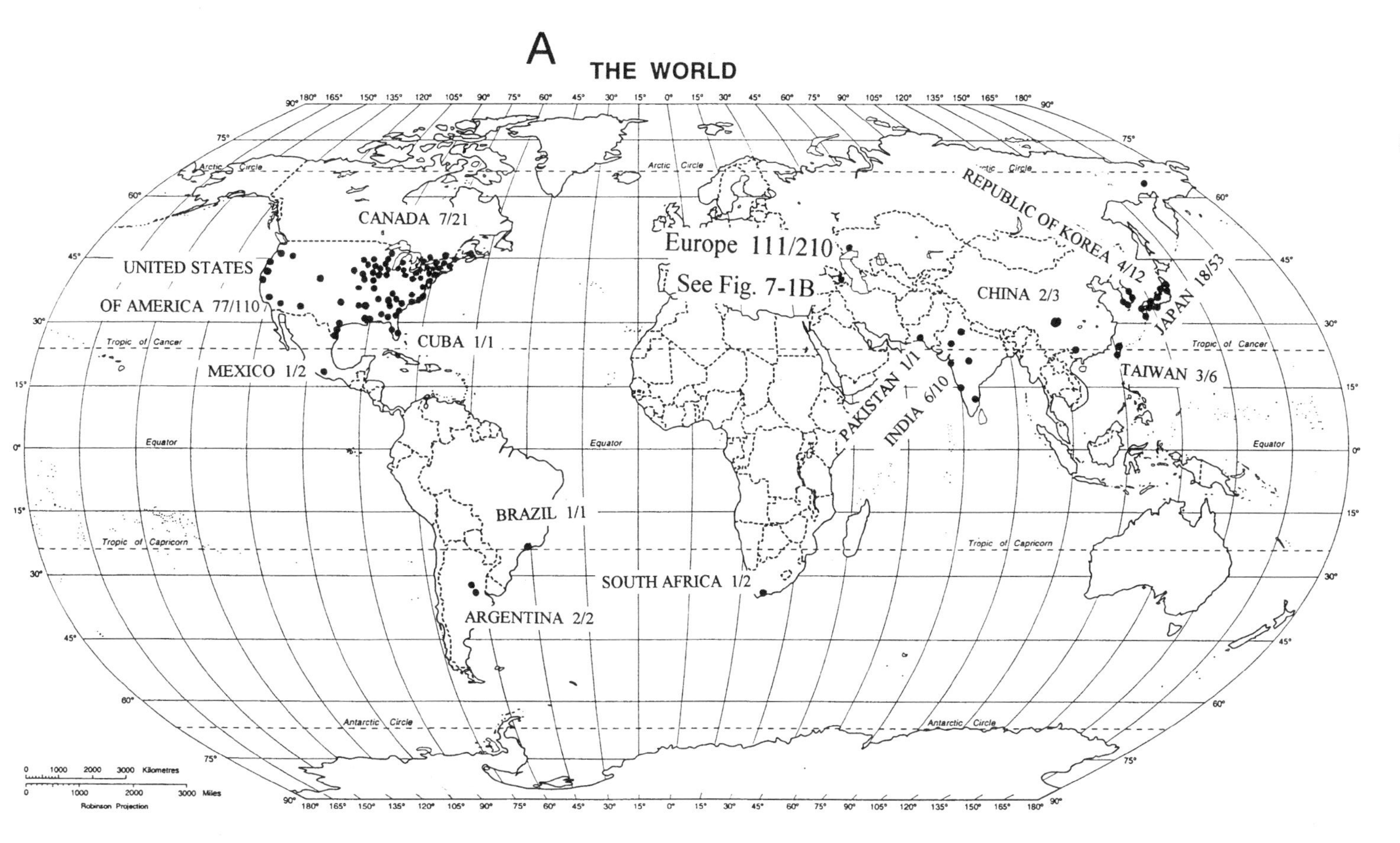

FIGURE 7.1. Maps showing the location of sites of nuclear power stations with the number of sites/total number of operating power stations as of 1996. (A) World, (B) Europe.

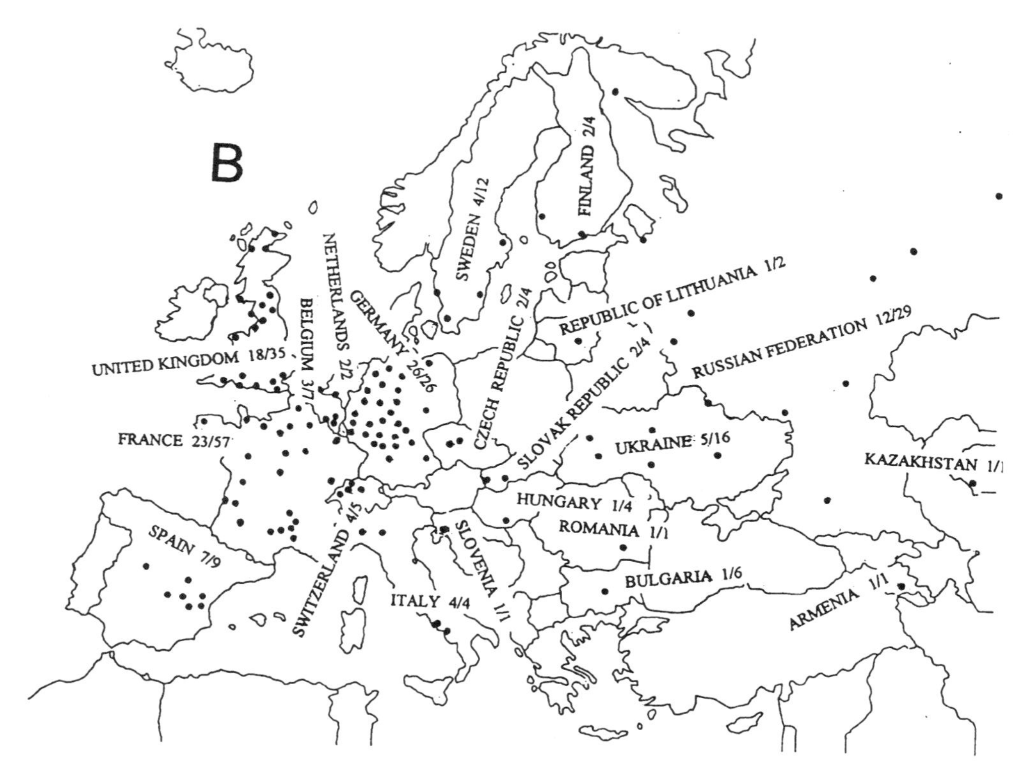

FIGURE 7.1. Continued.

The notation used to represent the nucleus of an isotope is in terms of the symbol of these elements, E, which has the appropriate number of protons, Z and the mass number A where $A = Z + N$ and N is the number of neutrons in the nucleus e.g., ${}^{A}_{Z}E$, ${}^{238}_{92}U$, ${}^{87}_{38}Sr$, ${}^{39}_{18}Ar$, ${}^{94}_{40}Zr$.

The mass of the components of the nucleus is invariably (except for H) greater than

TABLE 7.2
Nuclear Components

		Mass		Charge	
Component	Symbol	kg	amu[a]	Coulombs	Atomic
Electron	e, β^-	9.10939×10^{-31}	5.48580×10^{-4}	-1.60218×10^{-19}	-1
Proton	p	1.67262×10^{-27}	1.007276	$+1.60218 \times 10^{-19}$	$+1$
Neutron	n	1.67493×10^{-27}	1.008665	0	0
Positron	β^+	9.10939×10^{-31}	5.48580×10^{-4}	$+1.60218 \times 10^{-19}$	$+1$
Neutrino	v	0	0	0	0

[a] 1 amu = $1.6605402 \times 10^{-27}$ kg = 931.4874 MeV.

TABLE 7.3
A Isotopic Composition of Natural Uranium and Zirconium

(*a*) *Uranium*

Mass No.	Atomic mass	Abundance (%)
234	234.040946	0.006
235	235.043924	0.720
238	238.050784	99.274

(*b*) *Zirconium (Zr)*

Mass No.	Natural Abundance (%)	Thermal cross section Barns[a]
90	51.45	0.1 ± 0.07
91	11.27	1.58 ± 0.13
92	17.17	0.26 ± 0.08
94	17.33	0.08 ± 0.04
96	2.78	0.3 ± 0.1
(91.224) average		0.185

[a] 1 Barn = 10^{-24} cm^2.

the actual mass of the nucleus. The difference (a mass defect) represents the binding energy of the nucleus, ΔE, and represents the energy needed to break a nucleus into its individual components.

There are about 300 stable isotopes and over 1000 unstable (radioactive) isotopes which have been characterized. Nuclei with the same mass number A but having different nuclear charges Z are called *isobars*, there are 59 stable isobar pairs starting with $A = 36$, ${}^{36}_{16}S$, ${}^{36}_{18}Ar$, and ending with $A = 204$, ${}^{204}_{80}Hg$, ${}^{204}_{82}Pb$ as well as 5 isobaric triads such as $A = 50$, ${}^{50}_{22}Ti$, ${}^{50}_{23}V$, ${}^{50}_{24}Cr$.

The nucleus is very small, having a radius R which is approximated by the relation

$$R = R_0 A^{1/3} \tag{7.2}$$

where $R_0 = (1.5 \pm 0.2) \times 10^{-13}$ cm. The density of the nucleus is very high, approx. 10^{14} g/cm^3 or about 10^8 tonnes/cm^3.

The binding energy, ΔE_B, of a nucleus is the energy equivalent of the mass defect and is given by

$$\Delta E_B(A, Z) = \{(ZM_p + [A - Z]M_n) - M_N\}c^2 \tag{7.3}$$

where $\Delta E_B(A, Z)$ is the binding energy of a nucleus with A, Z values for the nucleons, expressed in MeV units of energy.

M_N is the isotopic mass of the particular nucleus
M_p is the mass of the proton

M_n is the mass of the neutron

$$\Delta E(A, Z) = 931.4874(1.007276Z + 1.008665)[A - Z] - M_N) \tag{7.4}$$

If the mass of the atom is utilized then $p + e = 1.007824$ is used instead of the simple proton mass.

The specific nuclear binding energy is $\Delta E_B/A$ and represents the average binding energy per nucleon. The binding energy of the helium nucleus (an alpha particle, He^{2+}) can be calculated from Eq. (7.4)

$$\begin{aligned}\Delta E &= 931.4874[(2 \times 1.007276 + 2 \times 1.008665) - 4.00151] \\ &= 931.4874(4.03188 - 4.00151) \\ &= 931.4874(0.03037) \\ &= 28.29 \text{ MeV} \quad \text{or} \quad 7.07 \text{ MeV/nucleon}\end{aligned} \tag{7.5}$$

The atomic mass of He is 4.00260 amu. The nuclear mass, M_N, is

$$4.00260 - 2 \times \text{Me} = 4.00260 - 2 \times 0.00054858$$
$$M_N = 4.00151 \text{ amu}$$

7.3. NUCLEAR MODEL AND NUCLEAR REACTIONS

Several models of the nucleus exist and attempt to account for the stability of various nuclei. The shell model predicts stable nuclei with N or Z equal to 2, 8, 20, 28, 50, 82, 126, and 152.

For nuclei with low values of Z, $n \simeq p$. However, as Z increases $n > p$ and for the largest nuclei $n = 1.5p$. Nuclei which are not in the stability region of the N/Z curve try to achieve stability by a change in the nuclear composition, i.e., they are radioactive and undergo a transformation which depends on whether the isotope is above or below the stability line.

Isotopes above the stability line can become stable by emitting a β^- particle thereby converting a neutron into a proton*

$$n \rightarrow p + \beta^- \tag{7.6}$$

$$^{294}_{90}\text{Th} \rightarrow ^{234}_{91}\text{Pa} + ^{0}_{-1}\text{e} \tag{7.7}$$

$$^{14}_{6}\text{C} \rightarrow ^{14}_{7}\text{N} + ^{0}_{-1}\text{e} \tag{7.8}$$

Another reaction which will move a nucleus closer to the stability line (see Fig. 7.2) would be the emission of a neutron. This occurs only for nuclei in excited states and therefore is primarily seen as a result of a reaction

$$^{9}_{4}\text{Be} + ^{4}_{2}\text{He} \rightarrow ^{12}_{6}\text{C} + ^{1}_{0}\text{n} \tag{7.9}$$

* The neutrino, υ, is produced in most of these reactions but is not discussed here.

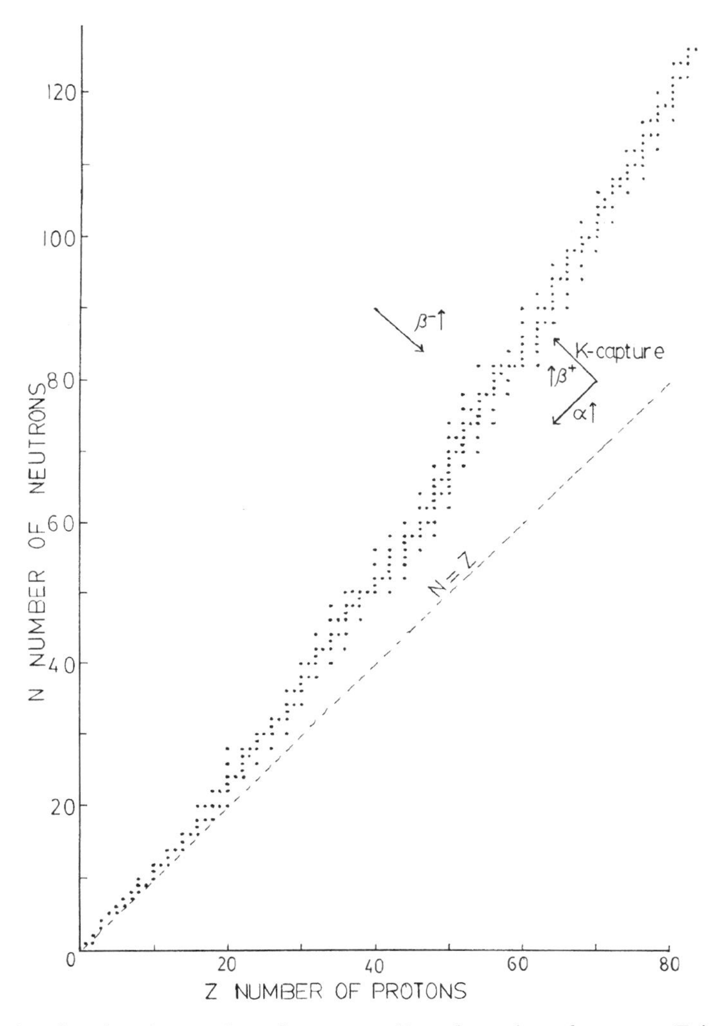

FIGURE 7.2. Plot showing the number of neutrons, N, and number of protons, Z, in the nucleus of naturally occurring isotopes of the elements from $Z = 1$ to $Z = 80$. The $N = Z$ line is shown to illustrate the deviation of the stable isotopes. Elements above the stability line can become more stable by β^- emission or neutron emission. Elements below the stability line can become more stable by β^+ emission, K-capture or α particle decay.

where the α-particles can be supplied by a radium source.

$$^{226}_{88}\mathrm{Ra} \rightarrow {}^{222}_{86}\mathrm{Rn} + {}^{4}_{2}\mathrm{He} \tag{7.10}$$

Isotopes which are below the stability line can achieve stability by (a) positron emission, β^+

$$^{11}_{6}\mathrm{C} \rightarrow {}^{11}_{5}\mathrm{B} + {}^{0}_{1}\mathrm{e} \tag{7.11}$$

$$^{39}_{19}\mathrm{K} \rightarrow {}^{39}_{18}\mathrm{Ar} + {}^{0}_{1}\mathrm{e} \tag{7.12}$$

TABLE 7.4
Characteristics of Some Natural Occurring Radioactive Isotopes

Isotope	Abundance (%)	$\lambda_{1/2}$ (Years)	Type of decay	Energy (MeV)
$^{14}_{6}C$	Trace	5730	β^-	0.156
$^{40}_{19}K$	0.0117	1.25×10^9	β^-	1.36
$^{50}_{23}V$	0.25	24×10^{17}	*e*-Cap	
$^{87}_{37}Rb$	27.83	4.5×10^{10}	β^-	0.273
$^{115}_{49}In$	95.77	6×10^{14}	β^-	0.5
$^{124}_{50}Sn$	6.1	1.5×10^{17}	β^-	
$^{147}_{62}Sm$	15.0	1×10^{11}	α	2.2
$^{187}_{75}Re$	62.6	4.5×10^{10}	β^-	0.0025
$^{238}_{92}U$	99.274	4.5×10^9	α	4.0
$^{235}_{92}U$	0.720	7×10^8	α	4.68

(b) by electron capture (K capture) in which the nucleus absorbs an electron from the external structure of the nucleus, the K shell of the atom ($n = 1$, s $= 0$, $m = 0$)

$$^{81}_{37}Rb + {}^{0}_{-1}e \rightarrow {}^{81}_{36}Kr \tag{7.13}$$

(c) by emitting an α-particle, $^{4}_{2}He$,

$$^{238}_{92}U \rightarrow {}^{234}_{90}Th + {}^{4}_{2}He \tag{7.14}$$

(d) by proton emission which seldom occurs.

7.4. RADIOACTIVE DECAY RATES

The radioactive nucleus decays spontaneously at a rate which is proportional to the number of active (unstable) nuclei

$$-\frac{dN}{dt} \propto N \qquad \text{or} \qquad -\frac{dN}{dt} = \lambda N \tag{7.15}$$

$$\frac{dN}{N} = \lambda\, dt; \qquad \ln N/N_0 = -\lambda t \tag{7.16}$$

where N_0 is the activity at $t = 0$

This is identical to a first order reaction with λ representing the decay constant. The half-life $t_{1/2}$ is given by $\ln 2/\lambda$ or $0.693/\lambda$ and is an important characteristic of each radioactive nucleus. Some values of half-lives of selected nuclei are given in Table 7.4.

The $^{14}_{6}C$ isotope is formed by the reaction of neutrons formed by cosmic radiation

$$^{1}_{0}n + {}^{14}_{7}N \rightarrow {}^{14}_{6}C + {}^{1}_{1}p \tag{7.17}$$

A shorthand notation for this reaction is $^{14}_{7}N(n, p)^{14}_{6}C$.

The $^{14}_{6}C$ reacts with oxygen and ends up as CO_2 which is incorporated into living matter. The rate at which the $^{14}_{6}C$ decays is balanced by the rate at which it is formed and thus an equilibrium is established and $^{14}_{6}CO_2$ is constantly replaced as it decays. When the living matter "dies" it ceases to absorb CO_2 and thus the $^{14}_{6}C$ begins to decay without being replaced.

Prior to 1945 when atomic bomb detonations began to contaminate the atmosphere, the rate of $^{14}_{6}C$ production and its decay had reached equilibrium. In 1960 Willard Libby received the Nobel Prize for his work showing that it was possible to date when natural materials ceased to grow by incorporating CO_2 into its matter. The following example will illustrate the application of C14 dating.

In 1950 a sample of linen in which some of the Dead Sea Scrolls were wrapped was burned to CO_2 and analyzed for ^{14}C. It was found that the activity was 12.05 Bq/g. Living matter had a ^{14}C activity of 15.30 Bq g^{-1}. What is the age of the linen? From Eq. (7.16),

$$In\,\frac{N_0}{N} = kt = In\,\frac{15.30}{12.05} = kt \qquad \textbf{(7.18)}$$

since $t_{1/2} = 5730$,

$$k = \frac{0.693}{t_{1/2}} = \frac{0.693}{5730}\ \text{yeas}^{-1}$$

$$\text{In}\,\frac{15.30}{12.05} = \frac{0.693}{5730}\,t \qquad t = 1974 \pm 200 \text{ years (with estimated error).}$$

Hence the linen was made in 1950–1974 = 24 B.C. ±200 years.

Radiation is measured by various methods which depend on the type of radiation, γ, β, α and the activity of the source. Early measurements were made using photographic films or electroscopes. Modern methods include ionization chambers and fluorescing plastics (scintillation counters) and solutions in which the emitting photons are counted individually. The Geiger Counter is one of the simplest means of detecting and measuring β and α particles of sufficient energy to penetrate the counter's thin window. A diagram of a Geiger tube is shown in Fig. 7.3.

7.5. RADIOACTIVITY UNITS

The radioactive nucleus decays by a spontaneous process. The activity is measured by several methods: (1) by the activity of the material. The basic unit was the Curie (Ci) which was the equivalent of 3.7×10^{10} disintegration/sec for 1 gram of radium. The present unit is the Becquerel (Bq) which is one disintegration per sec. (2) The energy of the radiation in terms of the absorbed energy. The Roentgen (r) is 83.8 ergs of absorbed energy defined as the quantity of x- or γ (as well as α and β) radiation which on passage through 1 mL of dry air at 0°C and 1 atm (STP) (0.001293 g/mL) will produce ions carrying 1 esu of charge (either positive or negative). This is limited to radiation with energy less than 3 MeV. A dose of 1 r produces 2.08×10^9 ion pairs corresponding to

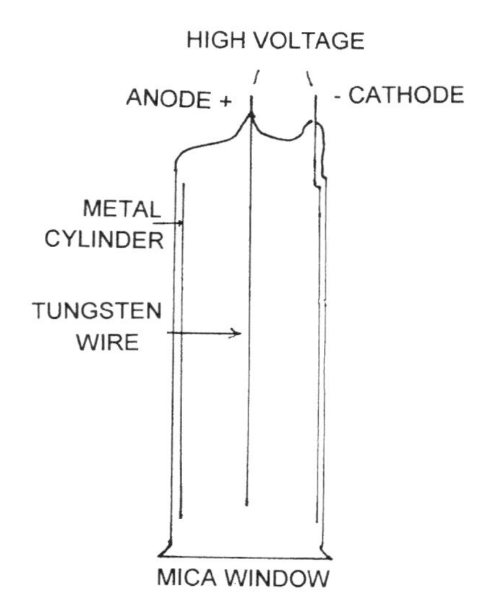

FIGURE 7.3. A Geiger Counter Tube. Radiation which passes through the thin window ionizes the gas in the tube. The electrons are accelerated by the high voltage (500–1200 V) creating a pulse which is detected electronically.

the absorption of 0.114 ergs. (2.58×10^{-4} coulombs of negative charges in 1 kg of dry air at STP).

The dose of radiation absorbed by a unit of mass per unit of time is called the *absorbed* dose rate in units of joules/kg. The rad (radiation absorbed dose) is 10^{-2}J/kg. A rad of α-rays is more harmful than a rad of β-rays. Hence the RBE (relative biological effectiveness) of radiation normalize the damage to tissue by different types of radiation. This is illustrated by the equation

$$\text{Rem} = (\text{Rad})(\text{RBE}) \tag{7.19}$$

where Rem is the roentgen equivalent to man. The values of some of the parameters are given in Table 7.5. Radiation units are given in Table 7.6. Radiation damage at about 400 to 500 Rem is fatal to humans (see Table 7.7). Some evidence exists which

TABLE 7.5
The Approximate Values of
RBE for Various Types of Radiation

Radiation	RBE
β^-, γ	1
α, p	10
n (thermal)	5
n (fast, $E < 400$ MeV)	10
Heay ions (cosmic ray)	20

TABLE 7.6
Units of Radiation

1 Curie (Ci) = 3.7×10^{10} disintegrations/sec[a]
Becquerel (Bq) = 1 disintegration/sec
1 Gray (Gy) = 100 Rads = 1 J/kg
1 Sievert (Sv) = 100 Rem

[a]The activity of 1 g of pure radium.

TABLE 7.7
Radiation Damage for Whole Body Exposure

0–25 Ram	No clinical effects
25–50 Ram	Decrease in white cells in blood
100–200 Rem	Nausea, fewer white blood cells
500 Rem	LD_{50}

indicates that small doses may be beneficial (called *hormesis*) and implies there is a repair mechanism in living man and animals exposed to small doses of radiation.

The human body contains some naturally occurring radioactive isotopes besides ^{14}C. These are all listed in Table 7.8. The energy of the radiation is very low and the biological damage due to these isotopes is of minor importance. The normal exposure levels to man are as given in Table 7.9.

7.6. NUCLEAR REACTORS

The first nuclear chain reaction occurred naturally about 2 billion years ago in Gabon, Africa where a uranium deposit moderated by water spontaneously became critical. In 1942 as a result of the war efforts, a sustained chain reaction was achieved by E. Fermi in Chicago working on the Manhattan Project which eventually led to the Atomic Bomb.

Of the naturally occurring isotopes only ^{235}U is fissionable with a neutron absorption cross section (σ) of 582 barns with thermal neutrons at 0.025 eV. This cross section decreases as the energy of the neutrons increases ($\sigma \propto E^{-1/2}$) and in the MeV

TABLE 7.8
The Naturally Occurring Radioactive Isotopes in a 70 kg Human Body

Source	Amount of isotope (g)	No. of atoms	Total activity (Bq)	Particle energy (MeV)	γ
$^{3}_{1}H$	8.4×10^{-15}	1.7×10^{9}	3	0.018	
$^{14}_{6}C$	1.9×10^{-8}	8.1×10^{14}	3.1×10^{3}	0.158	
$^{40}_{19}K$	8.3×10^{-2}	1.2×10^{21}	4×10^{7}	1.36	1.46

TABLE 7.9
Yearly Radiation Exposure

Natural Radiation	
Cosmic rays sea level	40 m Rem
at 2500 m (Banff AB)	90 m Rem
Terrestrial exposure due to Radium and other isotopes in ground and buildings	40 m Rem
Internal radiation mainly due to Potassium 40	18 m Rem
Cosmic radiation during 10,000 km flight at 10 km altitude	4 m Rem
Man-made Radiation	
1 Chest X-ray	40 m Rem
1 Dental X-ray	20 m Rem
Fall-out*	3 m Rem
Misc.–T.V., Monitors etc.	2 m Rem

*After atmospheric tests of atomic bomb.

range σ is about 2 barns. The ^{238}U reacts with neutrons ($\sigma \simeq 0$ for $E < 0.9$ MeV) to form plutonium (Pu) by the reaction sequence.

$$^{238}_{92}U(n, \gamma)\, ^{239}_{92}U \xrightarrow[23.5\text{ min}]{\beta^-} {}^{239}_{91}Np \xrightarrow[2.33\text{ days}]{\beta^-} {}^{239}_{90}Pu \tag{7.20}$$

The plutonium is radioactive with $t_{1/2} = 2.4 \times 10^4$ years and is fissionable like ^{235}U. It is also possible to convert ^{232}Th to fissionable ^{233}U which has a half-life of 1.63×10^5 years by the reaction sequence.

$$^{232}_{94}Th(n, \gamma)^{233}_{94}Th \xrightarrow[23.6\text{ min}]{\beta^-} {}^{233}_{93}Pa \xrightarrow[27.4\text{ days}]{\beta^-} {}^{233}_{92}U \tag{7.21}$$

The process of fission in ^{235}U proceeds by the absorption of a thermal neutron and formation of the compound nucleus ^{236}U in an excited state with about 6.5 MeV. During the fission process the compound nucleus is distorted and splits into two fission fragments, which by virtue of coulombic repulsion achieve a kinetic energy equivalent to about 80–90% of the fission energy (200 MeV).

The fragment nuclei decay by neutron and γ emission, leading to an average of 2.42 neutrons released for each neutron captured. The stabilization of the fragments and the radiation results in the thermal energy produced by the chain reaction.

The neutron flux can also be controlled by control rods made of cadmium or boron, which are strong absorbers of neutrons, and which are raised or lowered into the reactor to maintain the chain reaction at the appropriate level.

The enrichment of $^{235}_{92}U$ from natural uranium was first performed by the differential diffusion of gaseous UF_6 through porous barriers. This very energy intensive method was replaced by the simpler gas centrifuge. With the development of the tunable laser it became possible to photochemically excite one isomeric uranium atom by virtue of its hyperfine splitting of the isotopic lines, and thus enrich one isotope.

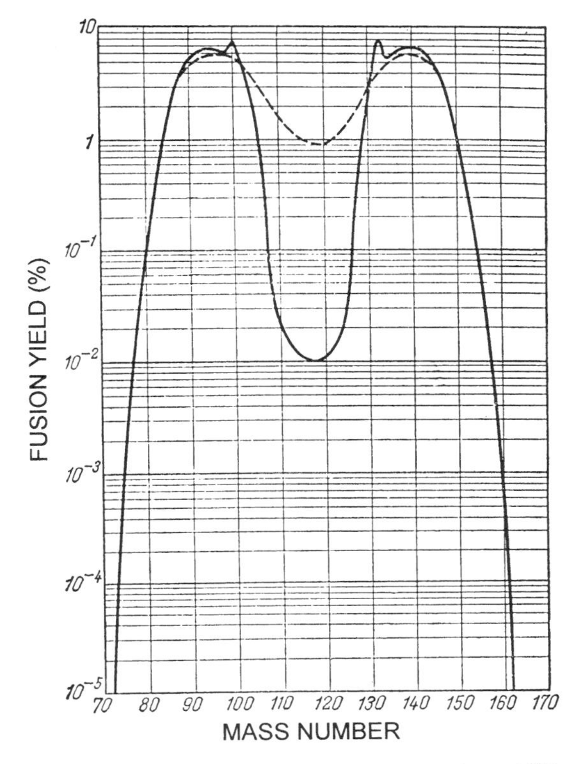

FIGURE 7.4. The distribution of fission products from the reaction of $^{235}_{92}U$ with slow neutrons.

Though the cost of the enrichment process has thus been greatly reduced, ^{235}U is still a very expensive fuel.

The yield of the fission products as a function of the atomic mass is shown in Fig. 7.4 and indicates that the major elements formed are in the range $A = 85–105$ and 130–150. The major elements formed and their fission yield are listed in Table 7.10.

TABLE 7.10
Most Important products of Thermal-Neutron Fission of Uranium-235

Isotope	Half-life	Fission yield, %	Isotope	Half-life	Fission yield, %
^{137}Cs	33 years	6	^{95}Zr	65 days	6
^{90}Sr	19.9 years	5	^{95}Nb	38.7 days	
^{140}Ba	13.4 days	5.7	^{99}Tc	2.1×10^5 years	6.2
^{91}Y	61 days	5.9	^{129}Te	35.5 days	0.2
^{140}La	1.65 days	5.7	^{131}I	8.1 days	3
^{141}Ce	33 days	5	^{129}I	1.7×10^7 years	—
^{144}Ce	282 days	3.6	^{103}Ru	39.8 days	3.7
^{143}Pr	13.5 days	5.3	^{106}Ru	290 days	0.5
^{147}Nd	11.9 days	2.6	^{105}Rh	1.54 days	0.5
^{147}Pm	2.26 years		^{133}Xe	5.3 days	6
^{155}Eu	1.7 years	0.03			

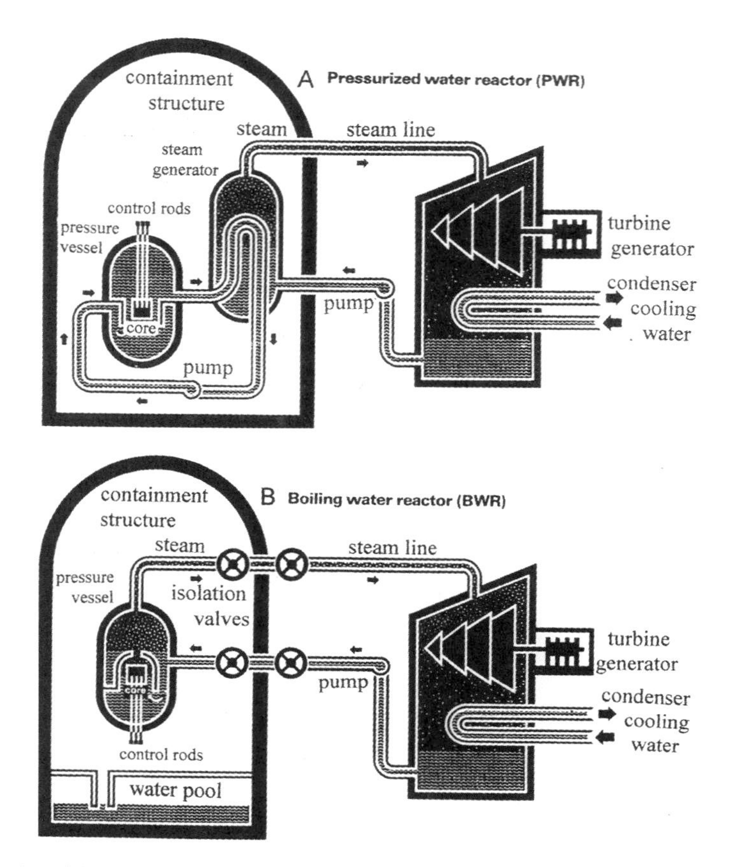

FIGURE 7.5. As with the boiler in a coal, oil or gas burning power plant, a nuclear power reactor produces steam to drive a turbine which turns an electric generator. Instead of burning fossil fuel, a reactor fissions nuclear fuel to produce heat to make steam. (a) The PWR shown here is a type of reactor fueled by slightly enriched uranium in the form of uranium oxide pellets held in zirconium alloy tubes in the core. Water is pumped through the core to transfer heat to the steam generator. This coolant water is kept under pressure in the core to prevent boiling and transfers heat to the water in the steam generator to make the steam. (b) The BWR shown here is a type of reactor fuelled by slightly enriched uranium in the form of uranium oxide pellets held in zirconium alloy tubes in the core. Water is pumped through the core, boils and produces steam that is piped to the turbine.

Reactors can be classified in terms of the fuel (^{235}U, ^{233}U, ^{239}Pu), and its containment alloy, zircaloy, the moderator which is needed to slow down the neutrons, and the heat exchanger used to generate steam. The most common type of classification is in terms of the coolant which can be either a liquid or gas.

The four categories are (1) the light water reactor (LWR) of which are 81% of all reactors, (2) gas cooled reactors (GR) make up 11%, (3) 7% use heavy water (HWR), and (4) the remaining reactors use molten metals (such as sodium) (see Fig. 7.5).

The LWR is further classified into the pressurized water reactor (PWR) which operates at about 150 atm and 318°C with a thermal efficiency of about 34%. The other

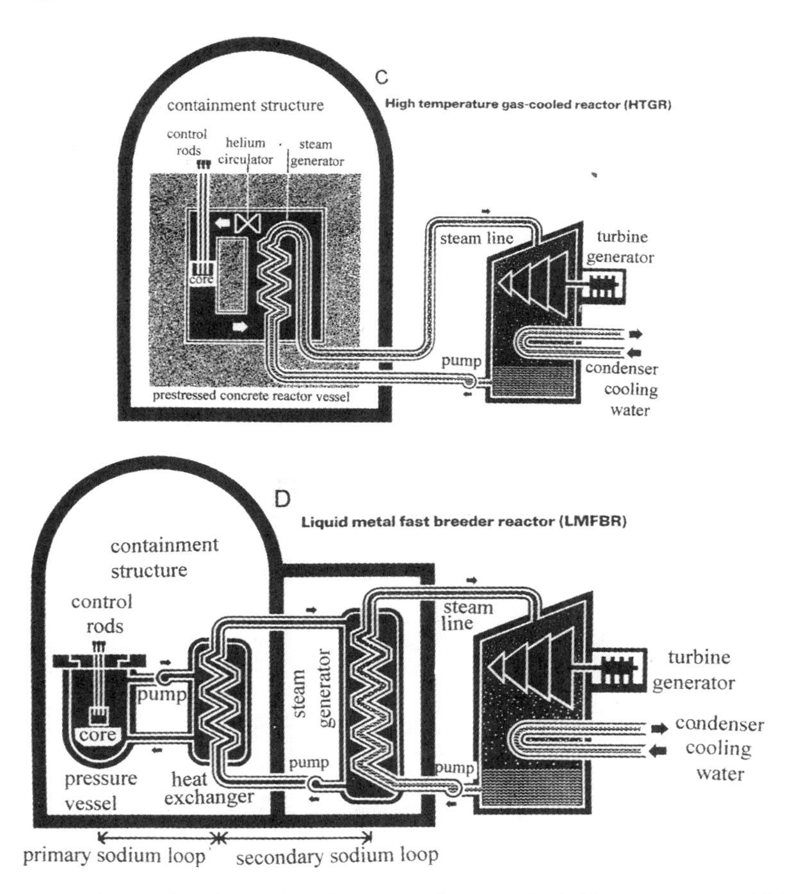

FIGURE 7.5. (c) The HTGR shown here is a type of reactor fueled by uranium carbide particles distributed in graphite in the core. Helium gas is used as a coolant to transfer the heat from the core to the steam generator. (d) In a LMFBR, molten sodium in the primary loop is pumped through the reactor core containing the fuel. This sodium collects the heat and transfers it to a secondary sodium loop in the heat exchanger from which it is carried to the steam generator. In addition to producing electricity, this reactor also produces more fissionable material than it consumes, which is why it is called a "breeder reactor." In the reactor uranium-238 is transmuted to fissionable plutonium-239 which is extracted periodically and fabricated into new fuel.

type of reactor is the boiling water reactor (BWR) which operates at 70 atm pressure and 278°C with a thermal efficiency of 33%. These reactors require fuel with enriched ^{235}U to about 3% to have a sufficient neutron flux for the chain reaction. The fuel, as UO_2, is in the form of pellets enclosed in a zirconium alloy, Zircaloy-2.

The gas cooled reactors were developed in Great Britain using CO_2 as the gas coolant and graphite as the moderator with natural uranium metal as the fuel. The thermal efficiency is about 25%. With uranium enriched to 2.2% as the oxide (UO_2) fuel, the thermal efficiency increased to 41% and is called the Advanced Gas Reactor (AGR). Helium is also used in a high temperature version.

TABLE 7.11
The Relative Moderating Characteristics of Various Materials

Substance	Moderating ratio	Average slow down length from "fast" to thermal (cm)
Light water	72	5.3
Organic liquids	60–90	
Beryllium	159	
Graphite	160	19.1
Heavy water (99.8%)	2300	
Pure D_2O	12,000	11.2

The heavy water reactor was developed in Canada and is known as the CANDU reactor. The D_2O is used as both coolant and moderator. The relative moderating efficiency of various materials is given in Table 7.11. Because of the superior moderating property of D_2O, it is possible to use natural uranium as the fuel in the form of UO_2 pellets in Zircaloy tubes. This makes the CANDU one of the best designed reactors in the world. The coolant cycle and the moderator are separate flow circuits as shown in Fig. 7.6. The fuel elements in the pressure tubes and the D_2O flow is shown in Fig. 7.7 where the coolant is at about 293°C and 100 atm pressure. The moderator is at lower pressure. The efficiency is rated at 29%.

The average composition of deuterium in water (H_2O) is about 150 ppm but varies from place to place as shown in Fig. 7.8 for Canada.

Heavy water is usually extracted from water in two steps; the first being a dual temperature deuterium transfer process, and the second through vacuum distillation.

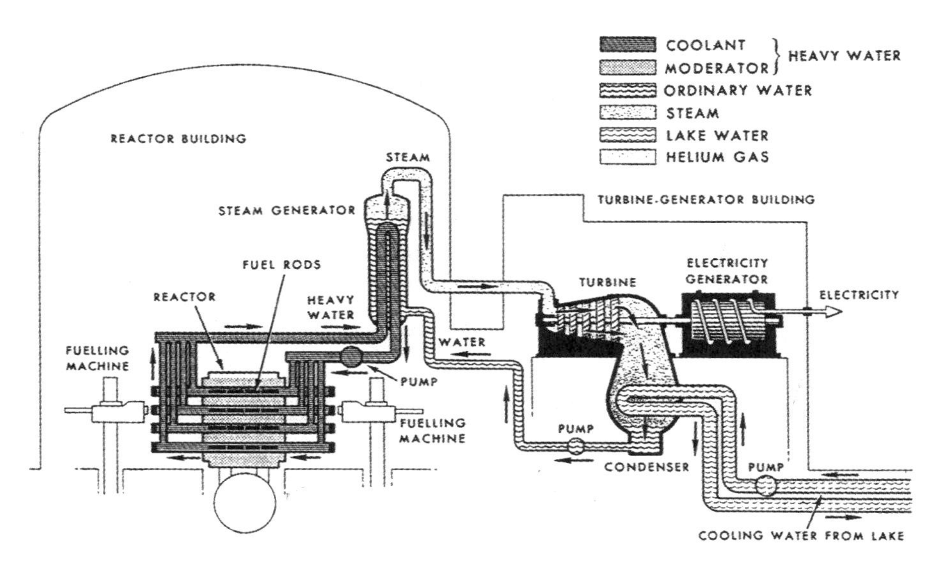

FIGURE 7.6. The simplified flow diagram for the CANDU reactor.

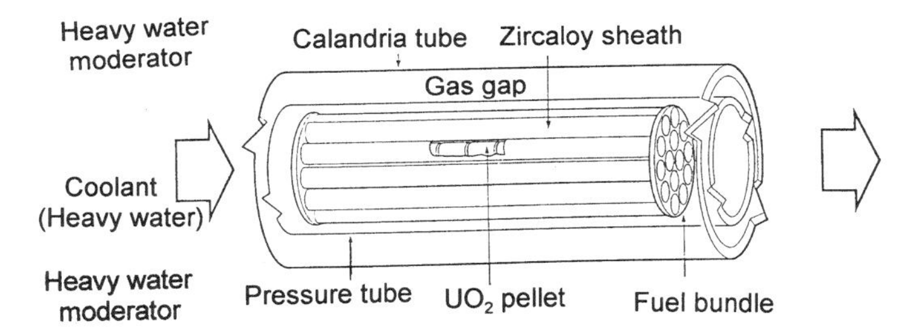

FIGURE 7.7. The coolant around the fuel elements in the Calandria tubes is D_2O.

The dual temperature process is based on the atomic exchange of hydrogen and deuterium between hydrogen sulphide gas (H_2S) and fresh water with a deuterium concentration of approx. 148 ppm.

When hot water is in contact with hot H_2S gas, the deuterium atom migrates from the water into the gas; when both are cold, it migrates in the reverse direction.

$$HOD + H_2S \rightleftharpoons H_2O + HSD \qquad K = 0.99 \quad (25^{\circ}C) \tag{7.22}$$

By repeating this process in successive stages, the deuterium oxide is increased to a concentration of 15%.

The ten extraction towers of the plant contains sieve trays. Water flows downward across each tray in succession. H_2S gas is forced upwards, bubbling through the sieve holes for contact with the water. This is shown in Fig. 7.9.

The 15% enriched heavy water is delivered to a finishing unit where, through vacuum distillation, the concentration is upgraded to 99.8% (reactor grade) deuterium oxide. The distillation unit consists of two towers containing special packing.

Deuterium is also enriched during the electrolysis of water since ${}^{1}_{1}H$ is more readily liberated as H_2 than ${}^{2}_{1}D$ as HD. Hence the first samples of heavy water were obtained from Norway where water electrolysis is the method of producing H_2 and O_2 since electricity is inexpensive there—even today it is the chosen method of heating homes.

It is possible to increase the fuel efficiency by selecting the zirconium isotope ${}^{94}_{40}Zr$ (which has the lowest neutron absorption cross section of 0.08 barns (see Table 7.3) for the fabrication of the Zircaloy alloy. This would allow the fuel to be useful for longer times or less fuel would be required for the reactor. Several attempts have been made to separate the heavier isotopes since the high cost of an enrichment process is a major capital investment which could be of continuous benefit as the ${}^{94}_{40}Zr$ can be recycled from spent fuel and reused with little reprocessing.

7.7. THE HAZARDS OF NUCLEAR ENERGY

It is at present somewhat uncertain if very low levels of radiation are harmful. We cannot avoid all radiation since there is a natural radiation background (with approximate yearly exposure) due to, for example, the cosmic rays (40 millirem at sea level, 250 Rem at 500 m elevation) radium and radon in ground and building material (40 Rem), potassium 40 (18 Rem). In addition, we can add some man-made radiation

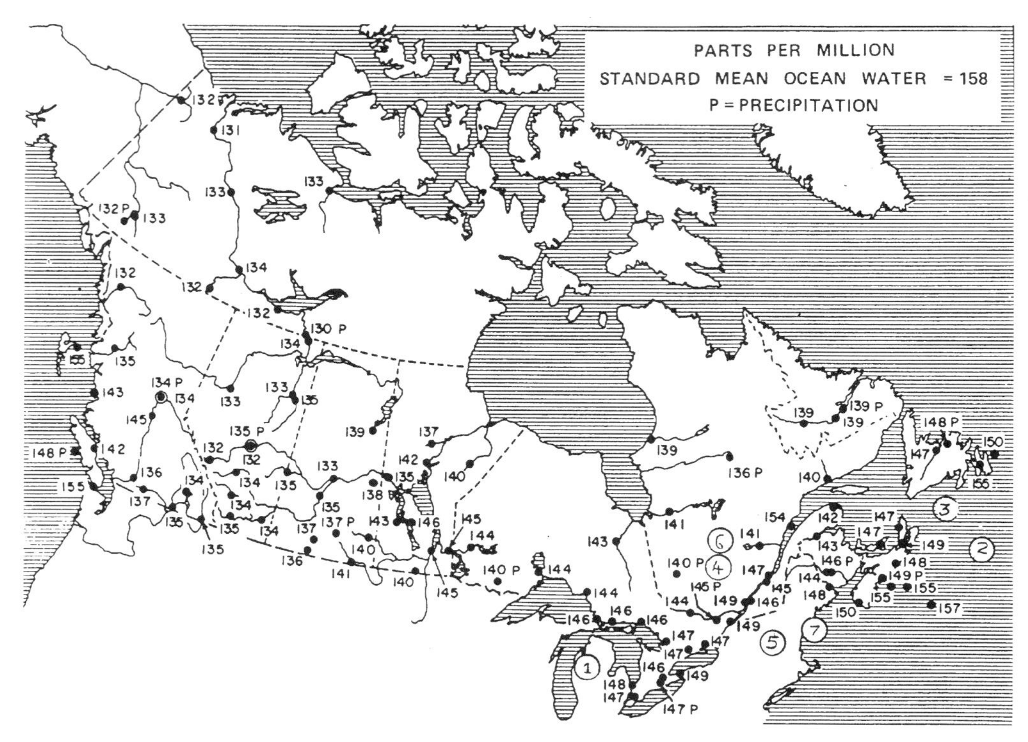

FIGURE 7.8. Map of Canada showing the deuterium concentration in ppm ($\mu g/g$) in various parts of the country. Also shown are the sites of heavy water plants and nuclear power plants.

Heavy water plants	Nuclear power plants[a]
1. Bruce	1. Bruce A, 4 × 769 MWe
2. Port Hawkesbury	B, 4 × 860 MWe
3. Glace Bay	Douglas Pt. 1 × 208 MWe
4. La Prade	5. Pickering A, 4 × 515 MWe
	B, 4 × 518 MWe
	Darlington 4 × 881 MWe
	6. Gentily 1 × 635 MWe
	7. Point Lepreau 1 × 640 MWe

*With rated electrical generating power.
Several plants have been shut down. Only 14 are in operation.

sources such as one chest X-ray (40 Rem), one dental X-ray (20 Rem), fallout from nuclear explosions (5 Rem) as well as miscellaneous sources such as T.V., CRT, etc., all of which total to 163 Rem/year (for see level). The average annual radiation dose to a nuclear reactor worker in Ontario is 0.68 Rem with an annual limit of 5 Rem set by radiation protection regulations.

Thus, a nuclear power reactor worker receives about as much additional radiation as that of an office worker who moves from New York to Denver. Hence, nuclear

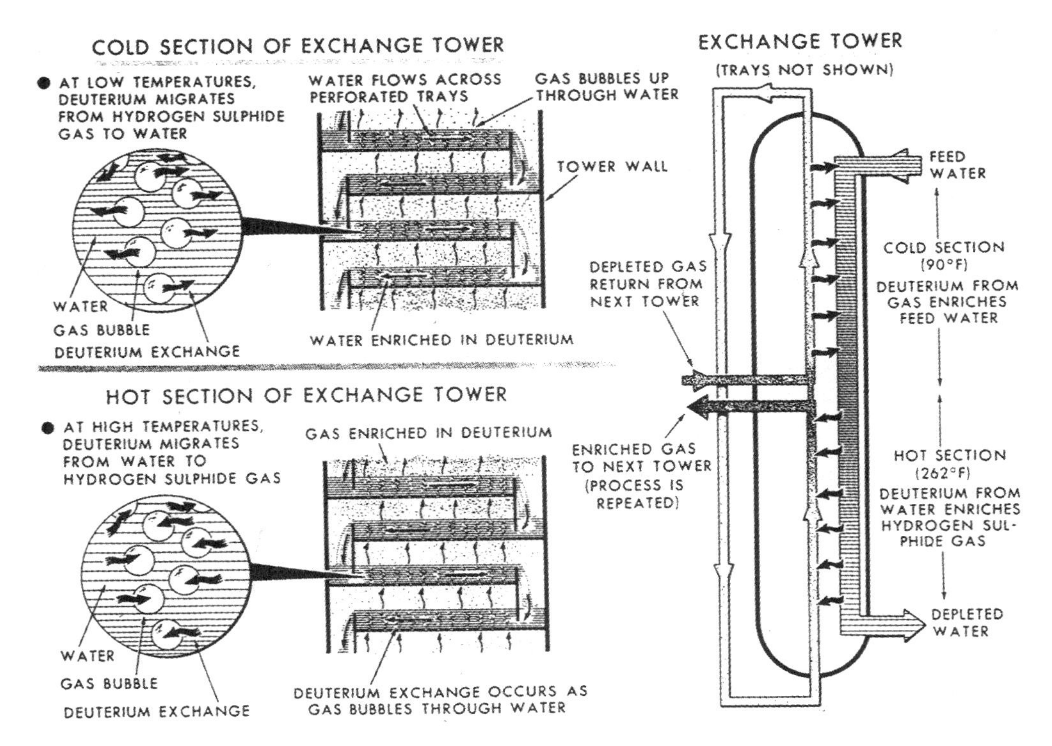

FIGURE 7.9. Basic principles of heavy water production by the $HDO + H_2S \rightarrow HDS + H_2O$ equilibrium reaction.

energy presents an insignificant risk to its workers. It can be argued that the risk of a nuclear accident (such as the Three Mile Island accident, March 1979, or the Chernobyl accident, April 1986) is not worth the benefits. This must be viewed in a perspective which is tempered by history and time. The Chernobyl nuclear accident on Saturday, April 26, 1986, has been called the worst accident in the world and will probably slow down and, in some countries, end the development of nuclear energy. The Russian reactor was a graphite moderated water cooled enriched uranium reactor which was idling at 7% power, generating 200 MW of heat while maintenance work was being done. Human error appears to have caused a sudden increase in power output, 7–50%, in 10 sec due to part of the reactor going critical. The resulting runaway high temperature converted water to hydrogen which exploded, setting the reactor on fire and releasing enormous amounts of radioactivity into the atmosphere. The radioactive cloud drifted across Europe and was first reported in Sweden, Monday morning, April 28. Based on the prevalence of radioactive cobalt, iodine and cesium, it was concluded that a nuclear bomb test was not the source of the activity but that a nuclear accident had occurred. This was acknowledged at 9:00 p.m. by Moscow T.V. news.

The radioactivity contaminated the vegetables and meats of central Europe. Millions of dollars were used to compensate farmers for the loss they suffered. Though about 31 deaths were associated with acute radiation exposure and fire, it is estimated that thousands of delayed cancers will result from the fallout.

TABLE 7.12
Loss of Lives due to Dam Disasters

Date	Place	Lives lost
1923	Santa Paula, CA, USA	450
1923	Gleno, Italy	600
1926	St. Francis, USA	430
1959	Malpaset, France	412
1961	Kiev, USSR	145
1963	Vaiont, Italy	2000
1967	Kayna, India	180
1972	Buffalo Creek, WV, USA	118

The accident at Three Mile Island released 15 curies of radiation and caused no direct fatalities. It has been estimated that not more than 1 delayed cancer victim will result.

However, we accept the construction of dams, some of which have failed and others, if they do fail, would cause hundreds of thousands of deaths. A list of some hydroelectric dam disasters is given in Table 7.12.

We accept the occasional mine disaster and continue to mine coal. Since a tonne of uranium ore yields 300 times more energy than a tonne of coal, it is obvious that mining uranium leads to fewer mining deaths per energy unit (1/300) than mining coal. Again it could be argued that the radon and radiation exposure in uranium mining is an additional hazard not encountered in coal mining. However, black lung disease kills thousands of miners every year. It has been estimated that for every 10^{15} We (Watts of electricity) generated there are 1000 deaths by black lung among coal miners and 20 deaths by lung cancer among uranium miners.

An indication of relative average risk of fatality by various causes is given in Table 7.13. The automobile remains the major cause of accidental deaths in our society, yet we accept this with minor token complaints. It is possible to predict that there will be 500 deaths in USA. due to automobile accidents (50 in Canada) during the Labor Day Weekend holiday in September.

A comparison of man-days lost per megawatt-year output by various energy sources has been given by Inhaber and is shown in Fig. 7.10. Coal and oil show the highest losses whereas natural gas shows the lowest loss. It may be surprising to note that solar energy sources are relatively hazardous. Edward Teller — the father of the H-bomb — once pointed out that solar energy is not free of dangers because the ladder, which would be required to clean solar cells on the roof of our homes, causes more accidents in the home than any other device.

7.8. NUCLEAR WASTE

The opponents of nuclear energy, besides being concerned with the hazards of accidents during power generation, are also anxious about the handling and storage of nuclear waste — the end-product of nuclear fuel. This waste consists of a multitude of

TABLE 7.13
Average Annual Risk of Fatality by Various Causes (U.S.A.).

Accident Type	Total Number	Individual Chance per year
Motor Vehicle	55,791	1 in 4,000
Falls	17,287	1 in 10,000
Fuels and Hot Substances	7 451	1 in 25,000
Drowning	6 181	1 in 30,000
Firearms	2 309	1 in 100,000
Air Travel	1 778	1 in 100,000
Falling Objects	1 271	1 in 160,000
Electrocution	1 148	1 in 160,000
Lightning	160	1 in 2,000,000
Tornadoes	93	1 in 2,500,000
Hurricanes	93	1 in 2,500,000
All Accidents	111,992	1 in 1,600
Total Nuclear Reactor Accidents (100 plants)	5 000*	1 in 200,000

*The British accident at Windscale in October 1957—a fire in the atomic pile—resulted in the release of radioactive gas $^{131}_{50}I$ (20,000 Curies), $^{210}_{82}Po$ (37 Ci) as well as $^{103}_{44}Ru$, $^{106}_{44}Ru$, $^{95}_{40}Zr$, $^{95}_{41}Nb$, $^{137}_{55}Cs$ (600 Ci), $^{89}_{38}Sr$ (80 Ci), $^{90}_{38}Sr$ (9 Ci), $^{144}_{58}Ce$, and $^{132}_{52}Te$. It has only recently been estimated that the number of deaths from leukemia and cancer which could be directly due to this accident is from 1000 to 2000 with the predominant effect on children. Hence, we must distinguish between direct and indirect or more remote fatalities. If we add the direct deaths and the leukemia and cancer deaths for the Three Mile Island and Chernobyl nuclear accidents we would estimate a total of about 5000 deaths. This gives a risk factor of 1 in 200,000 which is still better than air travel.

radioactive nuclei having a wide range in activity as well as lifetime. In general, about 2000 m^3 of uranium ore must be mined to produce 1 tonne of nuclear fuel grade material which yields 91 L of UO_2 reactor fuel. The spent fuel is stored in water pools to permit the radioactive decay of short-lived fission products. The discharged fuel is reprocessed, resulting in about 1200 L of high-level waste (HLW) which is stored and evaporated to about 570 L and calcined (heated) to 80 L of oxides. When fixed in glass the waste occupies 70 L or 1/3 of a canister 30 cm in diameter $\times$ 3 m long which is destined for disposal and storage underground. A plot of waste activity as a function of time is shown in Fig. 7.11.

After 450 years the residual activity is equivalent to that of natural uranium and presents the same radiation hazards. A list of the major waste components in spent fuel is given in Table 7.14. As illustrated in Fig. 7.11, the major activity is due to plutonium and strontium.

The storage and handling of nuclear waste by various countries all include an immobilization of either the reprocessed fuel waste or the unreprocessed spent fuel by glass or mineral formation followed by encapsulation. Storage varies due to facilities and geological formation within a country. Thus, salt mines have been used in Germany, USA, and the Netherlands. Salt mines by their existence testify to a guaranteed isolation from ground water in the environment. The abandoned salt mine Asse near Brunswick in Germany has since 1967 stored more than 100,000 special drums of nuclear waste. However, this has been stopped since 1979 due to political

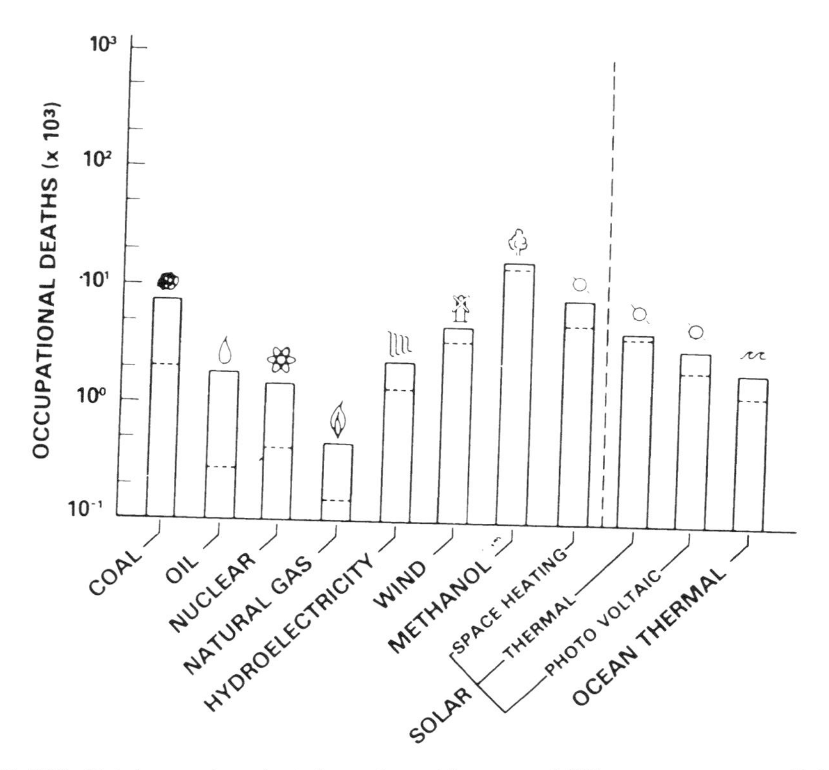

FIGURE 7.10. Total man-days lost through accidents per MW-year net output. Public and occupational deaths are combined. The top of the bars and the dotted lines indicate the upper and lower part of the range, respectively. Bars to the right of the vertical dotted lines indicate technologies less applicable to Canada. Note the logarithmic scale. Deaths are assumed to contribute 6000 man-days lost per fatality, and are combined with accidents and illness on that basis. Solid and dashed jagged lines on three of the bars indicate maximum and minimum values, respectively, when no back-up (or low-risk back-up) energy is assumed.

pressure. An alternate salt mine at Gorleben has been under consideration and testing. Germany's nuclear energy program is frozen and being phased out. It will be interesting to see what alternatives will be selected to replace the reactors shut down.

Some alternate proposals for waste disposal include: (a) the space option, (b) the Antarctic Ice Shelf, (c) Ocean dumping, (d) Ocean burial, and (e) Nuclear transformation.

(a) The soft-landing approach for the storage of nuclear waste on the moon has been evaluated but until such flights become routine it must be considered too risky.

(b) Disposal under the Antarctic ice is also too risky due to the unknown stability of the underlying shelf which requires predictable stability for thousands of years.

(c) Ocean dumping has been carried out since 1943 with disastrous results. Evidence of leaking drums has been obtained, and long term storage is not feasible for high-level waste.

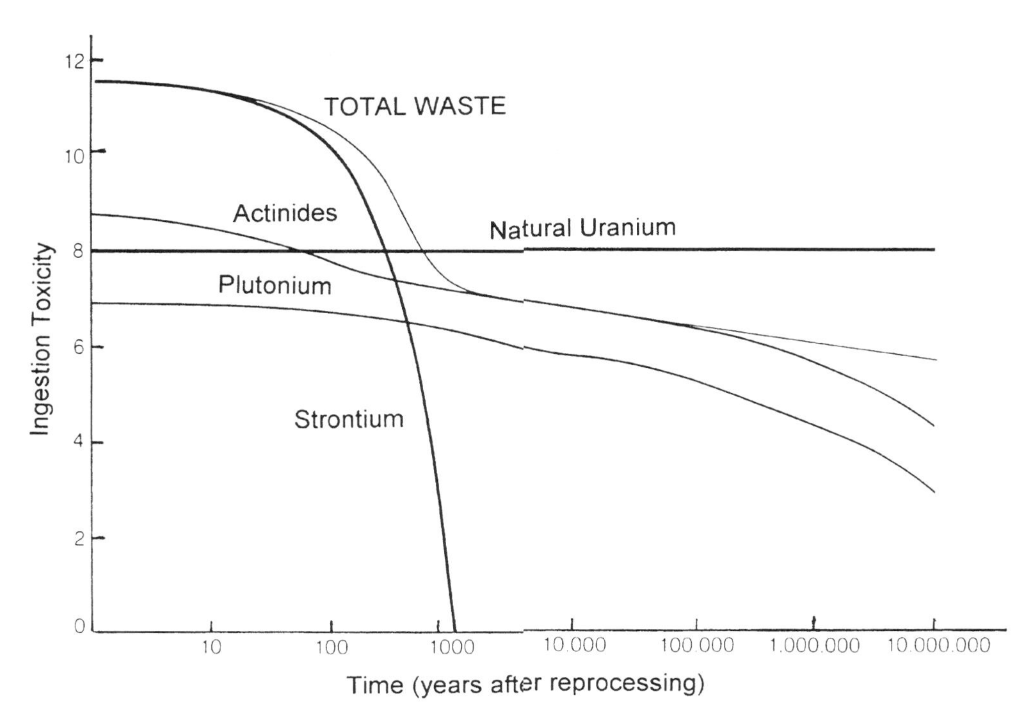

FIGURE 7.11. Within a finite period, the hazardous components of nuclear waste decay to a radioactive toxicity level lower than that of the natural uranium from which the waste was derived. The strontium in waste becomes less toxic to humans than natural uranium ore in 450 years; the total waste including plutonium, becomes less toxic in 500–1000 years, depending on fuel history and reprocessing-plant characteristics.

(d) Burial in the ocean floor is attractive if deep holes can be drilled and sealed with the sediment. This is being studied actively and the environmental impact evaluated.

(e) The conversion of radioactive nuclei by nuclear reactions and transformation into stable nuclei is possible. The cost, however, is prohibitive and no concerted effort is being made to make this a viable alternative.

Canada is very fortunate in having the Canadian Shield of Precambrian rock formation in Central Canada around Hudson's Bay. Some hard rock formations, called *Pluton*, within the Shield are ideal sites for the location of an underground nuclear waste storage facility.

Studies now in progress in an underground laboratory will establish the permeability and water movement in such formations.

It is interesting to note that the natural nuclear reactor discovered in Oklo, Gabon, Africa, operated for about 7×10^5 years and released about 15 GW years of energy about 1.8×10^9 years ago. The study showed that the fission products could lie immobile in the ground at the original reactor site during these many years. It is evident from this natural example that responsible disposal of nuclear waste is feasible and will eventually be accomplished.

TABLE 7.14
Activities of Selected Fission Products and Actinides in Irradiated Fuel[a]

Radionuclide	Mode of decay	Half-life (years)	Activity (Curies/kg U)		
			At discharge	After 1 year	After 10 years
Fission Products					
Tritium (H-3)	β	12.3	0.17	0.16	0.10
Krypton-85	β, γ	10.7	2.22	2.19	1.23
Strontium-89	β, γ	0.14	443	3.95	9.8×10^{-20}
Strontium-90	β	29	17.5	16.0	12.9
Yttrium-91	β, γ	0.16	578	77	1.1×10^{-16}
Zirconium-95	β, γ	0.18	825	17.3	1.3×10^{-14}
Niobium-95	β, γ	0.10	802	36.6	2.9×10^{-14}
Technetium-99	β, γ	2.1×10^{5}	3.4×10^{-3}	3.4×10^{-3}	3.4×10^{-3}
Ruthenium-106	β	1.0	182	101	0.21
Iodine-131	β, γ	1.6×10^{7}	7×10^{-6}	7.9×10^{-6}	7.9×10^{-6}
Iodine	β, γ	0.02	525	1.2×10^{-11}	0
Cesium-134	β, γ	2.17	16.9	11.3	0.55
Cesium-135	β	2.3×10^{6}	4.5×10^{-5}	3.8×10^{-5}	3.8×10^{-5}
Cesium-137	β, γ	30.2	25.3	24.8	20.2
Cerium-144	β, γ	0.78	424	181	0.06
Promethium-147	β, γ	2.6	58.9	50.7	4.7
Actinides					
Neptunium-237	$\alpha. \gamma$	2.1×10^{6}	2.1×10^{-5}	2.1×10^{-5}	2.2×10^{-5}
Plutonium-238	α, γ	87.7	7.2×10^{-2}	8.3×10^{-2}	8.0×10^{-2}
Plutonium-239	α, γ	2.4×10^{4}	0.15	0.15	0.15
Plutonium-240	α, γ	0.24	0.24	0.24	0.24
Plutonium-241	β, γ	14.7	22.9	21.8	14.2
Americium-241	α, γ	432	11.5×10^{-3}	4.7×10^{-2}	0.3
Americium-243	α, γ	7380	5.3×10^{-3}	5.3×10^{-4}	5.3×10^{-4}
Curium-242	α, γ	0.45	2.58	0.44	8.9×10^{-6}
Curium-244	α, γ	18.1	1.6×10^{-2}	1.5×10^{-2}	1.1×10^{-2}

[a]Based on Pickering fuel irradiated to 7.5 MWd/kg U.

7.9. NUCLEAR FUSION

In contrast with nuclear fission where a large nucleus is split into two more stable nuclei, fusion relies on the formation of larger stable nuclei from small nuclei. The main difference is that fusion requires an initial high temperature of millions of degrees to overcome the energy repulsion barrier of the nuclei. In the fusion H-bomb the high temperature (3×10^{8} K) is achieved by a fission bomb.

Various systems have been designed and tested to achieve these high temperatures and to initiate the fusion reaction. Such reaction have been maintained for very short time intervals (about 1 sec) which is slowly being extended to longer times.

Some fusion reactions, and the corresponding energy liberated, are given next:

1. DT: Deuterium—Tritium Reaction

$$^{2}_{1}D + ^{3}_{1}T \rightarrow ^{4}_{2}He\ (3.5\ MeV) + ^{1}_{0}n\ (14.1\ MeV) \quad (7.23)$$

This reaction can occur at the lowest temperature and produce the highest fusion–power density. The highly energetic neutrons present technical problems regarding material of construction.

2. DD: Deuterium—Deuterium Reactions

$$^{2}_{1}D + ^{2}_{1}D \nearrow ^{3}_{2}He\ (0.82\ MeV) + ^{1}_{0}n\ (2.45\ MeV) \quad (7.24)$$

$$^{2}_{1}D + ^{2}_{1}D \searrow ^{3}_{1}T\ (1.01\ MeV) + ^{1}_{1}H\ (3.02\ MeV) \quad (7.25)$$

The tritium and $^{3}_{2}He$ can react further to give

$$3^{2}_{1}D \rightarrow ^{4}_{2}He + ^{1}_{1}H + ^{1}_{0}n + 21.6\ MeV \quad (7.26)$$

In 1996 experiments on the electrolysis of LiOD in D_2O on a palladium cathode have been claimed to result in "cold" fusion. The detection of helium, neutrons and even tritium has been reported. However, there is considerable doubt about the validity of the claims and "cold" fusion like polywater* will soon be buried and its obituary published.

3. Other Reactions

The following reactions do not involve neutrons

$$^{1}_{1}H + ^{6}_{3}Li \rightarrow ^{3}_{2}He + ^{4}_{2}He \quad (4.0\ MeV) \quad (7.27)$$

$$^{3}_{2}He + ^{6}_{3}Li \rightarrow ^{1}_{1}H + 2^{4}_{2}He \quad (16.8\ MeV) \quad (7.28)$$

$$^{3}_{2}He + ^{3}_{2}He \rightarrow 2^{1}_{1}H + ^{4}_{2}He \quad (12.9\ MeV) \quad (7.29)$$

$$^{1}_{1}H + ^{11}_{5}B \rightarrow ^{12}_{6}C \rightarrow ^{8}_{4}Be + ^{4}_{2}He \quad (7.30)$$

$$\hookrightarrow 2^{4}_{2}He$$

The overall reaction is

$$^{1}_{1}H + ^{11}_{5}B \rightarrow 3^{4}_{2}He \quad (7.31)$$

Reactions involving $^{3}_{2}He$ are not really suitable because $^{3}_{2}He$ must be artificially produced. The amount of deuterium in the oceans which is available for fusion can supply enough energy to satisfy earth's energy requirements for many centuries.

* Polywater was a "new" form of water which was described in the literature in the early 1970s. Polymerized water was reported to have a freezing point of −40°C and a viscosity of several times that of normal water. It was later shown that the anomalous properties of polywater were due to impurities.

7.10. SUMMARY

Radioactivity can be harmful or beneficial. As a tracer isotope it can be used to map the fate of a chemical or the location of a malfunctioning organ. It is used in radiation therapy for cancer patients and in the irradiation of food to prevent or reduce spoilage. It is also dangerous in significant dosage, causing cancer and death. Nevertheless it is an efficient source of energy which does not contribute to the greenhouse effect and thus it helps preserve our climate. The need to deal with the problem of waste management remains but appears to be solvable. The real question is "can we survive without nuclear energy?"

EXERCISES

1. Uranium is present in seawater at about 3.3 mg/tonne. Calculate the amount of uranium in the oceans, ($V = 1.5 \times 10^9$ km^3).
2. Deuterium is present in water to about 150 ppm. Calculate the amount of deuterium present in earth's water.
3. Calculate the binding energy in the nitrogen nucleus.
4. Calculate the energy change in reactions (7.9) and (7.17).
5. A sample of wood from a cross was tested for ^{14}C and found to have 13.2 Bq/g of carbon. What was the age of the cross?
6. A sample of a radioactive mineral had an activity of 40,000 Bq. Three months later the activity of the same sample was 32,000 Bq. What is the half-life of the active component in the mineral?
7. A radioactive tracer (10 mL) with a half-life of 12 days was administered to a patent weighing 75 kg. Its initial activity was 1600 Bq/mL of solution. When testing the patient's blood a week later, what would be the expected activity of the blood (assuming the volume of blood in the patient is 5 L and that all the activity remains in the blood).
8. Explain how the breeder reactor can be used to extend the energy convertibility of nuclear fuel.
9. One of the major concerns about nuclear energy is the disposal of radioactive waste from the reactors. Discuss the problem and the various solutions which have been proposed.
10. From Table 7.3 calculate the average atomic mass of uranium.
11. The expected value of an event (EV) is equal to the risk (R) times the benefit (B), $EV = RB$. Calculate the expected value for each of the following.
    (a) Throw a 6-in cubic dice with 1–6 on the sides and win \$400.
    (b) Pick heads on the toss of a coin and win \$100.
    (c) Pick the ace of spades out of a full deck of cards (52) and win \$3500.
    (d) Pick any ace out of a full deck of cards (52) and win \$950.
    (e) Pick the month of birth of a total stranger and win \$800.
12. Calculate the energy liberated in the fusion reaction (7.31).
13. Write the shorthand notation for the reaction $^6_3Li + ^1_1H \rightarrow ^3_2He + ^4_2He$.
14. Write the nuclear reaction for $^{63}_{29}Cu(\alpha, n)^{66}_{31}Ga$.

15. Calculate the energy liberated when 1 g of U-235 is split by a slow neutron to form 3 neutrons, Ba-141 and Kr-92.

FURTHER READING

R. Alcraft, *Nuclear Disasters*, Heinemann Lib. (1999).
C. R. Hill, *Nuclear Energy: Promise or Peril*, WSC Inst., MA Studies (1999).
D. Bodansky, *Nuclear Energy, Principles, Practices, Practices and Prospects*, American Institute of Physics, Woodbury, New York (1996).
A. Walter, *America the Powerless: Facing Our Nuclear Energy Dilemma*, Medical Physic Pub., Madison, Wisconsin (1995).
J. Byrne and S. Hoffman, Editors, *Governing the Atom, The Politics of Risk*, Vol. 7, Transactions Pub., New Brunswick, New Jersey (1995).
F. L. Bouquet, *Nuclear Energy Simplified*, 2nd Ed. Systems Co., P.O. Box 339, Carlsborg, Washington (1994).
T. Lowinger and G. W. Hinnan, Editors, *Nuclear Power at the Crossroads: Challenges and Prospects for the Twenty-First Century*, Intl. Research Ctr. Energy. Boulder, Colorado (1994).
J. Lenihan, *The Good News about Radiation*, Medical Physics Publ., Madison, Wisconsin (1993).
J. Van Der Pligt, *Nuclear Energy and the Public*, Blackwell Publs., Cambridge, Massachusetts (1992).
R. Wolfson, *Nuclear Choices*, MIT Press, Cambridge, Massachusetts (1991).
C. K. Ebinger et al., *Nuclear Power: The Promise of New Technologies*, Center for Strategic and International Studies, Washington, DC (1991).
P. R. Mounfield, *World Nuclear Power: A Geographical Appraisal*, Routledge, Ndew York (1991).
B. L. Cohen, *The Nuclear Energy Option: An Alternative for '90s*, Plenum Press, New York (1990).
International Nuclear Energy Guide, 15th Ed., Editions Technip., Paris, France (1987).
T. A. Heppenheimer, *The Man-Made Sun*, Omni Press, Boston, Massachusetts (1984).
B. L. Cohen, *Before it's too Late*, Plenum Press, New York (1983).
G. Kessler, *Nuclear Fission Reactors*, Springer-Verlag, New York (1983).
S. E. Hunt, *Fission, Fusion and the Energy Crisis*, 2nd Ed., Pergamon Press, Oxford (1980).
J. J. Berger, *Nuclear Power—The Unviable Question*, Ramparts Press, Pala Alto, California (1976).
B. Orensen, *Renewable Energy*, Academic Press, New York (1974).
Nuclear Energy Institute, http://www.nei.org/
International Atomic Energy Agency, http://www.iaea.org
Westinghouse, http://www.westinghouse.com/c.asp
U.S. Nuclear Waste Technical Review Board, http://www.nwtrb.gov/
Atomic Energy of Canada Ltd., http://www.aecl.ca
South Africa Nuclear Energy, http://www.aec.co.za/
Ontario power generation, Nuclear fuel plants, http://www.opg.com/
Essential info, http://www.essential.org
Anti-nuclear Movement Links, http://www.neravt.com/left/
Greenpeace, http://www.greenpeace.org/
Chem instruction, http://www.chemcases.com/nuclear/index.htm
Nuclear waste, http://radwaste.org/ngo.htm
Anti-nukes, http://nuclearnukes.com/

8

Lubrication and Lubricants

8.1. AN INTRODUCTION TO TRIBOLOGY

Tribology, from the Greek word *tribos* meaning rubbing, is the science of friction, lubrication, and wear. The basic laws of tribology are concerned with the general behavior of surfaces interacting during sliding and to a first approximation are given by

$$F_{kin} \propto W \qquad \text{or} \qquad F_{kin} = \mu W \tag{8.1}$$

$$Z \propto W \qquad \text{or} \qquad Z = \kappa W \tag{8.2}$$

where F_{kin} is the sliding force of friction, W is the load applied normal to the sliding direction, Z is the wear rate (volume of surface removed per unit sliding distance), μ is the coefficient of friction and κ is the wear factor. F_{kin} and Z are independent of the apparent area of contact.

The factors of proportionality in laws (1) and (2) above are not necessarily related to the same operating parameters and it is possible to obtain heavy wear rates associated with low frictional forces.

Friction is not always undesirable. For example, the stopping of an automobile relies on friction between the tires and the road and on friction within the braking mechanism.

The function of a lubricant is (1) to prevent the moving interacting surfaces from coming into direct contact, (2) to provide an easily sheared interfacial film, (3) to remove the heat evolved in the process, and (4) to reduce wear of the surfaces. Solid lubricants can only satisfy (1), (2), and (4) but only liquids and gas can also satisfy (3). Lubricants can conveniently be classified into gases, liquids, and solids.

8.2. GASEOUS LUBRICANTS

Gas-lubricated bearings have special applications and their use is often neglected. The gases used include air, helium, nitrogen, and hydrogen—though air is the most commonly employed. It must be pointed out that in gas lubricated bearings, the coefficient of friction approaches zero. The viscosity of air under atmospheric conditions is 0.018 cP* (centipoise) and its temperature coefficient is small allowing both low

*A Poise = $1\ \mathrm{g\,cm^{-1}\,s^{-1}}$.

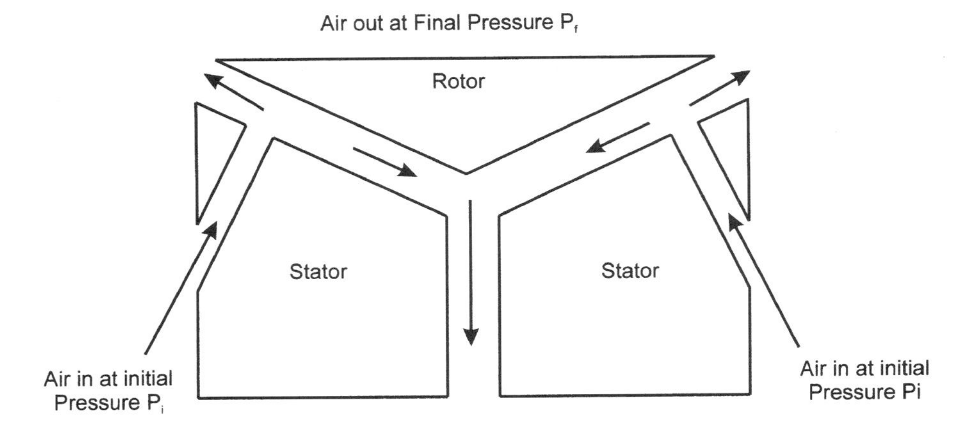

FIGURE 8.1. Schematic cross section of a hydrostatic (gas lubricated) cone step.

temperature and high temperature operation. A schematic representation of a hydrostatically lubricated gas thrust bearing is shown in Fig. 8.1.

For hydrodynamic conditions, the minimum film thickness for a gas thrust bearing, *ho*, is given by the relation:

$$ho = \sqrt{0.66\eta Vl/P_{av}} \tag{8.3}$$

where η = the viscosity, V = mean runner velocity, l = the length of shoe in rotation direction, P_{av} = average thrust pressure. The viscosity of a gas, η, is given by:

$$\eta = \frac{N'mc\lambda}{3} \tag{8.4}$$

where N' = the concentration of molecules in the gas, molecules per unit volume; m = mass of the gas molecule, c = average molecular speed, and λ = mean free path.

The mean free path of a gas λ is the average distance a molecule moves between collisions and is given by:

$$\lambda = \frac{1}{\sqrt{2}\,\pi\sigma^2 N'} \tag{8.5}$$

where σ is the molecular diameter and the average molecular speed c, is given by

$$c = \sqrt{\frac{8RT}{\pi Nm}} \tag{8.6}$$

where R = the gas constant, T = the absolute temperature, and N = Avogadro's number (6.022×10^{23}).

Thus, by combining Eqs. (8.2)–(8.4) we get

$$\eta = \frac{2}{3\pi^{3/2}\sigma^2}\sqrt{mkT} \tag{8.7}$$

where k is Boltzmann's constant,

$$k = R/N \tag{8.8}$$

Thus, $\eta \propto \sqrt{T}$ to a first approximation. Fluidity, φ, is the reciprocal of viscosity:

$$\text{fluidity } \varphi = 1/\eta \tag{8.9}$$

The ratio λ/ho is known as Knudsen number (K_n) and defines the conditions under which the gas can be considered a continuum. When $K_n < 0.01$ the gas can be considered as a fluid. When, however, $0.01 < K_n < 1.0$ the gas is a continuum if slip at the boundary is reduced or corrected. For air at 0°C and 1 atm, $\lambda \simeq 6 \times 10^{-6}$ cm, hence for film thicknesses of 6×10^{-4} cm or more, $K_n < 0.01$ and proper lubrication without slip can be expected.

Gas-lubrication is best suited to continuous operation at extreme temperatures as well as in radiation fields which would normally degrade oil. It is noncontaminating and "free." Starting and stopping gas-lubricated bearings can be harmful unless special design and precautions are taken. Other disadvantages include limited load capacity, susceptible to instability and requirement for precise machining.

8.3. LIQUID LUBRICANTS

Like gas lubrication, liquids lubricate by keeping the two moving surfaces apart. The bearing is one of the most common lubricated systems. There are four types of bearing: the journal bearing, the thrust bearing, the slider bearing and the ball bearing.

8.3.1. Journal Bearing

This consists of a rotating axle or journal which contacts the bearing for support. Rotations and the small angle between the two surfaces causes a buildup of pressure in the oil layer which supports the load during rotation. This is called hydrodynamic lubrication and is shown in Fig. 8.2. The important characteristic of the bearing is the dimensionless number C where

$$C = \frac{\eta N}{P} \tag{8.10}$$

η = the viscosity of the lubricant, N = the speed of the shaft, and P = the pressure or load divided by the projected area. Values of $C \geqslant 35$ implies that hydrodynamic lubrication is effective. When $C < 35$ the oil wedge responsible for hy-

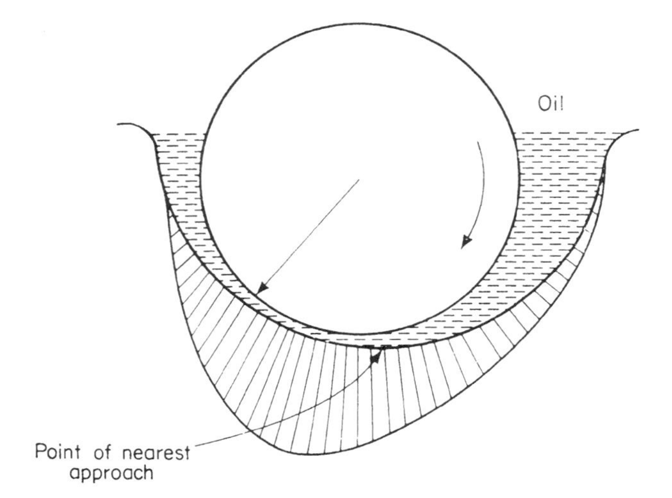

FIGURE 8.2. The journal bearing showing the pressure distribution and oil forces supporting the journal. The parallel lines indicate the oil pressure.

drodynamic lubrication is too thin, leading to an increase in friction, i.e., a reduction in lubrication.

The true surface area of a polished surface can be many times the nominal area. Thus, for example, the true area of a polished plate glass surface is about 10 times the nominal area. This implies the presence of many hills and valleys which can interact when two such surfaces are brought together (see Fig. 8.3). If the speed of rotation N and/or the viscosity of the lubricant is reduced or if the pressure is increased, then although the coefficient of friction may increase, the distance between the journal and the bearing will eventually decrease to less than the height of the largest surface irregularity and wear will result. There are thus three possible causes of lubrication failure: high pressures, low speeds, and of course, high temperature which reduces the viscosity of the lubricant oil.

The variation of viscosity (η) with temperature (T), in its simplest form, is given by the Reynolds relation:

$$\eta = ke^{-at} \tag{8.11}$$

where k and a are constants depending on the liquid. This can also be expressed as

$$\eta = ke^{E_{\text{vis}}/RT} \tag{8.12}$$

where E_{vis} is the activation energy for viscose flow and R is the gas constant. A more exact relationship is given by

$$\log_{10} \log_{10}(v + 0.8) = n \log_{10} T + C \tag{8.13}$$

where $v = \eta/\rho$ (kinematic viscosity with units of stokes) in which n and C are constants for a given oil or lubricant and ρ is the density of the liquid. The effect of temperature on the viscosity of a liquid is discussed in Appendix B.

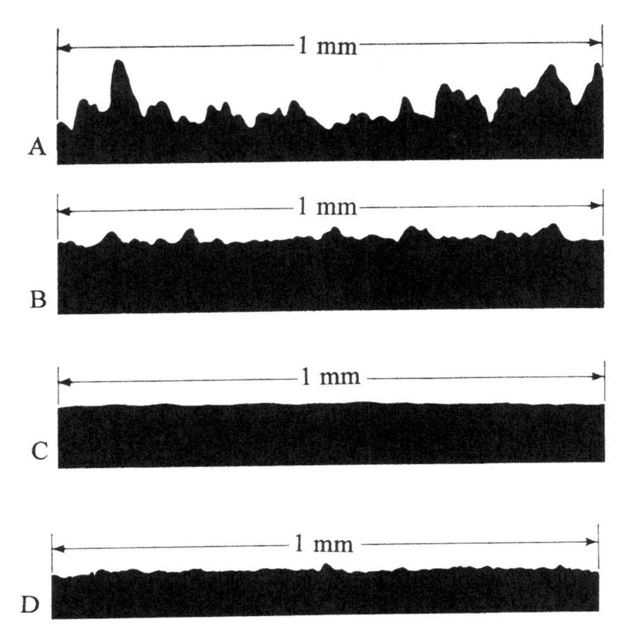

FIGURE 8.3. Microsections of various steel surfaces showing roughness depths (asperites). Vertical scale has been magnified.

Surface	Depth (μm)	Load bearing fraction (% area)
A. Smooth turned	10–15	0.5
B. Fine ground	1–10	25
C. Polished	0.1–0.6	50–80
D. Cold extruded	0.3–3.5	20–65

To reduce wear during low speeds, long chain fatty acids or soaps are added to the oil. These form chemical bonds to the metal surfaces and thus prevent direct metal–metal contact or welding of the two surfaces. This type of lubrication is called *boundary* lubrication and is illustrated in Fig. 8.4.

Two surfaces which are in relative motion with an oil film between them are hydrodynamically lubricated. The viscosity of the oil determines the friction of motion. When the thickness of oil film is less than the surface irregularities, the asperites, then surface wear will eventually bring the two surfaces sufficiently close so as to rely on boundary layer lubrication. This is illustrated in Fig. 8.5 for a moving piston in a car engine.

8.3.2. Thrust Bearings

Thrust bearings are meant to keep a rotating shaft from motion parallel to the axis of rotation. Various configurations are possible but the most common is one in which the load is supported by the bearing in a vertical shaft system.

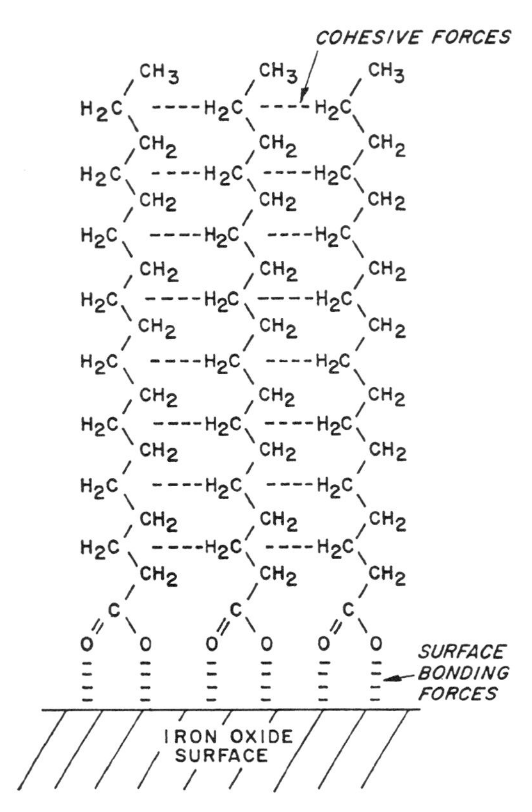

FIGURE 8.4. A schematic illustration of a monomolecular iron stearate boundary film. Cohesive forces between the iron stearate molecules result in a closely packed and difficult to penetrate film.

8.3.3. Slider Bearings

Slider bearings are also called *ways* or *guide* bearings and are meant for rectilinear and curvilinear motion.

8.3.4. Ball Bearings

Elastohydrodynamic Lubrication (EHL). In journal bearings or other surface area interactions under lubricating conditions, deformation of the materials or the influence of pressure on viscosity of the lubricant can be ignored. However, in roller and ball bearings the point of contact is deformed elastically under the high load pressure. This also increases the pressure on the film lubricant which increases in viscosity according to the equation

$$\eta_p = \eta_o e^{\alpha P} \tag{8.14}$$

where η_o is the viscosity of a liquid at 1 atm pressure, η_p is the viscosity of the liquid at pressure P and α is the pressure coefficient of viscosity.

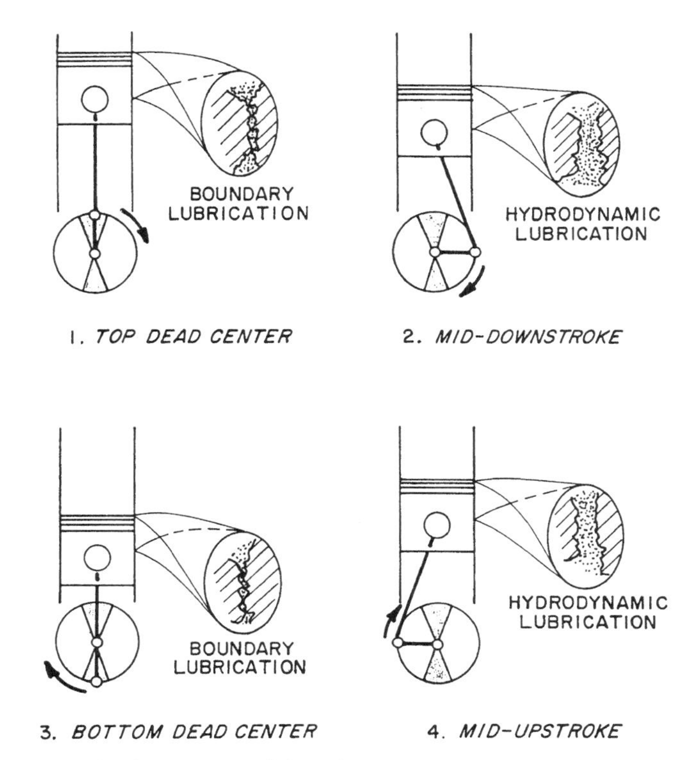

FIGURE 8.5. Schematic diagram describing piston-to-cylinder wall and piston skirt-to-cylinder wall lubrication regime changes during an engine revolution.

The film thickness is about 10–50 μm and increases as the speed increases. An increase in load has the effect of deforming the metal surfaces more, causing an increase in contact area rather than a decrease in film thickness.

An important characteristic of EHL is the specific film thickness (λ) which is the ratio of the film thickness (h) to the composite roughness (σ) of the two surfaces, i.e.,

$$\lambda = h/\sigma \tag{8.15}$$

The composite roughness is determined from

$$\sigma = \sqrt{\sigma_1^2 + \sigma_2^2} \tag{8.16}$$

where σ_1 and σ_2 are the root mean square roughness

$$h = B\sqrt{\eta_0 VR} \tag{8.17}$$

where η_0 = viscosity of oil at 1 atm pressure

$V = \frac{1}{2}(V_1 + V_2)$, mean speed
R = mutual radius of curvature of the two contacting surfaces

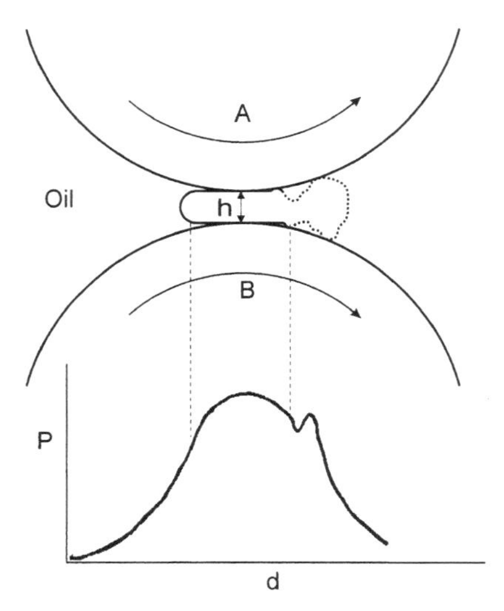

FIGURE 8.6. Schematic representation of elastic deformation of touching curved surfaces with the accompanying pressure increase on the lubricating oil.

and where B is a constant depending on the units used, e.g., $B = 8 \times 10^{-6}$ cm for cgs units. The elastic deformation and oil pressure profile is shown in Fig. 8.6.

8.4. EXTREME PRESSURE LUBRICATION (EP)

At higher pressures and higher sliding speeds or at temperatures above about 150°C, the surface boundary layer breaks down or "melts" and boundary lubrication ceases to operate. Such conditions can occur in hypoid and wormgears, and special oil-soluble substances called *extreme pressure* (EP) additives are used. These materials react with the metal surface at a specific temperature and pressure and form a layer of low shear strength solid; they are normally sulfides, chlorides or phosphorus compounds in the form of sulfurized fats, chlorinated esters and tricresyl phosphates of the base metal–iron. The reaction is a form of controlled corrosive attack which is preferred to uncontrolled abrasive wear. The solids formed by EP lubricants have high fusion temperatures, making them more effective than boundary lubricants. Though corrosive wear is slow, it cannot be eliminated by EP lubricants and the alternate use of solid lubricants such as graphite or molybdenum disulfide is employed. These lubricants will be discussed later.

8.5. WEAR

Not too many years ago a new car owner was required to "run the car in" by driving at lower speeds for a few months to wear off the high spots. The oil and lubricants were then changed and the car was allowed to run normally. The initial operation removes the asperites of the bearings and pistons and results in wear.

TABLE 8.1
Sources of Trace Elements in Oil

Element	Wear or contamination source	Typical values[a] (μg/g)
Si	Dirt due to faulty air filter, also found in antifoamers	5
Fe	Wear from engine block, gears, rings, camshaft, oil pump and/or crankshaft	20
Cu	Wear of bushings, valve guides, connecting rods or piston pins	6
Ni	Wear of plating on gears and some bearings	3
Sn	Wear of some bearings and coated surfaces	2
Pb	Wear of bearings in diesel engines	5
Cr	Wear of rings and cooling system if chromates are used as corrosion inhibitors	4
Al	Wear of pistons and some bearings	4
Mo	Wear of some alloy bearings and oil coolers	3
Na	Possible leakage of antifreeze	30
B	Coolant leakage if boron inhibitors are used	2
Ba	Normally used in rust and corrosion inhibition and in oil	2
Zn	Detergents	<2
P	Normally used as an antiwear additive in oil	<2
Ca	Normally used in detergents and antirusting additives	3000
	Normally found in detergents and dispersants	

[a]Under normal running conditions.

The analysis of lubricating oils for trace metals can give an early indication of excessive wear. More than 40 years ago the New York Central Railway routinely analyzed locomotive lubricating oils for trace metals and frequently predicted, and eventually prevented, engine failure. The development of modern automatic-computer controlled instrumental methods can routinely analyze as many as 20 elements at a rate of about 200 samples per hour. The methods include Atomic Absorption Spectrophotometry (AAS), Inductively Coupled Argon Plasma Spectrometry (ICAPS), and Atomic Fluorescence Spectrophotometry (AFS). The method is commonly referred to as SOAP (Spectrographic Oil Analysis Program). Table 8.1 shows the elements commonly found in oil from a diesel engine, the typical values and the source. The data in Table 8.2 show the results of consecutive analysis for several elements in diesel engines after different periods of operation. The high levels of lead in Engine B clearly indicates a potential problem.

This approach (SOAP) is now used for fleet vehicles, aircraft engines, and locomotives, and can prevent breakdowns by indicating maintenance in relation to wear of predictable parts of an engine.

8.6. OIL ADDITIVES

Straight chain aliphatic (alkane) hydrocarbons do not have optimum lubricating properties. However the addition of certain substances can improve a paraffin oil.

TABLE 8.2
Typical Concentrations (μg/g) of Trace Elements in Engine Oil as a Function of Time of Use

	Weeks	Al	B	Cr	Cu	Fe	Pb	Si	Sn
Engine A	1	3	2	2	4	10	4	4	2
	6	3	3	2	4	21	6	5	3
	12	2	2	2	5	22	12	5	2
	13	4	3	2	11	31	22	6	4
			Ni						
	0	<5	1.0	<1	1	31	4		<20
Engine B	25	12	1.7	<1	751	11	1200		<20
Engine C	25	<5	1.0	<1	6	1	18		<20

The general purpose of lubricant additives are:

(*a*) *To Extend the Useful Range of Conditions of the Lubricant by Increasing the Viscosity Index* (*VI*). One of the major factors affecting the conditions of a lubricant is its temperature, which usually starts low (cold) and increases with running time. Thus, it is most desirable for the temperature effect on viscosity to be a minimum. This is determined by the VI which is related to E_{vis} and determined by a comparison with standard oils. A high VI (80) signifies a low value in E_{vis} and a small dependence of viscosity on temperature, whereas the converse is true for a lubricant with a low VI (30).

Common VI improvers which are added to lubricants are high molecular weight ($\sim$20,000 g/mol) polymers. These tend to increase the viscosity of the oil to a greater extent at high temperatures than at low temperatures. Examples of polymers added include polymethacrylates, polyacrylates, polyisobutylenes, and alkylated styrenes. VI improvers are used in engine oils and automatic transmission fluids. The effect of a VI additive on an oil is shown in Fig. 8.7.

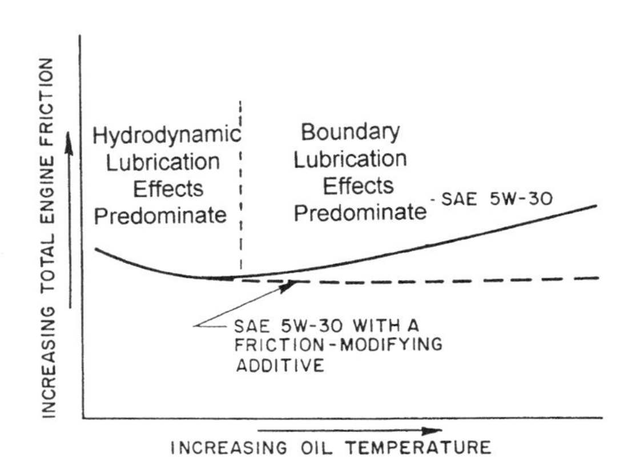

FIGURE 8.7. The effect of a friction-modifying additive on the total engine friction response to oil temperature for an SAE 5W-30 engine oil.

(*b*) *To Prevent or Retard the Deterioration of the Lubricant.* The most common additive designed to preserve the function of the lubricant is the antioxidant which reduces or inhibits the rate of oxidation of the oil. In oxidation the oil can form peroxides which eventually decompose to acids causing corrosion of the metallic components. Zinc dithiophosphate acts as both an antioxidant and a corrosion inhibitor. Other additives include sulfur compounds, amines, and metal salts such as calcium salts of alkylphenol formaldehyde.

(*c*) *To Protect the Surfaces in Contact with the Lubricant.* Surface protectors include boundary layer additives and rust inhibitors such as sulfonates, phosphates, and fatty acids salts. Extreme pressure additives also preserve metallic surfaces at lower temperatures.

(*d*) *To Depress the Pour Point and Break Foam.* The pour point (PP) is the temperature below which the oil will not flow due to the crystallization of the heavier wax components. Paraffinic type oils have higher pour points than the naphthenic oils. Hence the reduction in the PP temperature is achieved by either removing the wax components or adding PP depressants. Such depressants include the additions of high molecular weight polymers of polymethacrylate, polyalkyl-naphthalenes, and polyalkyl-phenol esters. An oil can have its PP depressed by 30° upon the addition of 1% depressant. The cloud point, the temperature at which the oil becomes cloudy on cooling, is due to the separation of wax or moisture. Thus low cloud points are a desirable property of a lubricant.

Foam and froth in the oil is caused by the entrapment of air during the pumping and movement of oil through various parts of the engine. Foam reduces the effectiveness of the lubricant and thus can result in increased friction and wear. Antifoaming agents are usually silicones which at 5 ppm or less can lower the surface tension of the oil and destroys the tendency to form bubbles.

(*e*) *To Improve the Wetting Properties of the Lubricant.* A good lubricant must be able to wet the metal surface that require lubrication. A discussion of wetting is given in Appendix C. Additives which lower the contact angle of oil on steel include fluorinated polymers. Other additives include dispersants and detergents which tend to keep solid particles (that pass through the filter) in suspension, preventing them from forming larger agglomerates.

The modern lubricating oil has many additives and further research and developments continue to improve oils, extending their useful life and effectiveness.

8.7. SYNTHETIC LUBRICANTS

Lubricants which are not based on petroleum are classed as synthetic and include silicones, polyglycols, polyphenyl ethers, fluorocompounds, chlorofluorocarbon polymers, and phosphate esters. The major advantage of these fluids is that they are tailor made for a specific function and usually require few additives.

Silicone oils are especially useful for high temperature applications and can be polymerized to give any desired viscosity. However their major disadvantage is that when they decompose they invariably form solid silicates and silica (SiO_2) which can damage lubricated joints and bearings. Their low surface tension make them less desirable for thin film lubrication and boundary conditions. However, the high viscosity index, good chemical stability, high shear resistance, low volatility, and antifoam characteristics make silicone oils good lubricants in special situations such as torque converters, ball bearings etc., despite the high cost.

Polyglycols are water insoluble but by incorporating ethylene oxide into the polymer a water soluble ethylene oxide–glycol polymer is obtained. They have very high viscosity index (160), low volatility, and good compatibility with rubber. The water soluble type, with as much as 40% water content, is considered to be fire-resistant and therefore used in mines and ships. Their degradation products are soluble in the fluid or volatile, thus leading to very low sludge formation.

Fluorolub consists of fluorinated hydrocarbons and polyethers. They can function effectively from −90 to over 250°C, have a high density, are thermally stable, have low volatility, and are nonexplosive. They are usually used in highly corrosive environments such as in the preparation of hypochlorite (ClO^-). The cost of such specialized lubricants is very high (about $2000/kg) but since they do not degrade they almost last forever.

8.8. SOLID LUBRICANTS

Solids in the form of soap has long been used as lubricants. More recently the three most prominent solids which have lubricating properties are graphite, molybdenum disulfide (MoS_2), and Teflon. A more complete list is given in Table 8.3. In graphite and MoS_2 the solid is a layer lattice (lamellar) in which the distance between the parallel planes is 3.40 Å for graphite and 3.49 Å for MoS_2. These large values are to be

TABLE 8.3
Selected Solid Lubricants

	Electrical resistivity (Ω cm)	Air stability (°C)
Teflon	10^{18}	320
MoS_2	8.5×10^2	350
$MoSe_2$	1.8×10^{-2}	400
WS_2	1.4×10^1	440
WSe_2	1.1×10^2	350
NbS_2	3.1×10^{-3}	420
$NbSe_2$	5.4×10^{-4}	350
TaS_2	3.3×10^{-3}	600
$TaSe_2$	2.2×10^{-3}	580
Graphite	1.4×10^{-3}	450
ReS_2		230
WS_2 + 10% Ga		800
WSe_2 + 10% Ga		800

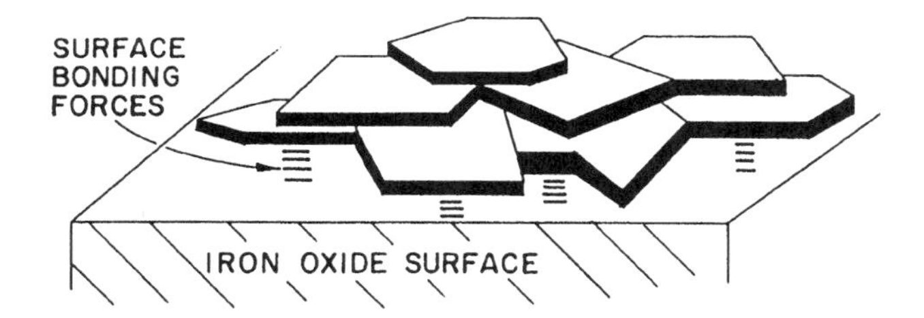

FIGURE 8.8. Schematic illustration of how microcrystalline plates of solid materials such as molybdenum disulfide can form strong surface coatings. The plates slide easily over each other thereby providing lubrication.

compared with 1.42 Å for the in-plane C–C distance and 2.41 Å for the in-plane Mo—S distance. Thus, the large interplanar distance means that the layers are bonded by relatively weak van der Waals forces which permit the planes to slide past each other.

Lamellar solid lubricants function by being interspersed between the asperites of the two moving surfaces. This is illustrated in Fig. 8.8 where the solid adhering to the surfaces cleave at the planes when the metal–lubricant bonding forces are greater than the interplanar forces. This is the case with MoS_2 even under vacuum conditions. However, in the presence of water vapor and to a lesser extent in O_2 or N_2 the lubricating effectiveness of MoS_2 is reduced due to a reduction in the metal–MoS_2 attractive forces. With graphite the converse applies, i.e., graphite is a poor lubricant in vacuum but much better in nitrogen and somewhat better in oxygen but best in the presence of water vapor. This is explained by the adsorption of O_2 and water to the dangling valences at the edge of the graphite planes, thereby reducing the shear strength required to move the interlaminar planes. With MoS_2 no such residual valency is present and the vapors adsorbed onto the metal surfaces reduces its adhesion to MoS_2. Other lamellar solids such as boron nitride (BN) or talc do not have similar lubricating properties because the adhesion to surfaces is less than the shear energy of the interlaminar planes. The mode of action of lamellar solid lubricants is shown in Fig. 8.9.

These solids, when dispersed in lubricating oils, increase the efficiency of lubricants and decrease wear during engine startup. Particle size must be less than 2 μm in order to pass through the filter and to remain in suspension.

A unique solid lubricant has been described which can function at over 800°C in air or other oxidizing environments and maintains its low coefficient of friction of 0.1–0.2. At very high temperatures normal solid lubricants such as MoS_2 or WSe_2 revert to abrasive metallic oxides while graphite is oxidized to CO and CO_2 thereby losing its lubricating surface. The lubricant is a blend of WSe_2 or WS_2 and gallium with 10–30% Ga. Under dynamic bearing pressures of 1500 psi (100 atm) a wear rate of less than 5 mg/h is obtained.

Teflon is a polymer of tetrafluoroethylene often referred to as PTFE. It is well known as a nonsticking surface on the frying pan and other cooking utensils. It has a low coefficient of friction and is chemically inert and thermally stable to over 300°C.

Teflon is a thermosetting plastic but it cannot be used as such because it is too viscous above its glass transition temperature of 325°C. It is therefore formed by the high pressure–high temperature compression of its powder. Teflon is often impregnated

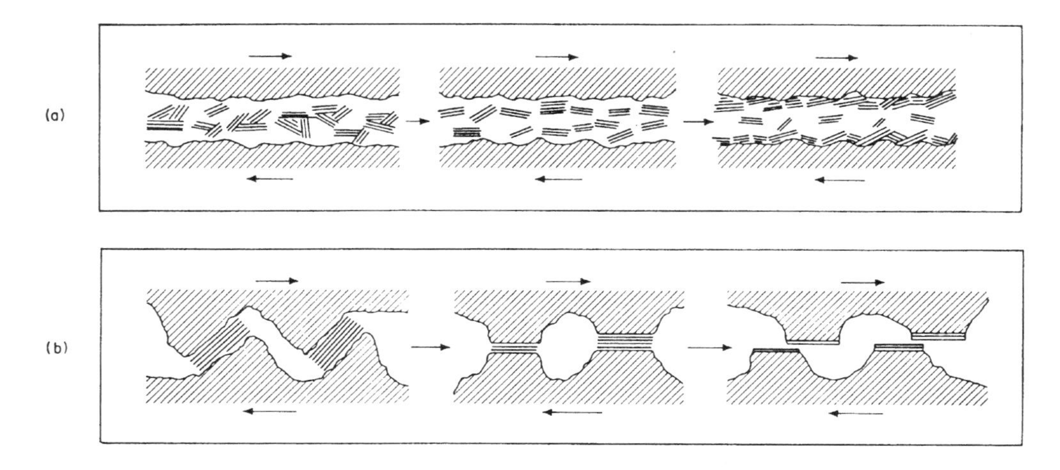

FIGURE 8.9. Schematic diagram of the mode of action of a lamellar solid lubricant when used in dispersion in a liquid such as oil. (a) Orientation of lubricant particles between sliding surfaces, (b) prevention of welding of asperites by the lubricant particle.

with powdered glass or other solids to increase its yield point since it tends to flow and deform under pressure. Other disadvantages of Teflon are that it is a poor conductor of heat and has a high coefficient of expansion. During wear the polymer breaks down, resulting in fluorocarbon free radicals and C_1 and C_2 fluorocarbons. The sliding friction (60 cm/sec, loaded at 0.7 kPa) of steel on Teflon (filled with 25% carbon) in a high vacuum ($P = 5 \times 10^{-8}$ torr, coefficient of friction = 0.17) was shown to give rise to fluorocarbon gaseous fragments which were identified in a mass spectrometer to be CF, CF_2, CF_3, C_2F_2, C_2F_3, C_2F_4. Thus, the wear of Teflon can give rise to dangerous toxic gases and its use in spacecraft is not recommended.

The effectiveness of Teflon as a means of reducing friction of a ski was compared to a conventionally waxed ski on crystalline snow at 0°C over a 213 m slope. With a 76 kg weight on the skis, the normal ski took 63 sec compared to 42 sec for a Teflon coated ski. When the weight was reduced to 64 kg the relative times for the descent of the skis were 83 sec and 54 sec, respectively. Teflon has many uses but its application is limited by the difficulty of bonding it to other surfaces and to itself (see Chapter 12). Another fluorocarbon polymer, perfluoropolypropylene (FEP) is similar to Teflon in its lubricating properties. However, its lower melting point (290°C) allows it to be fabricated by conventional methods when copolymerized with PTFE.

These polymers are more expensive than other plastics such as polyethylene but in special applications they fulfill exceptional needs.

8.9. GREASES

Grease consists of an oil which is thickened by the addition of a finely ground, highly dispersed solid and which is used over a temperature range of −70 to 300°C. The ancient Egyptians (~1400 B.C.) prepared lubricating grease by mixing olive oil with lime. In modern greases the solids include metallic soaps such as aluminum, barium, and lithium. Greases serve to (a) act as a barrier to water and abrasive materials in bearings, (b) protect surfaces against corrosion and wear, (c) lubricate bearings by virtue of the oil present in the grease, and (d) greases usually have a smaller temperature coefficient of viscosity than the base oil.

As much as 14% lithium soap is added to the base oil in a lithium grease. The oil is held in the gel structure by molecular and polar forces, capillarity and by mechanical entrapment. Greases are thixotropic gels which flow under shearing forces such as stirring or motion. The oil in the grease thus lubricates the bearing during motion.

Lithium soap greases with added MoS_2 are generally superior to the base lithium greases in wear reduction, lubrication, and rust prevention. In systems where the lithium soap acts as an EP lubricant the MoS_2 has little effect on the grease. The amounts of MoS_2 added vary from 0.25 up to 25%.

Silicone greases are readily made by adding finely ground or fumed silica (SiO_2) to silicone oil. The amount added determines the thickening effect. Greases are used in bearings and are often preferred to lubricating oils in cases where explosions and fires are possible from the vapors. For example, coal mining is reported to use 1 kg of grease for every 15 tonnes of coal mined.

EXERCISES

1. Gas under pressure can act as a lubricant. What factors affect the lubrication qualities of a gas?
2. What are the advantages and disadvantages of gas lubrication compared to normal liquid lubricants?
3. The analysis of trace metals in oils can be an indication of excessive wear. Explain.
4. Boron nitride forms a layer lattice—would you expect this substance to have lubricating properties?
5. Explain why Teflon-coated surfaces are considered to be self-lubricated in some cases.
6. Determine the mean free path of CO_2 at 2 atmospheres pressure and 300°C. The diameter of the CO_2 molecule is 520 pm.
7. The value of E_{vis} of a liquid is from 20 to 50% of the enthalpy of vaporization of the liquid. Why is this not unexpected?
8. The boiling point of an oil is 325°C. Using Trouton's Rule ($\Delta H_{vap}/T_{bp} = 88$ kJ/mol) and the ratio $\Delta H_{vap}/E_{vis} = 5$, determine E_{vis} for the lubricant.
9. Suggest a lubricant which is compatible with (a) liquid oxygen, (b) concentrated HNO_3 (70% by wt.), (c) concentrated HF (36% by wt.), (d) 50% NaOH solution, and (e) molten sodium at 250°C.
10. Based on the results in Table 8.2 what faults would you expect in the 3 engines, A, B, and C?
11. What lubricants would you use for operating in a spacecraft intended for a journey of 6 months?
12. Distinguish between hydrodynamic lubrication, boundary lubrication, and extreme pressure lubrication.
13. How can a viscous lubricant be converted into a grease?

FURTHER READING

R. M. Mortier and S. T. Orzulik, *Chemistry and Technology of Lubricants*, 2nd Ed., Kluwer Academic, New York (1996).

K. E. Bannister, *Lubrication for Industry*, Industrial Pr. Inc., New York (1996).

R. Miller, *Lubricants and Their Applications*, McGraw-Hill, NY (1993).

M. J. Neale, *Lubrication: A Tribology Handbook*, Soc. Auto. Engineers, Warrendale, Pennsylvania (1993).

Friction, *Lubrication, and Wear Technology: ASM Handbook*, ASM, Materials Park, Ohio (1992).

B. O. Jacobson, *Rheology and Elastohydrodynamic Lubrication*, Elsevier, New York (1991).

M. Ash and J. Ash, Editors, *Conditioners, Emollients, and Lubricants*, Chem. Pub., New York (1990).

J. G. Wills, *Lubrication Fundamentals*, Marcel Dekker Inc., New York (1980).

J. Halling, *Introduction to Tribology*, Wykeham Publ. Ltd., London (1976).

A. Cameron, *Basic Lubrication Theory*, J. Wiley and Sons, New York (1976).

A. D. Sarkar, *Wear of Metals*, Pergamon Press, New York (1976).

D. F. Moore, *Principles and Applications of Tribology*, Pergamon Press, New York (1975).

R. C. Gunther, *Lubrication*, Chilten Book Co., New York (1971).

T. F. J. Quinn, *The Application of Modern Physical Techniques to Tribology*, Van Nostrand Reinhold Co., New York (1971).

B. P. Trading Ltd., *Lubrication Theory and its Applications*, London (1969).

F. Bowden and D. Tabor, *Friction and Lubrication of Solids I and II*, Oxford (1950) and (1964).

Lubrizol--Theory of Lubrication, http://www.lubrizol.com/LubeTheory/default.htm

Soc. of Tribologists and Lubrication Engineers, http://www.stle.org

Oil Analysis Dictionary, http://www.noria.com/

Advanced Lubrication Technology Inc., http://www.altboron.com/

Dry Lubrication, http://www.dicronite.com/

RIV Bearings, http://www.riv.org/bearing.htm

9

Electrochemistry, Batteries and Fuel Cells

9.1. INTRODUCTION

Electrochemistry is concerned with the effect of electrical voltages and currents on chemical reactions (*ionics*) and chemical changes which produce the voltages and currents (*electrodics*). This is illustrated in Table 9.1 where ionics is governed by Faraday's laws whereas electrodics is determined by Nernst's equation.

9.2. IONICS

Faraday's laws of electrolysis are:

1. The mass of material formed at the electrodes is proportional to the quantity of electricity passed through the solution.
2. For a fixed quantity of electricity passed through a solution, the masses of different materials formed (or dissolved) at the electrode is proportional to their equivalent weight, which is the atomic mass/electron charge, Z.

The charge on the electron is 1.60219×10^{-19} coulombs. Hence one mole of electrons (6.02214×10^{23}) represents $1.60219 \times 10^{-19} \times 6.02214 \times 10^{23} = 9.6486 \times 10^{4}$ coulombs/mol which is called a Faraday ($\mathscr{F}$) and usually rounded off to 96,500 coulombs. A coulomb, Q, is equal to 1 amp $\times$ 1 sec, and 26.80 amp hr = 1 $\mathscr{F}$.

$$Q = It \tag{9.1}$$

where I is the current in amperes and t is the time in seconds.

Example 9.1

How much silver would be electrodeposited by a current of 6.0 amp for 3 hr from a solution of $AgNO_3$?

Answer: The atomic mass of silver is 107.87 g/mol. The charge on silver is +1 ($Z = 1$) and this means that 1 $\mathscr{F}$ would deposit 107.87 g. We must now determine the

TABLE 9.1
The Classification of Electrochemical Systems

Ionics	Electrodics
Faraday's law	Nernst equation
Current flow causes reaction to occur	Reactions cause voltage to develop and current to flow
Electrolysis	Batteries
Electrodeposition	Fuel cells
Electrochemical machining	Corrosion protection
Battery charging	

number of coulombs, Q, which were passed through the solution.

$$Q = It = 6.0 \text{ amp} \times 3 \text{ hr} \times 60 \text{ min/hr} \times 60 \text{ s/min} = 64{,}800 \text{ amp.sec} \tag{9.2}$$

96,500 C would deposit 107.87 g of Ag and 1 C would deposit 107.87/96,500 g of Ag. Hence 64,800 C would deposit 64,800 × 107.87/96,500 g of Ag or 72.43 g of silver would be deposited.

Example 9.2

During the 4.0 hr electodeposition of copper from a copper sulfate ($CuSO_4$) solution, 140.0 g of Cu was deposited. What was the current flowing through the cell?

Answer: The atomic mass of copper is 63.55 g/mol. Since the charge on the copper in the solution, Cu^{2+}, is +2 then the equivalent weight is 63.55/2 g. Hence 1 $\mathscr{F}$ would deposit 63.55/2 g.Cu. Since 140.0 g of Cu were deposited the quantity of current which flowed through the cell is 140.0/63.55/2 or 4.406 $\mathscr{F}$.

$$Q = It$$

$$I = \frac{Q}{t} = \frac{4.406 \times 96{,}500}{4 \times 3600} \text{ amp} = 29.53 \text{ amp} \tag{9.3}$$

Example 9.3

An alloy of tin and lead is deposited from a solution of $Sn(NO_3)_2$ and $Pb(NO_3)_2$. What is the percentage of tin in the alloy if 35.00 g of alloy are deposited when 3.00 amp of current are passed through the solution for 4.00 hr?

Answer: The quantity of electricity Q passed through the solution is given by:

$$Q = It = 3.00 \times 4.00 \times 3600 = 43{,}200 \text{ C}$$

96,486 C liberates 1 equivalent weight, 1 C liberates 1/648 equivalent weights, and 43,200 C liberates 43,200/96,486 equivalent weights; that is, a total of 0.448 equiv.

If M is the mass of Sn in the deposit then $35.00 - M$ is the mass of Pb.

The equivalent weight of Sn is

$$\frac{118.7}{2} = 59.35 \text{ g/eq.}$$

The equivalent weight of Pb is

$$\frac{207.2}{2} = 103.6 \text{ g/eq.}$$

Hence the number of equivalents of Sn in the deposit is

$$\frac{M}{59.35}$$

and the number of equivalents of Pb in the deposit is

$$\frac{35.00 - M}{103.6}$$

Thus

$$\frac{M}{59.35} + \frac{35.00 - M}{103.6} = 0.448$$

Solving for M

$$103.6M + 2077 - 59.35M = 2754$$

$$44.2M = 677$$

$$M = 15.3 \text{ g Sn}$$

$$35.00 - \text{M} = 19.7 \text{ g Pb}$$

%Sn in the alloy is

$$\frac{15.3}{35.00} \times 100 = 43.7\%$$

Example 9.4

High purity gases are often prepared by electrolysis. It is desired to produce O_2 at 25°C and 800 torr at a flow rate of 75.0 cm^3/min. What current would be required for the electrolysis cell which consists of aqueous NaOH solution and nickel electrodes?

Answer: Using PV = nRT we see that a flow rate of 75.0 cm^3/min corresponds to a flow rate of

$$n = \text{PV/RT} = \{800 \text{ torr}/760 \text{ torr/atm}\}$$
$$\times 75.0 \text{ mL/min} \times (82.06 \text{ mL.atm/K.mol})^{-1} \times [298 \text{ K}]^{-1}$$

$$n = 3.228 \times 10^{-3} \text{ mol/min}$$

For O_2 gas, 1 mole corresponds to 4 equivalents. Thus this flow rate corresponds to

$$4 \times 3.228 \times 10^{-3} \text{ equiv./min}$$

or

$$\frac{4 \times 3.228 \times 10^{-3}}{60} \text{ equiv./sec,}$$

i.e., $\mathscr{F}$/sec. Thus,

$$\text{current required} = \frac{4 \times 3.228 \times 10^{-3} \times 96{,}485}{60} \text{ C/sec}$$

$$I = 20.77 \text{ C/sec}$$

$$I = 20.77 \text{ amp}$$

One mole of H_2 prepared by the electrolysis of water would require $2\mathscr{F}$ whereas 1 mole of O_2 would require $4\mathscr{F}$. The electrolysis of Al_2O_3 in cryolite (Na_3AlF_6) requires $3\mathscr{F}$ for each mole of Al (27 g). This is very energy intensive and a tonne of aluminum uses over 15,000 kW. Some effort (though unsuccessful) has been made to convert Al^{3+} to Al^{+}* in order to save on electrical energy by electrolyzing Al^{+} instead of Al^{3+}.

9.3. ELECTROLYSIS AND ELECTRODEPOSITION OF METALS

When an increasing DC voltage is applied to a solution of metal ions the metal, in some cases, will begin to deposit the deposition material at the cathode at a minimum voltage. This is illustrated as D in Fig. 9.1. The deposition potential depends on the metal, the surface, the current density, the concentration of the metal, and other ions in solution. The electrodeposition or electroplating reaction

$$M^{+n} + ne^{-} \rightarrow M \tag{9.4}$$

is most common for metals such as Au, Ag, Pb, Cd, Zn, Cr, Ni, Cu, and Sn. Two or more metals can also be deposited to form alloys. If the voltage is too high then hydrogen will be evolved

$$2H^{+} + 2e^{-} \rightarrow H_2 \tag{9.5}$$

*The ground state electronic configuration of Al is $1s^2 2s^2 2p^6 3s^2 3p^1$ and it would be expected for Al to have a (+1) oxidation state. This is in fact the case and at high temperatures the reaction

$$AlCl_{3(g)} + 2Al_{(s)} = 3AlCl_{(g)}$$

occurs. However attempts to electrolyse AlCl have so far proved to be unsuccessful.

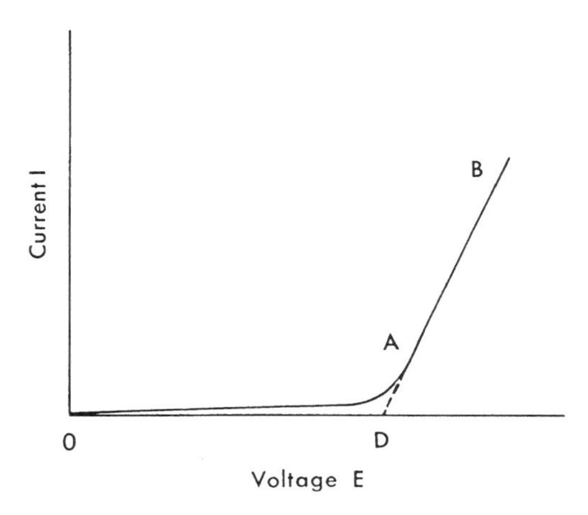

FIGURE 9.1. Typical current–voltage plot for an electrolyte solution. A: start of deposition, A–B: linear segment of plot, D: extrapolated deposition potential.

The theoretical minimum voltage required for the electrolysis of water is 1.23 V at 25°C. However, the process has an activation energy which is referred to as the overvoltage or polarization which depends on the current, temperature, and materials used for the electrodes. Some typical overvoltages are given in Table 9.2.

The industrial preparation of hydrogen by the electrolysis of water on nickel electrodes requires a voltage of more than 1.50 V (1.23 V + 0.210 V + 0.060 V) since it is necessary to add the RI drop (due to the internal resistance of the electrolyte). However, at very high current densities the polarization is much higher and higher temperatures are used to reduce the excess power requirements. The overvoltage encountered in the electrodeposition of a metal can be associated with the various steps by which the metal in solution, e.g., Cu^{2+} becomes the atom in a copper lattice.

The various steps in the overall mechanism are:

(1) Cu^{2+} (hydrated in solution) diffuses to the cathode.
(2) Cu^{2+} (hydrated) at electrode is transferred to the cathode surface.

TABLE 9.2
Overvoltage (mV) in Water Electrolysis, 25°C

Metal surface	Cathode polarization dilute H_2SO_4	Anode polarization dilute KOH
Platinized platinum	5	250
Smooth platinum	90	450
Palladium	~0	430
Gold	20	530
Silver	150	410
Nickel	210	60
Lead	640	310
Cadmium	480	530
Mercury	780	—

(3) Cu^{2+} (partially hydrated and adsorbed onto the surface) the ad ion—diffuses across the electrode surface to a crystal building site.
(4) Cu^{2+} (adsorbed at a crystal building site) becomes a part of the lattice.
(5) $[Cu^{2+} + 2e^-]$ occurs and Cu is part of the metal.

The sequence of relative importance is $2 > 3 > 4 > 1 > 5$ with step 2 being the slowest step or rate controlling process.

9.4. ELECTROCHEMICAL MACHINING

Electrolysis, with an anode that dissolves under controlled conditions, is the basis of electrochemical machining (ECM). High-strength metals such as Nimonic alloys (nickel with Al, Ti, and Mo) used in the aircraft industry resist deformation even at high temperatures and are exceedingly difficult to machine by the normal cutting process because of the limitations and expense of tool materials. In ECM the metal alloy does not determine the rate or characteristics of the dissolution process and hard tough metals can be dissolved as readily as soft metals. Only the current density (amp/cm^2) determines the rate of machining where approx. 5×10^{-3} mm/min can be removed at a current density of 0.3 amp/cm^2. The additional advantage of ECM is the absence of mechanical and thermal stress usually associated with conventional machining.

A schematic representation of a typical ECM apparatus is shown in Fig. 9.2 and consists of a workpiece to be machined (the anode), a properly shaped cathode tool

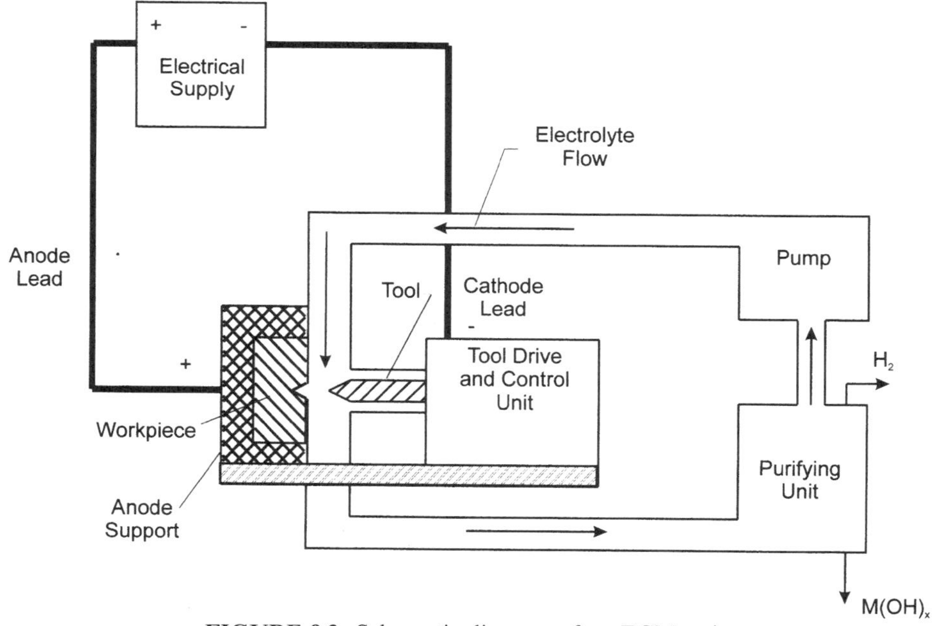

FIGURE 9.2. Schematic diagram of an ECM unit.

which is movable and maintains a constant gap with the workpiece. The electrolyte flows between the two electrodes, removing the products of electrolysis as well as heat. The power supply furnishes the high currents necessary to dissolve the anode.

9.4.1. The Cathode

The cathode is a tool shaped to conform to the desired cut in the workpiece. The tool is usually made from copper, steel, or alloy and insulated on the sides to give directed current lines. The tool is moved during electrolysis to maintain a constant small gap (about 0.25 mm) between the electrodes to reduce the voltage required.

9.4.2. The Electrolyte

The purpose of the electrolyte is to provide a conducting medium and at the same time it must not corrode the cathode tool. The cheapest material commonly used is sodium chloride (NaCl) at about 30% by weight. In some cases additives such as alcohols, amines, thiols, and aldehydes are used to inhibit stray currents, which results in overcuts. Other electrolytes such as $Na_2Cr_2O_7$, $NaNO_3$, and $NaClO_3$ at 50 to 250 g/L have also been used but the choice is limited primarily by cost.

The electrolyte is usually recirculated with the metal products removed or reduced before being reused. This minimizes cost and pollution and prevents the formation of a precipitate in the electrolysis gap.

The effect of electrolyte on surface roughness, H, was studied by Y. Sugie (1978) for 5 different iron alloys (characterized in Table 9.3) and is shown in Fig. 9.3. The roughness depends on electrolyte and alloy.

The accuracy of the machining, Figs. 9.4 and 9.5, was determined for the 5 alloys by measuring the overcut and machined angle (shown in Fig. 9.6) for the various electrolytes. Based on the results, it was concluded that $NaClO_3$ was most suitable for the low alloy steel and that Na_2SO_4 was best for the high nickel alloy steels.

The power supply and tool drive complete the apparatus. Current densities of 100–200 A/cm^2 at voltages from 10 to 50 V give cutting rates of about 1 mm/min with surface finishes of 5×10^{-4} cm. By pulsing the DC current it is possible to reduce the required flow rate of the electrolyte. However, one difficulty, sparking, common in some

TABLE 9.3
Composition of Various Iron Base Alloys[a]

Iron base alloy	Element (%)							
	C	Cr	Mo	Mn	Si	P	S	Ni
SCM 3	0.35	1.1	0.2	0.7	0.25	0.03	0.03	—
SKD 11	1.5	12.0	1.0	—	—	—	—	—
SNC 2	0.3	0.8	—	0.5	0.2	0.03	0.03	2.7
SUS 304	0.08	17.5	—	20	1.0	0.045	0.03	9.5
Inconel 718	0.06	18.26	2.93	0.11	0.13	0.0001	0.003	51.85

[a]Iron is the remaining element making up the alloy's 100%.

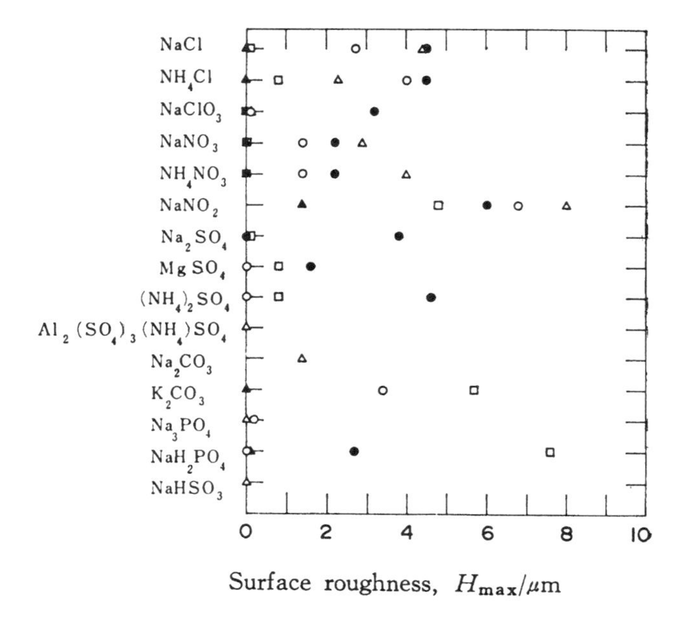

FIGURE 9.3. Surface roughness of steels (see Table 9.3) obtained in various electrolytes ○: SCM 3, ●: SKD 11, △: SNC 2, ▲: SUS 304, □: Inconel 718.

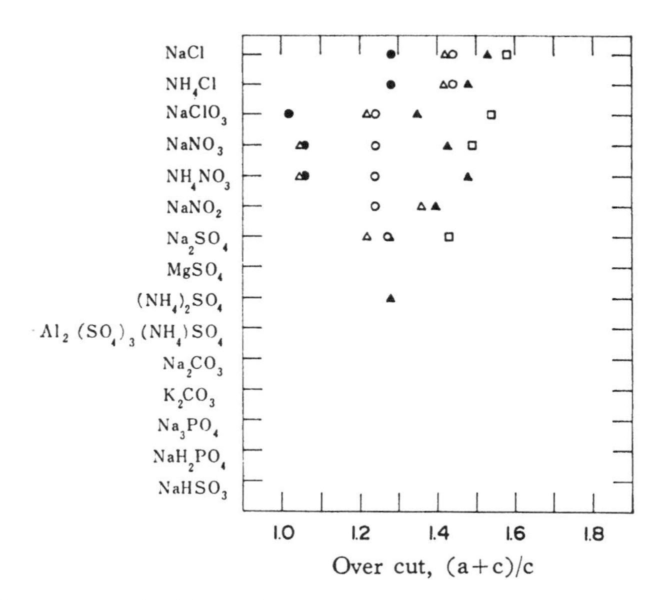

FIGURE 9.4. Over cut obtained for various steels (see Table 9.3) in various electrolytes. ○: SCM 3, ●: SKD 11, △: SNC 2, ▲: SUS 304, □: Inconel 718.

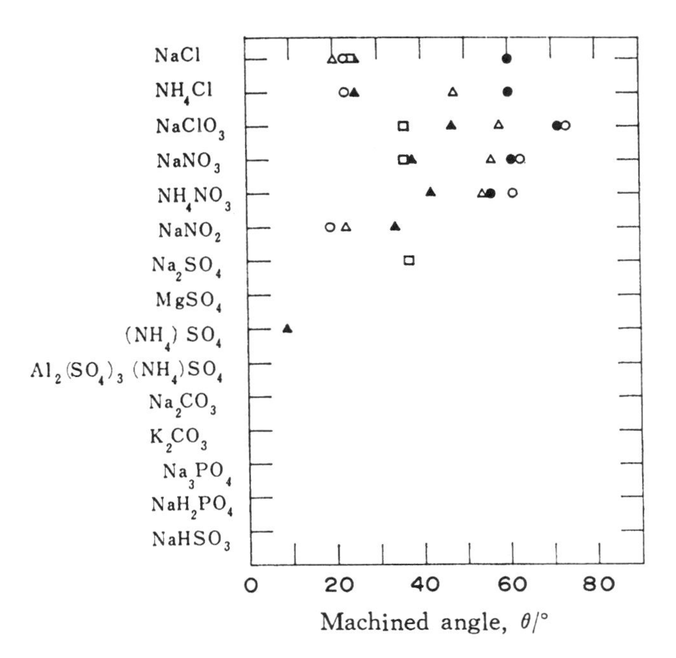

FIGURE 9.5. Machined angle obtained for various steels (see Table 9.3) in various electrolytes. ○: SCM 3, ●: SKD 11, △: SNC 2, ▲: SUS 304, □: Inconel 718.

continuous systems, occurs in intermittent electrolysis and is a decided disadvantage. Other disadvantages in ECM include high initial cost of the equipment, the hazards of hydrogen evolution at the cathode, sparking and "wild cutting" by stray current, and the need of a machinist who has some knowledge of chemistry.

Intergranular corrosion observed for some alloys can be reduced by using mixed electrolytes such as 15% NaCl with 20% $NaClO_3$.

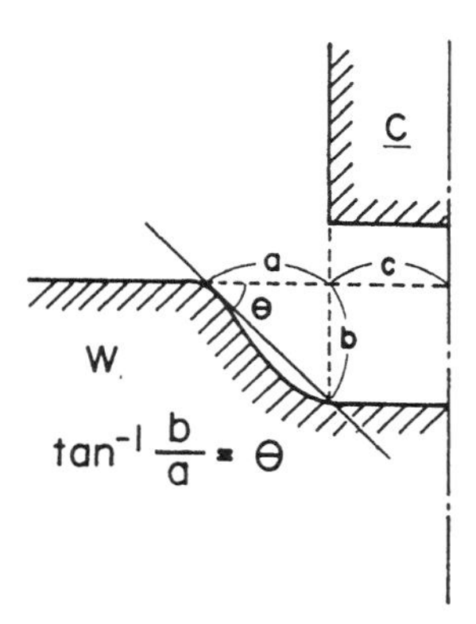

FIGURE 9.6. Schematic diagram of machined surface (see Figs. 9.3, 9.4, and 9.5). W: working electrode (15 mmφ), C: counter electrode (5 mmφ), θ: machined angle.

It is possible to effect electrochemical grinding and polishing by using a cathodic wheel with an anodic workpiece separated by a flowing electrolyte. Tungsten carbide tools are usually prepared by such a process. An electrochemical saw has also been produced.

Other refractor materials such as TiC, ZrC, TiB_2, ZrB_2, TiC/Ni, can also be subject to ECM using NaCl and KNO_3 as electrolytes. However, the number of equiv./mol (Z) is usually not a whole number because of multiple dissolution reactions, e.g., $Z = 6.6$ for TiC because

$$\mathrm{TiC} + 3\mathrm{H_2O} \rightarrow \mathrm{TiO_2} + \mathrm{CO} + 6\mathrm{H}^+ + 6\mathrm{e}^- \tag{9.6}$$

as well as

$$\mathrm{TiC} + 4\mathrm{H_2O} \rightarrow \mathrm{TiO_2} + \mathrm{CO_2} + 8\mathrm{H}^+ + 8\mathrm{e}^- \tag{9.7}$$

One interesting result of ECM is that the surface composition of an alloy may be significantly different from the bulk composition. Thus a TiC/Ni alloy of Ni, Ti, C, O of 11, 44, 39 and 6 at.%, respectively, had a surface composition of 47, 22, 10, and 21 at.% after ECM. This difference disappears within 1 μm from the surface and reflects the preferential dissolution of one or more elements in the metal.

The major disadvantages of ECM over conventional machining are the high cost of the ECM machine and its high maintenance cost, the technical expertise required of the operator and the lower accuracy achieved.

9.5. ELECTRODICS

A metal in equilibrium with its ions in solution will establish a potential difference which depends on the concentration of the metal ion in solution and the temperature. This potential is best determined by comparison with a standard which is arbitrarily set to zero—the hydrogen electrode

$$\mathrm{H^+(aq)} + \mathrm{e}^- \rightleftharpoons \tfrac{1}{2}\mathrm{H_2(g)} \tag{9.8}$$

The electrical energy, $\mathscr{E}$ is directly related to the free energy, ΔG, by the relation

$$\Delta G = -n\mathscr{F}\mathscr{E} \tag{9.9}$$

where n is the number of electrons transferred in the process and $\mathscr{F}$ is the Faraday.

For an equilibrium reaction

$$\Delta G = \Delta G^\circ + RT \ln Q \tag{9.10}$$

where ΔG° is the standard free energy and

$$\Delta G^\circ = -RT \ln K_{eq} \tag{9.11}$$

R is the gas constant equal to 8.314 J/K.mol; T is the absolute temperature; Q is the ratio of the concentration of products to concentration of reactants, and K_{eq} is the value of the equilibrium constant at the specified temperature.

For a chemical reaction, Eq. (9.9) becomes

$$\mathscr{E} = \mathscr{E}^{\circ} - RT \ln Q \tag{9.12}$$

which is called the Nernst equation. For example in the Daniell cell

$$\mathrm{Zn(s) + Cu^{2+}(qa) \rightarrow Zn^{2+}(aq) + Cu(s)} \tag{9.13}$$

$$\mathscr{E} = \mathscr{E}^{\circ} - \frac{RT}{n\mathscr{F}} \ln \frac{[\mathrm{Zn^{2+}}]}{[\mathrm{Cu^{2+}}]} \tag{9.14}$$

Reaction (9.13) can be considered to be composed of two reactions called *half-cell* reactions

$$\mathrm{Zn(s) \rightarrow Zn^{2+}(aq) + 2e^{-}} \tag{9.15}$$

and

$$\mathrm{Cu^{2+}(aq) + 2e^{-} \rightarrow Cu(s)} \tag{9.16}$$

The potential of these half-cell reactions can be determined by comparison with the hydrogen electrode* all under standard conditions, which is unit concentration for ions in solution and 1 atm pressure for gases at 25°C. The value of these standard electrode reduction potentials is given in Table 9.4. The standard cell potential of reaction (9.15) is

$$\mathscr{E}^{\circ}_{\mathrm{Zn/Zn^{2+}}} = -\mathscr{E}^{\circ}_{\mathrm{Zn^{2+}/Zn}} = 0.763\ \mathrm{V}, \qquad \mathscr{E}^{\circ}_{\mathrm{Cu^{2+}/Cu}} = 0.337\ \mathrm{V}$$

and

$$\mathscr{E}^{\circ}_{\mathrm{cell}} = 0.763 + 0.337 = 1.100\ \mathrm{V}$$

This is represented in shorthand notation as

$$\mathrm{Zn|Zn^{2+}(1M)\,\|\,Cu^{2+}(1M)|Cu} \tag{9.17}$$

where the single | line indicates a phase change and the double || line represents a salt bridge or connection between the two half-cells. This is illustrated in Fig. 9.7.

The simplest cell is one in which the electrode material is the same in each half-cell but the metal ion concentration is different in the two half-cells. This is called a *concentration cell*, an example of such a cell is

$$\mathrm{Cu|Cu^{2+}(0.01M)\,\|\,Cu^{2+}(0.10M)|Cu}$$

*The standard potential of the hydrogen electrode (SHE.) is defined as zero at all temperatures for $[H^{+}]aq = 1M$ and $P(H_2) = 1$ atm.

TABLE 9.4

Standard Reduction Potentials for Some Common Redox Systems at 25°C (↑ Signifies Gas State and ↓ Represents Solid State; in All Other Cases, State is Liquid or Solution)

Reduction half-reaction	$\mathscr{E}^\circ_{cell}$ (V)	Reduction half-reaction	$\mathscr{E}^\circ_{cell}$ (V)
$F_2\uparrow + 2e^- = 2F^-$	+2.87	$Cu^+ + e^- = Cu\downarrow$	+0.521
$O_3\uparrow + 2H^+ + 2e^- = O_2\uparrow + H_2O$	+2.07	$Fe(CN)_6^{3-} + e^- = Fe(CN)_6^{4-}$	+0.356
$Co^{3+} + e^- = Co^{2+}$	+1.82	$Cu^{2+} + 2e^- = Cu\downarrow$	+0.337
$H_2O_2 + 2H^+ + 2e^- = 2H_2O$	+1.77	$Hg_2Cl_2\downarrow + 2e^- = 2Hg + 2Cl^-$	+0.2680
$Ce^{4+} + e^- = Ce^{3+}$ (in $HClO_4$ solution)	+1.70	$AgCl\downarrow + e^- = Ag\downarrow + Cl^-$	+0.2224
$MnO_4^- + 4H^+ + 3e^- = MnO_2\downarrow + 2H_2O$	+1.69	$Cu^{2+} + e^- = Cu^+$	+0.153
$2HClO + 2H^+ + 2e^- = Cl_2\uparrow + 2H_2O$	+1.63	$Sn^{4+} + 2e^- = Sn^{2+}$ (in HCl solution)	+0.14
$Ce^{4+} + e = Ce^{3+}$ (in HNO_3 solution)	+1.60	$S\downarrow + 2H^+ + 2e^- = H_2S$	+0.14
$2HBrO + 2H^+ + 2e^- = Br_2 + 2H_2O$	+1.6	$S_4O_6^{2-} + 2e^- = 2S_2O_3^{2-}$	+0.09
$MnO_4^- + 8H^+ + 5e^- = Mn^{2+} + 4H_2O$	+1.51	$2H^+ + 2e^- = H_2\uparrow$	0.0000
$2BrO_3^- + 12H^+ + 10e^- = Br_2 + 6H_2O$	+1.5	$Pb^{2+} + 2e^- = Pb\downarrow$	−0.126
$Mn^{3+} + e^- = Mn^{2+}$	+1.49	$Sn^{2+} + 2e^- = Sn\downarrow$	−0.140
$Ce^{4+} + e^- = Ce^{3+}$ (in H_2SO_4 solution)	+1.44	$Ni^{2+} + 2e^- = Ni\downarrow$	−0.23
$Cl_2\uparrow + 2e^- = 2Cl^-$	+1.359	$Co^{2+} + 2e^- = Co\downarrow$	−0.28
$Cr_2O_7^{2-} + 14H^+ + 6e^- = 2Cr^{3+} + 7H_2O$	+1.33	$Cr^{3+} + e^- = Cr^{2+}$ (in HCl solution)	−0.38
$Ce^{4+} + e^- = Ce^{3+}$ (in HCl solution)	+1.28	$Cd^{2+} + 2e^- = Cd\downarrow$	−0.402
$MnO_2\downarrow + 4H^+ + 2e^- = Mn^{2+} + 2H_2O$	+1.23	$Fe^{2+} + 2e^- = Fe\downarrow$	−0.440
$O_2\uparrow + 4H^+ + 2e^- = 2H_2O$	+1.229	$2CO_2\uparrow + 2H^+ + 2e^- = H_2C_2O_4$	−0.49
$ClO_4^- + 2H^+ + 2e^- = ClO_3^- + H_2O$	+1.19	$Cr^{3+} + 3e^- = Cr\downarrow$	−0.74
$2IO_3^- + 12H^+ + 10e^- = I_2\downarrow + 6H_2O$	+1.19	$Zn^{2+} + 2e^- = Zn\downarrow$	−0.7628
$Br_2 + 2e^- = 2Br^-$	+1.087	$SO_4^{2-} + H_2O + 2e^- = SO_3^{2-} + 2OH^+$	−0.93
$N_2O_4\uparrow + 2H^+ + 2e^- = 2HNO_2$	+1.07	$Mn^{2+} + 2e^- = Mn\downarrow$	−1.190
$NO_3^- + 3H^+ + 2e^- = HNO_2 + H_2O$	+0.94	$Al^{3+} + 3e^- = Al\downarrow$	−1.66
$2Hg^{2+} + 2e^- = Hg_2^{2+}$	+0.907	$H_2\uparrow + 2e^- = 2H^-$	−2.25
$2NO_3^- + 4H^+ + 2e^- = N_2O_4\uparrow + 2H_2O$	+0.80	$Mg^{2+} + 2e^- = Mg\downarrow$	−2.37
$Ag^+ + e^- = Ag\downarrow$	+0.7994	$Na^+ + e^- = Na\downarrow$	−2.7
$Hg_2^{2+} + 2e^- = 2Hg$	+0.792	$Ca^{2+} + 2e^- = Ca\downarrow$	−2.87
$Fe^{3+} + e^- = Fe^{2+}$	+0.771	$Sr^{2+} + 2e^- = Sr\downarrow$	−2.89
$O_2\uparrow + 2H^+ + 2e^- = H_2O_2$	+0.69	$Ba^{2+} + 2e^- = Ba\downarrow$	−2.90
$H_3AsO_4 + 2H^+ + 2e^- = HAsO_2 + 2H_2O$	+0.56	$K^+ + e^- = K\downarrow$	−2.925
$I_3^- + 2e^- = 3I^-$	+0.545	$Rb^+ + e^- = Rb\downarrow$	−2.93
$I_2\downarrow + 2e^- = 2I^-$	+0.536	$Li^+ + e^- = Li\downarrow$	−3.03

The cell potential is

$$\mathscr{E}_{cell} = \mathscr{E}^\circ_{cell} - \frac{RT}{n\mathscr{F}} \ln \frac{0.01}{0.10}, \qquad \mathscr{E}_{cell} = 0 - \frac{0.02568}{2} \ln 0.10, \qquad \mathscr{E}_{cell} = 0.0296 \text{ V}$$

The cell will continue to develop a potential until the concentration of Cu^{2+} in the two half-cells become equal.

The concentration inequality adjacent to metals is the basis of the corrosion of metals which is discussed in the next chapter.

All redox reactions can be divided into two or more half cells which can be combined into a full cell. The voltage generated and the current which can be drawn determines its usefulness as a battery.

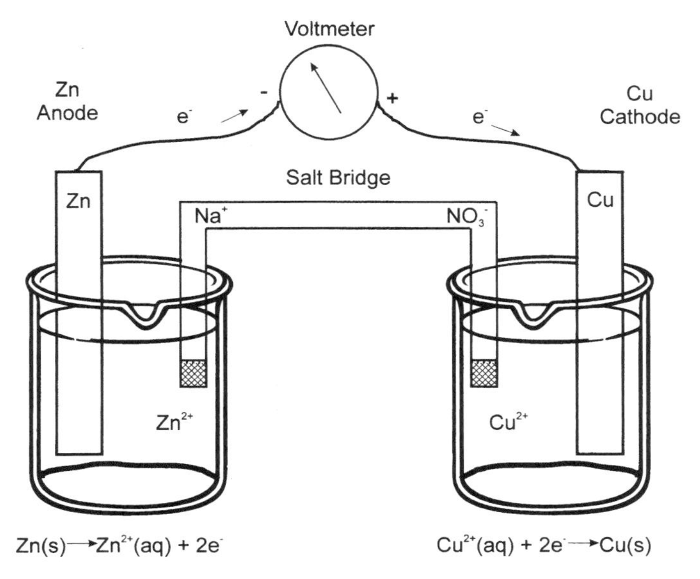

FIGURE 9.7. The zinc–copper Daniell cell where the zinc dissolves at the anode (−) and copper is plated out at the cathode (+).

9.6. BATTERIES AND CELLS

The means by which chemical energy is stored and converted into electrical energy is called a *battery* or *cell*. We saw how the Daniell cell composed of a zinc electrode immersed in a 1M zinc sulfate solution and a copper electrode dipping into a 1M copper sulfate

$$Zn|ZnSO_4(1M)\|CuSO_4(1M)|Cu \tag{9.18}$$

will develop 1.10 V.

This cell, however, has limited use because as soon as current is drawn from the cell the voltage drops because of polarization which is primarily caused by the buildup of hydrogen at the copper electrode. This polarization can be minimized by the addition of a depolarizer which supplies oxygen readily, and so removes the hydrogen from the electrode to form water. Such a cell, nonetheless, has a limited lifetime and restricted use.

Batteries are classified as primary or secondary. Batteries which are not rechargeable are referred to as primary batteries. Secondary batteries are rechargeable either by an electrical current or by a replacement of the electrode material (anode).

9.6.1. Primary Batteries

A dry cell, which has become a primary power source for transistorized electronic equipment, was developed over a hundred years ago in 1865, is called the *Leclanché* cell. It consists of a zinc negative electrode, which acts as the container and which is slowly oxidized; a porous carbon rod as the positive electrode, which takes no part in the overall reaction but can act as a gas vent for the cell and an electrolyte, which is ammonium chloride to which is added manganese dioxide. The MnO_2 acts as a depolarizer and is reduced at the carbon electrode by the following reaction:

$$MnO_2 + 4H^+ + 2e^- \rightarrow Mn^{2+} + 2H_2O \tag{9.17}$$

This is complicated by further reactions:

$$Mn^{2+} + MnO_2 + 2OH^- \rightarrow 2MnOOH \tag{9.18}$$

$$Mn^{2+} + MnO_2 + 4OH^- + Zn^{2+} \rightarrow ZnO{\cdot}Mn_2O_3 + 2H_2O \tag{9.19}$$

The zinc ion is then complexed by the ammonia as follows:

$$Zn^{2+} + 4NH_3 \rightleftharpoons [Zn(NH_3)_4]^{2+} \tag{9.20}$$

The cell voltage is between 1.5 and 1.6 V. A D-type battery, commonly used in flashlights, has a capacity of about 4 amp-hr. Although this type of cell contains about 20 g of zinc, only about 5 g is used. The overall reaction, although complex, can be represented as follows:

$$Zn + 4NH_4Cl + MnO_2 \rightarrow MnCl_2 + [Zn(NH_3)_4]Cl_2 + 2H_2O \tag{9.21}$$

Because the reduction reaction of MnO_2 is not a well-defined reaction, the Nernst equation cannot be applied successfully, and the cell voltage changes unpredictably with time and discharge condition. The shelf life of this battery is rather poor, as shown in Fig. 9.8, because of the slow loss of water and because of side reactions due to impurities in the MnO_2 ore commonly used. When specially purified MnO_2 is used, the

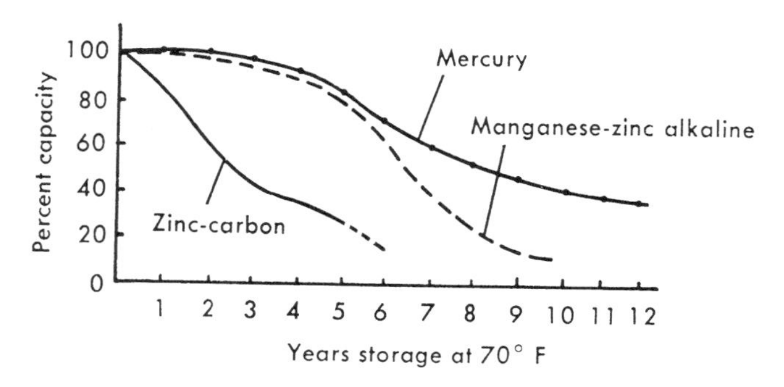

FIGURE 9.8. Plots showing the loss in capacity with storage time for the zinc–carbon, manganese–zinc–alkaline, and mercury batteries.

performance of the battery is greatly improved. The zinc-carbon dry cell is considered the workhorse of the battery industry: it provides power at a very low cost.

Another popular dry cell that is commonly associated with transistorized electronic components is the manganese-zinc alkaline cell. It utilizes refined MnO_2, as does the improved zinc–carbon cell, but in large excess. The electrolyte is 40% KOH presaturated with zinc (ZnO) to prevent the zinc electrode from dissolving while in storage. A steel can, instead of the zinc electrode, usually serves as the container; hence the cell is highly leak resistant. However, the cost of the cell is about twice that of the zinc–carbon cell.

A most efficient dry cell is the mercury cell, which has an excellent stabilized voltage. Developed in 1942 by Ruben, it consists of zinc, which dissolves and becomes the negative electrode and mercuric oxide, which is reduced to mercury at the positive electrode. The overall cell, which has no salt bridge, can be represented as follows:

$$\mathrm{Zn|ZN(OH)_2|40\%\ KOH|HgO|Hg} \tag{9.22}$$

One half-cell reaction is

$$\mathrm{Zn} \rightleftharpoons \mathrm{Zn^{2+}} + 2\mathrm{e^-}, \qquad \mathscr{E}^\circ_{\mathrm{Zn^{2+}/Zn}} = -0.763\ \mathrm{V} \tag{9.23}$$

This is followed by the reaction

$$\mathrm{Zn^{2+}} + 2\mathrm{OH^-} \rightleftharpoons \mathrm{Zn(OH)_2} \tag{9.24}$$

for which $K_{sp} = 4.5 \times 10^{-17}$.

The other half-cell reaction is

$$\mathrm{Hg^{2+}} + 2\mathrm{e^-} \rightleftharpoons \mathrm{Hg}, \qquad \mathscr{E}^\circ_{\mathrm{Hg^{2+}/Hg}} = 0.085\ \mathrm{V} \tag{9.25}$$

This is preceded by the reaction

$$\mathrm{HgO} + \mathrm{H_2O} \rightarrow \mathrm{Hg(OH)_2} \rightleftharpoons \mathrm{Hg^{2+}} + 2\mathrm{OH^-} \tag{9.26}$$

for which $K_{sp} = 1.7 \times 10^{-26}$. The cell emf is given as follows:

$$\begin{aligned}\mathscr{E}_{\mathrm{cell}} &= \mathscr{E}_{\mathrm{Zn/Zn^{2+}}} + \mathscr{E}_{\mathrm{Hg^{2+}/Hg}} = -(\mathscr{E}_{\mathrm{Zn^{2+}/Zn}}) + \mathscr{E}_{\mathrm{Hg^{2+}/Hg}} \\ \mathscr{E}_{\mathrm{cell}} &= -\left(\mathscr{E}^\circ_{\mathrm{Zn}} - \frac{RT}{n\mathscr{F}} \ln \frac{1}{[\mathrm{Zn^{2+}}]}\right) + \left(\mathscr{E}^\circ_{\mathrm{Hg}} - \frac{RT}{n\mathscr{F}} \ln \frac{1}{[\mathrm{Hg^{2+}}]}\right)\end{aligned} \tag{9.27}$$

However,

$$[\mathrm{Zn^{2+}}] = \frac{K_{sp}(\mathrm{Zn(OH)_2})}{[\mathrm{OH^-}]^2} \qquad \text{and} \qquad [\mathrm{Hg^{2+}}] = \frac{K_{sp}(\mathrm{Hg(OH)_2}}{[\mathrm{OH^-}]^2} \tag{9.28}$$

Therefore,

$$\mathscr{E}_{\text{cell}} = -\left(\mathscr{E}^{\circ}_{\text{Zn}} - \frac{0.02568}{2}\ln\frac{[\text{OH}^-]^2}{K_{\text{sp}}(\text{Zn(OH)}_2)}\right) + \left(\mathscr{E}^{\circ}_{\text{Hg}} - \frac{0.02568}{2}\ln\frac{[\text{OH}^-]^2}{K_{\text{sp}}(\text{Hg(OH)}_2)}\right)$$

$$= -\mathscr{E}^{\circ}_{\text{Zn}} + \mathscr{E}^{\circ}_{\text{Hg}} + \frac{0.02568}{2}\ln\frac{K_{\text{sp}}(\text{Hg(OH)}_2)}{K_{\text{sp}}(\text{Zn(OH)}_2)}$$

$$= 0.763 + 0.850 + \frac{0.02568}{2}\ln\frac{1.7\times10^{-26}}{4.5\times10^{-17}} \tag{9.29}$$

$$= 1.613 + \frac{0.02568}{2}\ln 3.8\times10^{-10}$$

$$\mathscr{E}_{\text{cell}} = 1.613 - 0.278 = 1.335\ \text{V}$$

This value compares favorably with the actual cell voltage of about 1.35 V. Thus, the cell potential is independent of the concentration of the electrolyte, $[OH^-]$. The cell has a low internal resistance and has a very long shelf life when compared to the other two dry cells discussed previously, as shown in Fig. 9.8. For example, after a three-year storage, a typical cell voltage changed from an initial value of 1.357 V to 1.344 V; i.e., there was about a 1% change. Thus, the use of the mercury dry cell as a reference voltage is widespread. A comparison of cell voltage of the three types of dry cells with time during constant current drain is shown in Fig. 9.9. The remarkable constancy in voltage of the mercury cell in contrast to the sharp voltage drop in the zinc–carbon and manganese–zinc alkaline cells is obvious. The mercury cell, although initially about three times more expensive than the zinc–carbon cell, has a lower operating cost per hour than either of the other two dry cells. Therefore, it is not too difficult to understand why the mercury cell is being used increasingly as a convenient source of power and reference voltage.

Also shown in Fig. 9.9 is the voltage curve for the zinc–air cell. The cell consists of an anode of amalgamated zinc powder in contact with the electrolyte, which is concentrated potassium hydroxide, and a cathode of metal mesh, which is a catalyst for

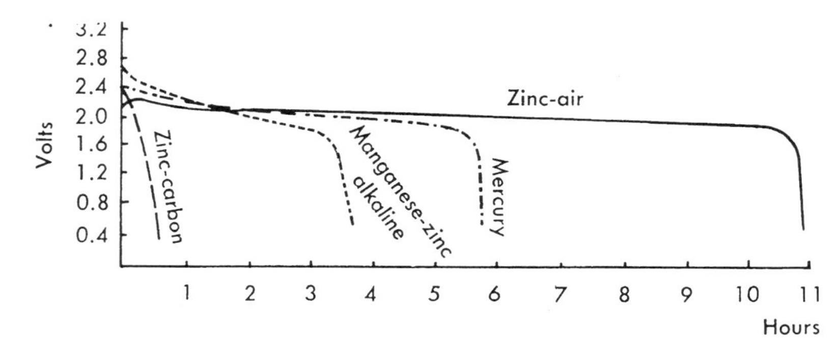

FIGURE 9.9. Typical discharge curves of voltage plotted against time for four different types of cells of comparable size (two AA penlight cells discharged at 250 mA).

the conversion of oxygen to the hydroxide ion. The half-reactions are

$$Zn \rightarrow Zn^{2+} + 2e^- \tag{9.30}$$

$$O_2 + 2H_2O + 4e^- \rightarrow 4OH^- \tag{9.31}$$

and the overall reaction is

$$2Zn + O_2 + 2H_2O + 4OH^- \rightarrow 2Zn(OH)_4^{2-} \tag{9.32}$$

The cell is encased in a porous polymer that allows oxygen from the air to diffuse to the cathode but does not allow the electrolyte to leak out. The shelf-life is almost indefinite when the cell is stored in an airtight container. The cell is used to best advantage when continuous high currents are required for a short period of time, since it cannot be left in contact with air without losing capacity. Resealing the cell or cutting off the air supply to the cell when it is not in use extends the life during intermittent use. The catalytic cathode for the zinc–air cell is a direct development from work on fuel cells, which are discussed later.

In contrast to the zinc–air battery the lithium–iodine solid LiI electrolyte battery will last for almost 15 years. A 120-mAh battery with an initial voltage of 2.8 V drops to 2.6 V when discharged continuously at about 1 μA. The cell is written as

$$Li(s)|LiI(s)|P2VP \cdot nI_2(s)$$

where $P2VP \cdot nI_2$ is a complex between poly-2-vinylpyridine (P2VP) and iodine. The reaction is given as

$$Li(s) + \tfrac{1}{2}I_2(s) \rightarrow LiI(s) \tag{9.33}$$

This type of cell is highly reliable and it is commonly used in cardiac pacemaker batteries which are implanted.

The temperature coefficient of a cell's potential is determined by the change in free energy, ΔG, with temperature and is given by

$$\frac{d\mathscr{E}^\circ}{dT} = \frac{\Delta S^\circ}{n\mathscr{F}} \tag{9.34}$$

where ΔS° is the standard entropy change for the reaction.

9.6.2. Secondary Batteries

The most common secondary battery is the lead storage battery, which has as an essential feature an ability to be recharged. The cell consists of a lead plate for the negative electrode, separated by a porous spacer from the positive electrode, which is composed of porous lead dioxide. The electrolyte is sulfuric acid—about 32% by weight. The electrode reactions are as follows:

Negative electrode

$$\begin{array}{r} Pb \rightleftharpoons Pb^{2+} + 2e^- \\ Pb^{2-} + SO_4^{2-} \rightleftharpoons PbSO_4 \\ \hline Pb + SP_4^{2-} \rightleftharpoons PbSO_4 + 2e^- \end{array} \tag{9.35}$$

Positive electrode

$$\begin{array}{c} PbO_2 + 4H^+ + 2e^- \rightleftharpoons Pb^{2+} + 2H_2O \\ Pb^{2+} + SO_4^{2-} \rightleftharpoons PbSO_4 \\ \hline PbO_2 + 4H^+ + SO_4^{2-} + 2e^- \rightleftharpoons PbSO_4 + 2H_2O \end{array} \tag{9.36}$$

The net overall reaction is as follows:

$$PbO_2 + Pb + 2H_2SO_4 \underset{\text{charge}}{\overset{\text{discharge}}{\underset{2\mathscr{F}}{\rightleftharpoons}}} 2PbSO_4 + 2H_2O \tag{9.37}$$

Polarization by hydrogen is minimized by the PbO_2 electrode, which is also a depolarizer. The discharge of the battery consumes acid and forms insoluble lead sulfate and water, i.e., the density of the solution decreases from about 1.28 g/cm^3 in the fully charged condition to about 1.1 g/cm^3 in the discharged state. The overall open cell voltage (when no current is being drawn) depends on the acid concentration (i.e., SO_4^{2-} ion concentration, which in turn controls the concentration of the Pb^{2+} ion via the K_{sp} for $PbSO_4$). The voltage varies from 1.88 V at 5% H_2SO_4 by weight to 2.15 V at 40% acid by weight. The conductivity of aqueous H_2SO_4 is at a maximum when H_2SO_4 is about 31.4% by weight at 30°C (or 27% at −20°C); it is best to control concentration in this range, since the internal resistance of the battery is at a minimum. Another factor influencing the choice of the acid concentration is the freezing point of the sulfuric acid solution; thus in cold climates a higher acid level (38% H_2SO_4 by weight, specific gravity 1.28) is required in order to minimize the possibility of the electrolyte freezing at the relatively common temperature of −40°C*. The amount of lead and lead dioxide incorporated into the electrodes is three to four times the amount used in the discharging process because of the construction of the electrodes and the need for a conducting system that makes possible the recharging of a "dead" battery.

In 1988 a collection of 8256 lead–acid batteries was used by a California electric power plant to store energy and to deliver it during peak power demands, i..e., load leveling. The batteries contained over 1800 tonnes of lead and could supply 10 MWe for 4 hr, enough to meet the electrical demands of 4000 homes. The efficiency of the system was rated at 75%.

The capacity of a battery is rated in terms of amp-hours and depends on the rate of discharge and, even more significantly, on the temperature. For example, a battery with a rating of 90 amp-hours at 25°C has a rating of about 45 amp-hours at −12°C

*This refers to winter in Winnipeg, Canada, where I live.

and about 36 amp-hours at $-18°C$. The lead–acid battery in the fully discharged state slowly loses capacity, since the lead sulfate recrystallizes and some of the larger crystals are then not available for the reverse charge reaction. When this happens, the battery is said to be sulfonated. This can be remedied by the process of removing the "insoluble" sulfate, recharging the battery, and reconstituting the acid to the appropriate specific gravity.

In respect to this property as well as others, the nickel alkaline battery is superior to—although about three times more costly than—the lead–acid battery. There are two types of nickel alkaline storage batteries: the Edison nickel–iron battery and the nickel–cadmium battery.

In the Edison battery the cell can be represented as follows:

$$\text{Steel} \mid Ni_2O_3;\ Ni(OH)_2 \mid KOH(aq\ 20\%) \mid Fe(OH)_2\ Fe \mid \text{Steel}$$

The overall reaction is as follows:

$$Ni_2O_3 + 3H_2O + Fe \underset{\text{charge}}{\overset{\text{discharge}}{\underset{\longleftarrow}{\overset{\longrightarrow}{\text{KOH}}}}} 2Ni(OH)_2 + Fe(OH)_2 \tag{9.38}$$

The cell potential has an average value of 1.25 V. Although the overall reaction does not apparently involve the electrolyte, the KOH does in fact participate in each of the half-cell reactions. Although the Edison battery is designed and suitable for regular cyclic service, the efficiency of charge is only 60%; thus it has now been almost completely replaced by the more efficient (72%) nickel–cadmium battery, which is itself inferior in energy efficiency to the lead–acid battery with an efficiency of about 80%.

In the nickel–cadmium alkaline storage battery, the iron of the Edison cell is replaced by cadmium to give the following equivalent reaction:

$$Ni_2O_3 + 3H_2O + Cd \underset{\text{charge}}{\overset{\text{discharge}}{\underset{\longleftarrow}{\overset{\longrightarrow}{\text{KOH}}}}} 2Ni(OH)_2 + Cd(OH)_2 \tag{9.39}$$

The average cell voltage of 1.2 V is slightly lower than that of the Edison cell. Cadmium is preferred to iron in the nickel alkaline cell because cadmium hydroxide is more conductive than iron hydroxide. The absence of higher oxidation states for cadmium minimizes side reactions, which occur in the Edison cell. The nickel–cadmium cell can also be charged at a lower voltage, since there is no over-voltage, as there is at the iron electrode.

One major disadvantage of the nickel alkaline battery is the alkaline electrolyte, which picks up CO_2 from the atmosphere and must therefore be replaced periodically. However, the advantages of the nickel–cadmium cell over the lead–acid battery are numerous; some of these are as follows:

(1) the freezing point of the KOH electrolyte is low (about $-30°C$) regardless of the state of charge,

(2) the capacity does not drop as sharply with drop in temperature,
(3) the cell can be charged and discharged more often and at higher rates (without gassing) and thus has a longer useful life.

The storage battery has become an accepted source of power in our modern technological world, and in a environment-conscious society the storage battery will play an ever increasing role.

9.7. FUEL CELLS

The discovery of the fuel cell followed soon after Faraday developed his laws of electrolysis. In 1839 Grove showed that the electrolysis of water was partially reversible. Hydrogen and oxygen formed by the electrolysis of water were allowed to recombine at the platinum electrodes to produce a current or what appeared to be "reverse electrolysis." Using the same fundamental principles but somewhat more advanced technology, Bacon in 1959—after about 20 years of intensive effort—produced a 6 kW power unit that could drive a small truck.

It was recognized early that the overall thermodynamic efficiency of steam engines is only about 15%. The efficiency of modern electrical generators is about 20–50%, whereas the efficiency of the fuel cell (in which there is direct conversion of chemical energy into electrical energy) does not have any thermodynamic limitation. Theoretically, the efficiency of the fuel cell can approach 100%, and in practice, efficiency of over 80% can be achieved.

Interest in the fuel cell has increased remarkably in the last decade primarily because of (1) the high efficiency associated with the energy conversion, (2) the low weight requirement essential for satellite and spacecraft power sources that is readily satisfied with hydrogen as a fuel, and (3) the recent requirement of a pollution-free power source.

Any redox system with a continuous supply of reagents is potentially a fuel cell. Some reactions that have been studied are given in Table 9.5 with the corre-

TABLE 9.5
Values of Standard Cell Voltages of Selected Fuel Cell Reactions at 25°C

Reaction	$\mathscr{E}^\circ_{cell}$ (V)
$2C + O_2 \rightarrow 2CO$	0.70
$C + O_2 \rightarrow CO_2$	1.02
$CH_4 + 2O_2 \rightarrow CO_2 + 2H_2O$	1.04
$C_3H_8 + 5O_2 \rightarrow 3CO_2 + 4H_2O$	1.10
$4NH_3 + 3O_2 \rightarrow 2N_2 + 6H_2O$	1.13
$CH_3OH + \frac{3}{2}O_2 \rightarrow CO_2 + 2H_2O$	1.21
$H_2 + \frac{1}{2}O_2 \rightarrow H_2O(l)$	1.23
$2CO + O_2 \rightarrow 2CO_2$	1.33
$N_2H_4 + O_2 \rightarrow N_2 + 2H_2O$	1.56
$2Na + H_2O + \frac{1}{2}O_2 \rightarrow 2NaOH$	3.14

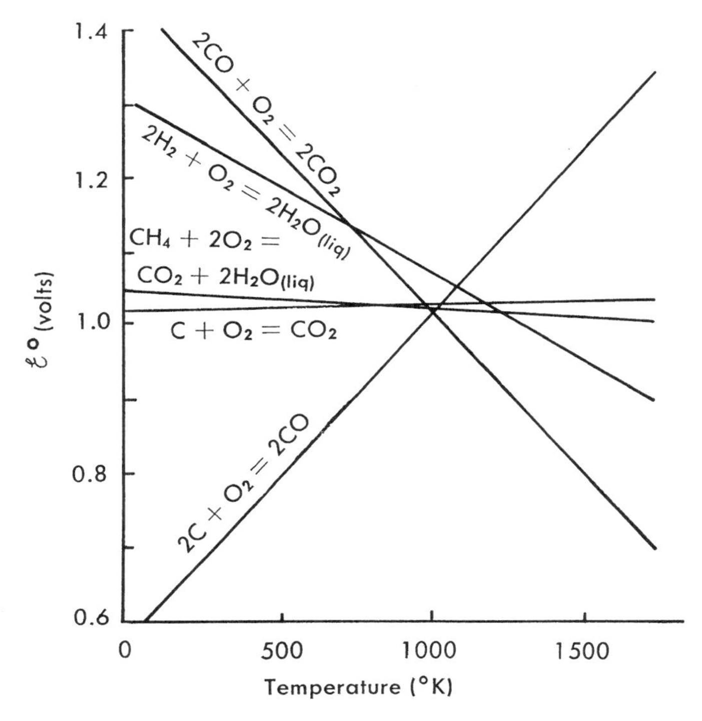

FIGURE 9.10. Effect of temperature on the cell voltage, $\mathscr{E}^\circ_{\text{cell}}$, for some fuel cell reactions. *Note:* Slope is related to the entropy change of the reaction.

sponding theoretical $\mathscr{E}^\circ_{\text{cell}}$ values, which are calculated from thermodynamic data ($\Delta G^\circ = -n\mathscr{F}\mathscr{E}^\circ$). The temperature coefficients of the $\mathscr{E}^\circ_{\text{cell}}$ values of some of the reactions in Table 9.5 are shown in Fig. 9.10.

In practice, the suitability of a reaction system is determined by the kinetics of the reaction, which depends on temperature, pressure of gases, electrode polarization, surface area of electrodes, and presence of a catalyst. A fuel cell that is thermodynamically and kinetically feasible must be considered from an economic viewpoint before it is accepted. Thus, since hydrogen, hydrazine, and methanol are too expensive for general application, their use in fuel cells has been limited to special cases. Hydrogen has been used for fuel cells in satellites and space vehicles, in which reliability and lightness are more important than cost. Hydrazine fuel cells have been used in portable-radio power supplies for the United States Army because of their truly silent operation. Methanol fuel cells have been used to power navigation buoys and remote alpine television repeater stations because such power systems are comparatively free from maintenance problems over periods of a year or more. The polarization at the electrodes of a fuel cell is the most important single factor that limits the usefulness of the cell. The various polarization characteristics for a typical fuel cell are plotted separately as a function of current density in Fig. 9.11.

The most successful fuel cell to date is the hydrogen–oxygen fuel cell, which deserves special attention since it has been used in the Apollo and Gemini space flights

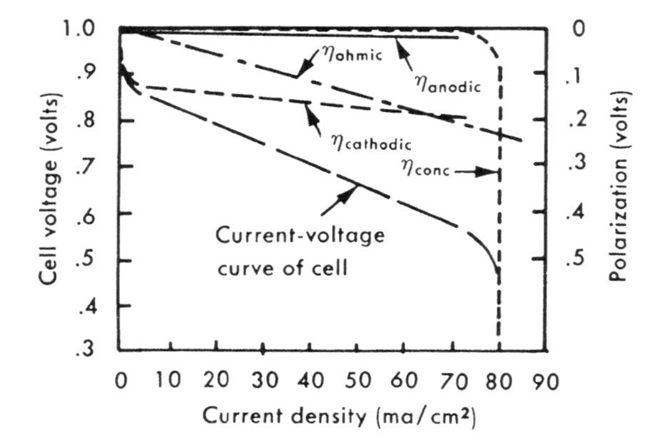

FIGURE 9.11. Operating characteristics of a typical fuel cell. Net polarization is given by $\eta_{total} = \eta_{anodie} + \eta_{cathodic} + \eta_{ohmic} + \eta_{conc}$.

and moon landings. The reaction

$$H_2(gas) \rightleftharpoons 2H(solid) \rightleftharpoons 2H^+(solution) + 2e^- \quad \textbf{(9.40)}$$

occurs at the $gas \rightleftharpoons solid$ interface. To facilitate the rapid attainment of equilibrium, a

$\downarrow\uparrow \quad \uparrow\downarrow$
liquid

liquid gas-diffusion electrode was developed whereby concentration polarization could be minimized. The ohmic polarization (the RI drop between the electrodes, which gives rise to an internal resistance) is also minimized when the anode-to-cathode separation is reduced. The apparatus of the hydrogen–oxygen fuel cell developed by Bacon with gas-diffusion electrodes is shown in Fig. 9.12. The operating temperature of 240°C is attained with an electrolyte concentration of about 80% KOH solution, which with the high pressures of about 600 psi for H_2 and O_2, allows high current densities to be drawn with relatively low polarization losses. Units such as these with power of 15 kW have been built and used successfully for long periods.

The fuel cells in general use today are still in the development stage, and much further work must be done before an efficient economical fuel cell is produced. The oxidation of coal or oil to CO_2 and H_2O has been achieved in a fuel cell, the system uses platinum as a catalyst and an acid electrolyte at high temperature, and thus the cost of materials for the cell construction is very high. The economic fuel cell-powered automobile, although a distinct possibility, is not to be expected in the immediate future.

9.8. HYBRID CELLS

The hybrid cell is one which is not rechargeable by simply reversing the voltage. Some of these use oxygen in air as the cathode material

$$O_2 + 4H_2O + 4e^- \rightarrow 4OH^- \quad \textbf{(9.41)}$$

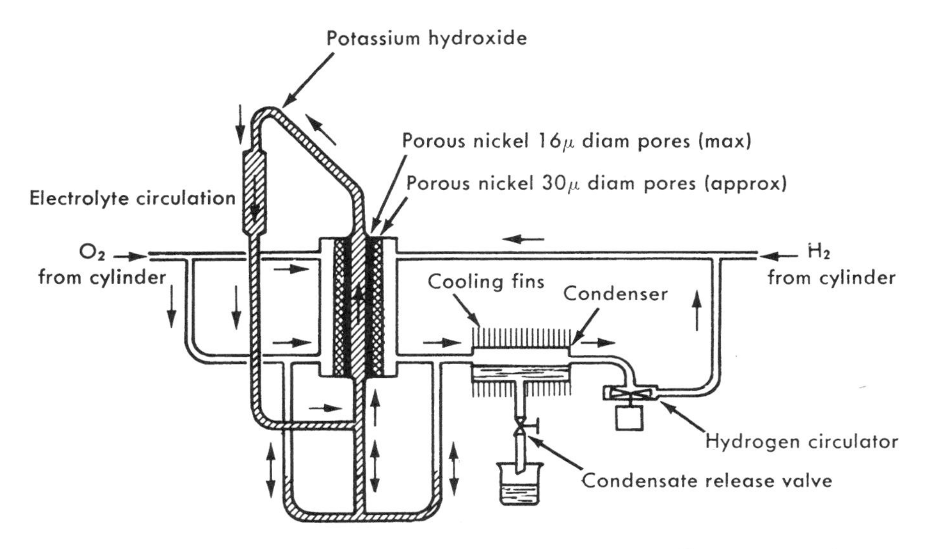

FIGURE 9.12. Bacon hydrogen–oxygen fuel cell with gas-diffusion electrodes.

and a metal, e.g., Al, as the anode material

$$Al \rightarrow Al^{3+} + 3e^- \quad (9.42)$$

Such systems are called *metal–air batteries* and are mechanically rechargeable (anode metal is replaced). Such batteries have only recently become practicable due to the developments of the O_2-electrode in fuel cells. Some characteristics of selected metal–air batteries are given in Table 9.6.

The aluminum air battery has recently received some attention as a result of work done by the Lawrence Livermore National Laboratory. It was estimated that a 60 cell system with 230 kg of aluminum can power a VW for 5000 km before requiring mechanical recharging. Periodic refill with water and removal of $Al(OH)_3$ would be required after 400 km. The conversion of the $Al(OH)_3$ back to Al at an electrolytic refinery completes the recycling process. In 1986 an Al–air battery producing 1680 W was shown to power an electric golf cart for 8 hr.

A battery where the active components are flowed past electrodes in a cell with two compartments separated by an appropriate membrane is called a *flow battery*. One such battery is the Fe/Cr redox system

cathode $\quad Cr^{2+} \rightarrow Cr^{3+} + e^- \qquad \mathcal{E}^\circ = -0.410\text{ V}$

and

anode $\quad Fe^{3+} + e^- \rightarrow Fe^{2+} \qquad \mathcal{E}^\circ = 0.771\text{ V}$

$$Cr^{2+} + Fe^{3+} \rightarrow Cr^{3+} + Fe^{2+} \qquad \mathcal{E}^\circ = 1.181\text{ V} \quad (9.43)$$

TABLE 9.6
Some Properties of Selected Metal–Air Batteries

Battery	Electrolytes	Cell volt (V)	Energy density[a] (Wh/kg) Theoretical	Energy density[a] (Wh/kg) Actual	Peak power (W/kg)	Cycle life	Comments
Lithium–air $2Li + \frac{1}{2}O_2 \rightarrow Li_2O$	LiOH	2.9	11,148	290		Mechanical	Unlikely to be developed for commercial use. High Li costs.
Aluminum–air $2Al + \frac{3}{2}O_2 \rightarrow Al_2O_3$	NaOH	2.71	8,081	440		Mechanical rechargeable	Prototype developed, and tested. Good energy density, low cost
Magnesium–air $2Mg + O_2 \rightarrow 2MgO$	NaCl	3.09	6,813			Mechanical rechargeable	No advantage over Al, not being considered seriously at present

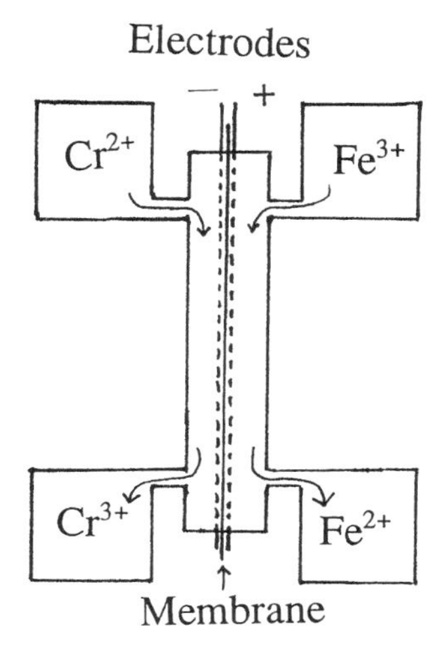

FIGURE 9.13. Schematic diagram of a redox cell (battery) using Cr^{2+} and Fe^{3+} aqueous solutions as reactants.

The overall voltage is given by

$$\mathscr{E} = \mathscr{E}^{\circ}_{\text{cell}} - 0.059 \log K \tag{9.44}$$

$$\mathscr{E} = 1.181 - 0.059 \log \frac{[Fe^{2+}][Cr^{3+}]}{[Fe^{3+}][Cr^{2+}]} \tag{9.45}$$

The reactant solutions Cr^{2+} and Fe^{3+} are reacted as shown in Fig. 9.13.

The product solutions are kept separate and the Fe^{2+} can be oxidized by air back to Fe^{3+} whereas the Cr^{3+} can be electrolytically reduced back to Cr^{2+}. An Australian redox flow battery has been described which uses vanadium both as oxidant and reductant in the following reactions:

$$\text{Negative} \qquad \text{V(III)} + \text{e} \underset{\text{discharge}}{\overset{\text{charge}}{\rightleftharpoons}} \text{V(II)} \tag{9.46}$$

$$\text{Positive} \qquad \text{V(IV)} \underset{\text{discharge}}{\overset{\text{charge}}{\rightleftharpoons}} \text{V(V)} + \text{e} \tag{9.47}$$

The electrolyte is 2M $VOSO_4$ in 2M H_2SO_4 with graphite plates acting as electrodes to collect the current. The open circuit voltage (OCV) is 1.45 V with a 95% charging efficiency and little or no H_2 or O_2 evolution. One of the major advantages of this battery is that if the membrane leaks, then the separation of the two flow streams is not necessary as in the Fe/Cr system. Several such redox cells are available and being studied primarily as potential power sources for the electric vehicle (EV) which will most assuredly be a reality in the near future.

9.9. ELECTRIC VEHICLE

The first EV was built in 1839 by Robert Anderson of Aberdeen, Scotland. The first practical one was a taxi introduced in England, 1886, which had 28 bulky batteries and a top speed of 12.8 km/hr. By 1904 the electric vehicle was common throughout the world but its production peaked at about 1910 when the self-starting gasoline powered internal combustion engine began to dominate. This was due to the availability of cheap gasoline and mass-produced cars.

However, the EV is due to make a come-back because of the rising cost of gasoline and diesel fuel, the pollution of the environment and the prevalence of a second small car in most families. Some of the major American automobile producers have been planning to have an EV on the market for the past 20 years. The major stumbling block is the batteries which must be reliable, lightweight, take hundreds of full discharges and recharges, and be inexpensive as well. The desirable features of an ideal battery are compared in Table 9.7 with the lead–acid battery still used at present in EV. A list of possible batteries and some of their properties are given in Table 9.8. The use of the fuel cell and hybrid fuel cell type power sources must also be included.

The choice of batteries available for an EV is both expanding in number as well as narrowing in type. Several batteries listed in Table 9.8 are being given commercial pilot production tests. It must be recognized that winter restricts the choice or design of a suitable system for cold climate regions. Recent tests in Winnipeg, Canada of a US-made EV using lead–acid batteries showed it to be appropriate in summer (about 80 km/charge) but in winter the lower capacity resulted in less than 8 km/charge. This could undoubtedly be corrected by an integrated design. Since the batteries are only about 70% efficient on charge, the excess energy (heat) could be stored by insulating the batteries or adding a heat storing medium such as Glauber's Salt (see Chapter 1) between the batteries and the insulation. This however adds both weight and volume to the system.

TABLE 9.7
A Comparison of Performance and Cost Goals for a Practical EV Battery with Those of Deep-Cycling Industrial Lead–Acid Battery

Parameter	Goal	Lead–acid
Cost ~$/kWh	45	90
Life (cycles)	1000	700
Life (years)	10	5
Energy efficiency $\left(\frac{\text{discharge energy}}{\text{charge energy}}\right)$	0.80	0.65
Charge time (h)	1–6	6–8
Discharge time (h)	2–4	2–4
Energy density (Wh/kg)	140	35
Power density—peak (Wh/kg)	200	80
Power density—sustained (Wh/kg)	70	30
Volume density (Wh/L)	200	50
Typical size (kWh)	20–50	20–40

TABLE 9.8
Evaluation and Characteristics of Electric Vehicles Batteries (Values Depend on Source)

Battery and reaction electrolyte				Energy density: Mass	Vol.								
Discharge	Electrolyte	Temp. (°C)	Cell voltage (V)	Theoretical (Wh/kg)	Actual (Wh/L)	Power peak (W/kg)	Cycle life	Depth charge (%)	Charge efficient (%)	Cost initial ($/kwh)	Projected	Advantage	Disadvantage
Improved lead–acid $^{\ominus}Pb + {}^{\oplus}PbO_2 + H_2SO_4 \rightarrow 2PbSO_4 + H_2O$	H_2SO_4	0–40	2.05	171/30	30/46	50–100	500	90–60	65	90	50	Now available	Low specific energy marginal peak power limited to about 100 miles (150 km)
Nickel–zinc $^{\ominus}Zn + {}^{\oplus}2NiO(OH) \rightarrow 2H_2O + 2Ni(OH)_2$	KOH $2H_2O$	−45+40	1.7	321/66	66/140	150	400	65	65	150	50	Excellent power and volume density. (UK prospect)	Solubility of Zn in KOH shortens shelf life. High Ni cost
Nickel–iron $^{\ominus}Fe + 2NiO(OH) + 2H_2O \rightarrow Zn(OH_2) + Fe(OH)_2$	KOH	−40+40	1.37	267/55	55/100	75	1500	90	60	120	65	Developed in 1901 by Edison, now mature and rugged	Low power density H_2 gasing during charge. Poor peak power at low temperature
Zinc–chlorine $^{\ominus}Zn + Cl_2 6H_2O \rightarrow ZnCl_2 + 6H_2O$	$ZnCl_2$	<9.0+40°	2.12	465/100	100/60	80	400	—	65	100	35	Inexpensive lightweight materials, high power density	Complex system requiring refrigeration and heating. Cl_2 hazard
Sodium–sulfur $^{\ominus}xNa + YS \rightarrow Na_xSy$ $x = 2, y = 5.3$	βAl_2O_3	350	1.76–2.08	664/150	150/160	200	300	60	85	50	35	Inexpensive light material, high energy density	Short life due to seals and corrision, Na hazard. High temperature requirement
Lithium–sulfur $^{\ominus}2Li + S \rightarrow Li_2S$	LiCl/KCl	450	2.2	2567								Improved with Li/Al alloy (~5 at.% Li)	Too reactive, Li cost high

FIGURE 9.14. Mercedes-Benz zero emission class A EV. The car uses a 40 kW (54 hp) — three phase induction motor developing a rated torque of 155 Nm which can accelerate the car to 100 km/h in 17 sec with a top speed of 120 km/h and a normal usage range of 150 km. Recharging can be made in 6–12 h using normal household sockets. The battery system is sodium/nickel chloride with an energy storage capacity of 100 Wh/kg and a life of over 100,000 km.

The modern design EV will be lightweight and have minimal aerodynamic drag and rolling resistance, efficient motor control system and transmission as well as regenerative (battery charging) braking. The usual goal of EV is a range of about 100 km, a maximum speed of 90 km/h and a cruise speed of 45 km/h, with a recharge time of 8–10 hr. It would be interesting to speculate that as the EV becomes common and recharging is performed at night, the resulting power drain may invert the peak load, i.e., the greater load would occur overnight. The low vehicle emissions set for California are readily met by the EV and major automobile manufacturers are striving to meet the demand.

One example is the Mercedes-Benz 5-seater 190 Electro car which develops up to 32 kW (44 hp), has a maximum speed of 115 km/h and an operating range of 150 km. The sodium–nickel chloride batteries were chosen over nickel–cadmium and sodium–sulfur alternatives. The car is shown in Fig. 9.14.

A second example is the use of a hydrogen fuel cell to run a bus. Using a Proton Exchange Membrane Fuel Cell (PEMFC), Ballard of Vancouver has built a prototype bus for Chicago Transit Authority. The bus stores hydrogen at high pressure in cylinders on the roof of the bus — enough to give the bus a 560 km range (see Fig. 9.15). Designs have been developed for a more compact (volume of 32 L) stack of fuel cells which delivers 32.3 kW and intended for a small passenger vehicle.

FIGURE 9.15. (A) The Ballard H_2 fuel cell powered bus and (B) the fuel cell unit at the rear of the bus.

General Motors has announced (February 1996) the start of a mass produced EV powered by lead–acid batteries and which will have a range of about 110 km per charge and a maximum speed of 100 km/hr. The estimated cost is expected to be US$35,000.

It is obvious that much experimental work remains to be done before a reliable and economical EV will be on the market. The incentive—lower fuel costs and a cleaner environment for the EV in comparison to the ICE—will not diminish with time.

EXERCISES

1. What is a concentration cell?
2. What is a redox cell?
3. Distinguish between primary and secondary batteries with examples.
4. What factors determine the voltage of a cell and its temperature coefficient?
5. Explain how the capacity of a storage battery is measured.
6. Explain why the capacity of the lead–acid battery drops so rapidly with decrease in temperature.
7. What factors influence the drop in cell voltage of a battery when current is withdrawn?
8. The voltage of the H_2–O_2 fuel cell is given by $\Delta G^\circ = -n\mathscr{F}\mathscr{E}^\circ$ where n is the number of electrons transferred in the reaction, $\mathscr{F}$ is the Faraday and ΔG is the standard free energy change in the reaction. Show that $\mathscr{E}^\circ = 1.23$ V from thermodynamic data.
9. Explain why in the electrolysis of water into H_2 and O_2, the voltage required is greater than 1.23 V.
10. (a) Calculate the voltage of the methanol–oxygen fuel cell.
    (b) Explain why the voltage will be lower if air was used instead of pure oxygen.
    (c) Calculate the standard voltage for the methanol–air fuel cell.
11. The approximate energy consumption for an EV is 0.16 kWh/km to 0.32 kWh/mile. If electrical energy is priced at 3¢/kWh, calculate the energy cost/km for an EV.
12. The minimum energy required to dissociate H_2O into H_2 and O_2 is 1.23 V. It is possible to use a carbon anode and produce CO_2 instead of O_2. Calculate the minimum cell potential for such a cell.

$$H_2O_{(1)} \rightarrow H_{2(g)} + \tfrac{1}{2}O_{2(g)} \qquad \Delta G^\circ = 237.2\ \text{KJ}$$

$$C_{(s)} + O_{2(g)} \rightarrow CO_{2(g)} \qquad \Delta G^\circ = -394.4\ \text{KJ}$$

    Note: The actual voltage required is from 0.85 to 1.0 V. Give some reasons for the discrepancy between the calculated and actual values. (Fuel **58** 705 (1979)).
13. Consider an automobile with an internal combustion engine running on hydrogen which can somehow be stored for a modest run of 50 km. The hydrogen is prepared by the electrolysis of water (in one cell at about 2.5 V) during the night (from 10:00 p.m. to 7:00 a.m.). What would be the required current? Make the following assumptions: (1) heat of combustion of gasoline and hydrogen are given in Table

6.8), (2) the automobile has the efficiency of the new generation of cars, *viz.* 14 km/L gasoline, (3) the efficiency of the hydrogen driven vehicle is related by its comparable heat of combustion to the gasoline efficiency.
Note: Do not be surprised by the large current that is necessary.

14. An ICE vehicle has been described running on H_2 produced on board the vehicle by the reaction of an Al wire with KOH/H_2O. Assume that the vehicle is the same one as in Exercise 9.14. Calculate the mass of Al required for the 50 km trip.
15. Calculate the standard cell potential for the Ti/Fe flow battery where

$$\mathscr{E}^{\circ}_{Ti^{4+}/Ti^{3+}} = 0.04 \text{ V}$$

$$\mathscr{E}^{\circ}_{Fe^{3+}/Fe^{2+}} = 0.771 \text{ V}$$

16. Calculate the RI drop of a lead–acid battery (in which the internal resistance is 0.01052 ohms) when it cranks an engine drawing 200A.
17. Estimate the amount of Glauber's salt needed to keep a lead–acid battery pack from cooling to 0°C if the outside temperature drops to −40°C. The excess heat for charging the 40 kW power supply will be insulated with 8 cm of Styrofoam. Assume that the volume of the battery pack is 0.3 m^3, its area is 2.5 m^2 and its heat capacity is 3 kJ/K. The thermal conduction of the polystyrene foam is 0.0003 J s^{-1} cm^{-1} K^{-1}. Assume that the EV is stored at work from 8:00 a.m. to 6:00 p.m. at −40°C.
18. From Eqs. (9.6) and (9.7), calculate the ratio of CO_2/CO produced when TiC is electrolytically machined.
Note: $Z = 6.6$.

FURTHER READING

H. Wendt and G. Kreysa, *Electrochemical Engineering*, Springer-Verlag, New York (1999).
Fuel Cell Power for Transportation, Soc. Auto. Engineers, Warrendale, PA (1999).
T. Koppel, *Powering the Future, The Ballard Fuel Cell and the Race to Change the World*, Wiley, New York (1999).
J. A. Bereny, Editor, *Battery Performance, Research and Development*, Bus. Tech. Bks., Orinda, California (1996).
T. Turrentine and K. Kurani, *The Household Market for Electric Vehicles*, Bus. Tech. Bks., Orinda, California (1996).
K. Kordesch and G. Simader, *Fuel Cells and Their Applications*, VCH Pubs., New York (1996).
D. Linden, *Handbook of Batteries*, 2nd Ed., McGraw-Hill, New York (1996).
D. Saxman and S. Grant, Editors, *Batteries and EV Industry Review*, BCC Norwalk, CT (1995).
L. J. Blomen and M. N. Mugerwa, Editors, *Fuel Cell Systems*, Plenum Press, New York (1994).
OECD Staff, *Electric Vehicles: Performance and Potential*, OECD, Washington, DC (1994).
D. Pletcher and F. C. Walsh, *Industrial Electrochemistry*, 2nd Ed., Chapman and Hall, London (1994).
?. B. Brant, *Build Your Own Electric Vehicle*, TAB Bks., New York (1993).
J. Koryta and J. Dvorak, *Principles of Electrochemistry*, Wiley, New York (1987).
P. H. Rieger, *Electrochemistry*, Prentice-Hall Inc., Englewood Cliffs, New Jersey (1987).
B. V. R. Chowdari and S. Radhakaishna, Editors, *Materials for Solid State Batteries*, World Scientific, Singapore (1986).

D. A. J. Rand and A. M. Bond, Editors, *Electrochemistry: The Interfacing Science*, Elsevier, New York (1984).
C. A. Vincent, *Modern Batteries*, Edward Arnold, London (1983).
L. E. Unnewehr and S. A. Nasar, *Electric Vehicle Technology*, Wiley, New York (1982).
M. Barak, Editor, *Electrochemical Power Sources*. Peter Peregrinus Ltd., New York (1980).
R. W. Graham, *Primary Batteries, Recent Advances*, Noyes Data Corp., Park Redge, New Jersey (1978).
L. I. Antropor, *Theoretical Electrochemistry*, MIR Publishers, Moscow (1977).
G. R. Palin, *Electrochemistry for Technologists*, Pergamon Press, Oxford (1969).
E. H. Lyons, Jr., *Introduction to Electrochemistry*. Heather and Co., Boston (1967).
E. C. Potter, *Electrochemistry—Principles and Application*, Clever-Hume Press, London (1956).
E.V. Association of America, http://www.evaa.org/
Canadian Electric Vehicle Ltd., http://www.canev.com
Electrochemistry lecture notes,
http://www.chem.ualberta.ca/courses/plambeck/p102/p0209x.htm
http://chemed.chem.purdue.edu/genchem/topicreview/bp/ch20/electroframe.html
http://edu.cprost.sfu.ca/~rhlogan/electrochem.html
http://www.chem1.com/CB1.htm
Help files for Electrochemistry, http://www.learn.chem.vt.edu/tutorials/electrochem/
Batteries, http://www.panasonic.com/
Batteries, http://www.varta.com/
All about batteries, http://www.howstuffworks.com/battery.htm

10

Corrosion

10.1. INTRODUCTION

Corrosion is the unwanted reaction or destruction of a metal component by the environment. The annual cost of corrosion to the US economy has been estimated to be over \$70 billion. Similar costs are associated with other industrialized countries. Many of the problems can be avoided if basic precautions and design processes are followed.

The mechanism of corrosion is electrochemical, and can be induced by the flow of current or will cause a current to flow. When a corroding metal is oxidized, the reaction

$$M \rightarrow M^{+n} + ne^- \tag{10.1}$$

must be accompanied by a reduction reaction which is usually the reduction of oxygen whether in the air or dissolved in water.

$$O_2 + 4H^+ + 4e^- \rightarrow 2H_2O \tag{10.2}$$

or

$$O_2 + 2H_2O + 4e^- \rightarrow 4OH^- \tag{10.3}$$

In some cases the reduction of hydrogen occurs

$$2H^+ + 2e^- \rightarrow H_2 \tag{10.4}$$

The usual classification of corrosion is according to the environment to which the metal is exposed or the actual reactions which occur. We have seen that the concentration cell is a simple cell in which a metal can corrode as dissolution takes place.

10.2. FACTORS AFFECTING THE RATE OF CORROSION

It is convenient to classify the corrosion of metals in terms of: (a) the metals; (b) the environment.

The reduction potential is the most important characteristic of a metal that determines its susceptibility to corrosion. This has been illustrated by Table 9.4. Thus the noble metals, gold and platinum, are resistant to corrosion and will only dissolve in strong oxidizing solutions which also contain complexing halides or other ions, e.g., (CN^-). For metals in seawater the relative order of the reduction potential of metals and alloys has been established. This is illustrated in Table 10.1 where distinction is made between active and passive surfaces for some metals. Magnesium is a most active metal whereas platinum and graphite are the least active materials. The voltages are

TABLE 10.1
Galvanic Metal and Alloy Potential V (vs. SCE) in Seawater

	−V (volts)		−V (volts)
Mg	1.6 ± 0.02	Cu/Ni 90/10	0.26 ± 0.04
Zn	1.00 ± 0.02	Cu/Ni 80/20	0.26 ± 0.04
Be	0.99 ± 0.01	430 stainless steel	0.24 ± 0.04
Al alloys	0.89 ± 0.11	Active potential	0.52 ± 0.06
Cd	0.71 ± 0.01	Pb	0.23 ± 0.03
Mild steel	0.65 ± 0.05	Cu/Ni, 70/30	0.21 ± 0.02
Cast iron	0.63 ± 0.05	Ni/Al bronze	0.20 ± 0.05
Low alloy steel	0.60 ± 0.02	Ni/Co 600 alloy	0.17 ± 0.02
Austenite Ni	0.50 ± 0.03	Active potential	0.41 ± 0.06
Bronze	0.36 ± 0.05	Ag bronze alloys	0.15 ± 0.05
Brass	0.35 ± 0.05	Ni 200	0.15 ± 0.05
Cu	0.34 ± 0.04	Ag	0.13 ± 0.03
Sn	0.32 ± 0.02	302, 304, 321, 347, SS	0.08 ± 0.02
Solder Pb–Sn	0.31 ± 0.03	Active potential	0.51 ± 0.05
Al brass	0.31 ± 0.03	Alloy 2C, stainless steel	0.00 ± 0.06
Manganese bronze	0.31 ± 0.02	Ni/Fe/Cr/Alloy 825	−0.08 ± 0.04
410, 416 stainless steel	0.31 ± 0.03	Ni/Cr/Mo/Cu/Si alloy	−0.07 ± 0.03
Active potential	0.51 ± 0.04	Ta	−0.09 ± 0.06
Silicon bronze	0.29 ± 0.02	Ni/Cr/Mo alloy C	−0.07 ± 0.07
Tin bronze	0.29 ± 0.03	Pt	−0.13 ± 0.10
Nickel silver	0.28 ± 0.02	Graphite	−0.14 ± 0.16

TABLE 10.2
Effect of O_2 Pressure on Corrosion of Iron in Seawater

$P(O_2)$ (atm)	Rate of corrosion (mm/year)
0.2	2.2
1	9.3
10	86.4
61	300

given with respect to the saturated calomel electrode (SCE).* The oxidation reaction (10.1) represents corrosion which must be accompanied by a reduction reaction (10.2), (10.3), or (10.4) as well as reactions such as

$$Fe^{3+} + e^- \rightarrow Fe^{2+} \tag{10.5}$$

and

$$3H^+ + NO_3^- + 2e^- \rightarrow HNO_2 + H_2O \tag{10.6}$$

The reaction which occurs depends on the solution in which the metal corrodes but in most cases the cathodic reaction involves O_2.

The corrosion rate will thus depend on the partial pressure of oxygen. This is shown in Table 10.2. Hence the removal of oxygen from water in steam boilers is one method of reducing corrosion.

If hydrogen evolution is the cathodic reaction (10.4) then it can be reduced by increasing the overvoltage. The overvoltage of H_2 on mercury is very high (see Table 9.2) and reaction (10.4) can be inhibited if mercury is used to coat the metal surface and to form an amalgam (see the zinc–air cell, Sec. 9.6). The overvoltage is dependent on current density which is determined by the area of the metals. Hence as the cathode area decreases the polarization can be expected to increase resulting in a decrease in rate of corrosion. In the case of iron (anode) on a large copper sheet (cathode) the large cathode/anode ratio favors corrosion of the iron. This is shown in Fig. 10.1.

The type and amount of impurities in a metal will affect the rate of corrosion. For example, a zinc sample which is 99.99% pure (referred to as 4n zinc) would corrode about 2000 times faster than a 5n sample. Even improperly annealed metals will show excessive corrosion rates.

Another factor which controls the rate of corrosion is the relative volume of the corrosion product (oxide) to the metal as well as the porosity of the oxide layer. For example, the volume ratio of oxide/metal for Al, Ni, Cr, and W is 1.24, 1.6, 2.0, and 3.6, respectively. The oxide layer on a metal can convert a metal from one that corrodes to one that is inert. Aluminum can react with water to form hydrogen by the reaction

$$2Al + 6H_2O \rightarrow 2Al(OH)_3 + 3H_2 \tag{10.7}$$

*The saturated calomel electrode is a convenient reference electrode often used instead of the standard hydrogen electrode: $\frac{1}{2}Hg_2Cl_2 + e^- \rightarrow Hg + Cl^-$, $\mathscr{E}° = 0.2224$ (25°C).

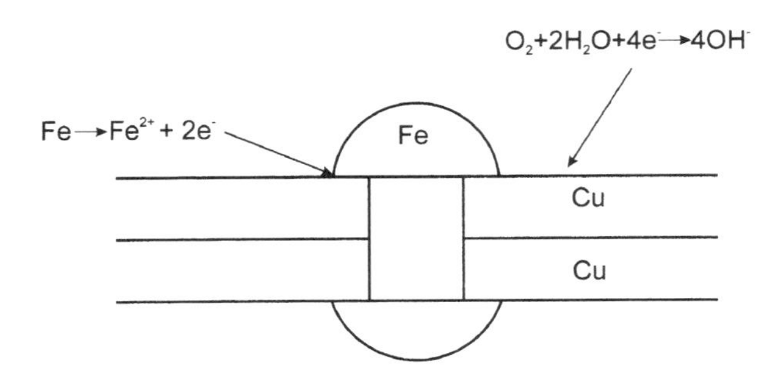

FIGURE 10.1. The corrosion of an iron rivet in a copper plate. The large copper surface results in a low O_2 overvoltage allowing the corrosion to proceed at a rate controlled by O_2 diffusion.

followed by

$$2Al(OH)_3 \rightarrow Al_2O_3 + 3H_2O \tag{10.8}$$

However, the oxide layer which forms prevents the water from contacting the aluminum surface. Only in acid or alkali is the Al_2O_3 solubilized and the aluminum reacts to liberate hydrogen.

An oxide layer is readily formed on many metals when they are made anodic in aqueous solutions. In the case of aluminum this process is called anodization. It is also referred to as a passive film which reduces the corrosion rate. Such passive films can be thin, from 0.01 μm, and fragile and easily broken. Thus when steel is immersed in nitric acid or chromic acid and then washed, the steel does not immediately tarnish nor will it displace copper from aqueous $CuSO_4$. The steel has become passive due to the formation of an adhering oxide film which can be readily destroyed by HCl which forms the strong acid $H^+FeCl_4^-$.

The factors influencing the rusting of iron can be illustrated by the electrochemical treatment of the overall reaction

$$\begin{array}{ll} 2Fe + O_2 + 4H^+ \rightarrow 2Fe^{2+} + 2H_2O & \\ \qquad Fe \rightarrow Fe^{2+} + 2e^- & \mathscr{E}^o = 0.440\text{ V} \\ O_2 + 4H^+ + 4e^- \rightarrow 2H_2O & \mathscr{E}^o = 1.23\text{ V} \\ & \mathscr{E}^o_{cell} = 1.67\text{ V} \end{array} \tag{10.9}$$

From the Nernst equation (9.12)

$$\mathscr{E}_{cell} = \mathscr{E}^o_{cell} - \frac{RT}{4\mathscr{F}} \ln \frac{[Fe^{2+}]^2}{P_{O_2}[H^+]^4}$$

The corrosion reaction (10.9) ceases when $\mathscr{E}_{cell} \leqslant 0$

$$0 = 1.67 - \frac{0.0591}{4} \log \frac{[Fe^{2+}]^2}{P_{O_2}[H^+]^4} \tag{10.10}$$

Hence $\mathscr{E}_{cell} \leqslant 0$ when

$$\log \frac{[Fe^{2+}]^2}{P_{O_2} \cdot [H^+]^4} \geqslant 113 \tag{10.11}$$

Let us consider extreme conditions where

$$P_{O_2} = 10^{-6} \text{ atm}, \qquad [Fe^{2+}] = 10 \text{ M}, \qquad [H^+] = 10^{-14} \text{ M}$$

$$\log \frac{(10)^2}{10^{-6} \cdot 10^{-56}} = \log 10^{64} = 64$$

Since $64 < 113$ corrosion will continue to occur. In strong NaOH solution rusting is reduced because the Fe_2O_3 forms a protective layer over the metal.

10.3. TYPES OF CORROSION

The various forms of corrosion can be classified by their various causes. These are: uniform corrosion attack (UC); bimetallic corrosion (BC); crevice corrosion (CC); pitting corrosion (PC); grain boundary corrosion (GBC); layer corrosion (LC); stress corrosion cracking (SCC); cavitation corrosion (CC); hydrogen embrittlement (HE).

10.3.1. Uniform Corrosion

Such corrosion is usually easy to detect and rectify. The slow corrosion of a metal in aqueous acidic solution is an example of such corrosion. Impurities in a metal can result in local cells which, in the presence of electrolyte, will show corrosive action.

10.3.2. Bimetallic Corrosion

This type of corrosion, also called *galvanic corrosion* is characterized by the rapid dissolution of a more reactive metal in contact with a less reactive more noble metal. For example, galvanized steel (Zn–Fe) in contact with copper (Cu) pipe is a common household error. A nonconducting plastic spacer would reduce the corrosion rate in the pipe. The rate of corrosion is partially determined by the difference in the standard cell potentials of the two metals in contact (see Table 9.4). The relative potential of metals in seawater is given in Table 10.1 and represents the driving force of the corrosion which includes the current, or more precisely, the current density, i.e., A/cm^2.

An electrochemical cell is formed and the anodic metal dissolves. This can be corrected by applying a counter current or voltage or by introducing a more reactive, sacrificial anode, e.g., adding a magnesium alloy to the above Zn–Fe—Cu system, a procedure commonly used for hot-water pipes in renovated buildings.

10.3.3. Crevice Corrosion

A nonuniform environment or concentration gradient due to material structure or design leads to concentration cells and corrosion. Differential aeration is, for example, the cause of corrosion at the waterline or at the edges of holes or flange joints. The size of a crevice can range from 25 to 100 μm in width—small enough to create an oxygen concentration cell between the crevice solution and that on the outer surface. Oxygen can form a thin oxide layer on metals which acts as a protective passive film.

10.3.4. Pitting Corrosion

Like CC, PC is due to differential aeration or film formation (due to dust particles). The breakdown of a protective oxide layer at a lattice defects is another common cause of pits. The mechanism of pitting of iron under a water drop is shown in Fig. 10.2 and as in a CC, a differential concentration of oxygen in the drop creates a concentration cell. Rust has the composition of Fe_3O_4 and FeO(OH). Fe_3O_4 is a mixed oxide of $FeO \cdot Fe_2O_3$ where iron is in the +2 and +3 oxidation state. The PC of various iron alloys induced by Cl^- in the presence of 0.5 M H_2SO_4 is given in Table 10.3. High chromium alloys are effective in reducing PC but a limit is reached at about 25% Cr whereas nickel seems to have little effect on corrosion resistance. Other salts in solution also can affect the pitting rate as well as the depth of the pits.

10.3.5. Grain Boundary Corrosion

Coarse crystalline rolled metals or alloys can corrode at the edge of the crystallites. Thus the iron impurity in aluminum is responsible for aluminum corrosion. Similarly, stainless steel (18/8 Cr/Ni) when heated (during welding) results in the precipitation of chromium carbide at the grain boundaries. This forms an enriched nickel layer anodic

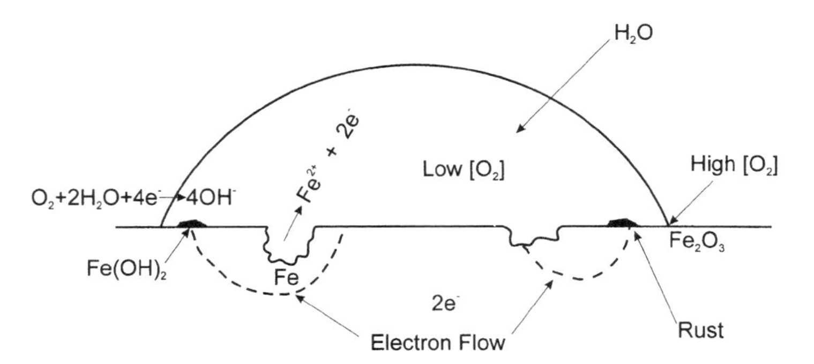

FIGURE 10.2. A gradient in O_2 concentration in the water drop makes the center portion of the iron anodic where $Fe \rightarrow Fe^{2+} + 2e^-$ while the edge is cathodic and oxygen is reduced, $O_2 + 2H_2O + 4e^- \rightarrow 4OH^-$. The basic OH^- reacts with the solubilized Fe^{2+} to form the insoluble $Fe(OH)_2$. This will oxidize and then dehydrate to form Fe_2O_3 (rust).

TABLE 10.3
Minimum Concentration of Chloride Ion Necessary for Starting Pitting in 0.5 M H_2SO_4

Alloy	$[Cl^-]$ (M)
Fe	0.0003
5.6 Cr–Fe	0.017
11.6 Cr–Fe	0.069
20 Cr–Fe	0.1
18.6 Cr, 9.9 Ni–Fe	0.1
24.5 Cr–Fe	1.0
29.4 Cr–Fe	1.0

to the bulk alloy, and severe intergranular corrosion and pitting results. The corrosion rate of stainless steel (18/8) in aqueous HCl solutions depends on the concentration of acid, temperature, and the oxygen pressure. In contrast, an equivalent metallic glass* ($Fe{-}Cr_{10}Ni_5P_{13}C_7$) showed no detectable corrosion. This is illustrated in Fig. 10.3 and clearly shows how important corrosion is along the grain boundaries in stainless steel. Similar results were obtained for immersion tests in 10% wt. of $FeCl_3 \cdot 6H_2O$ at 60°C

*Metallic glasses are amorphous noncrystalline solids which are usually prepared by rapidly cooling the molten metal. Such metals are devoid of grain boundaries.

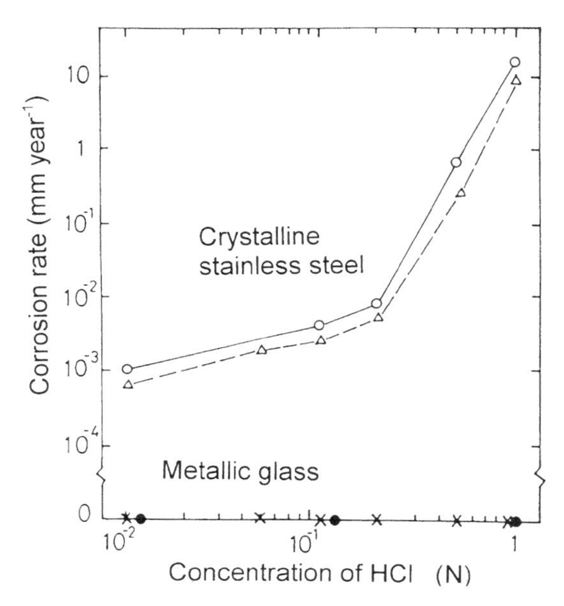

FIGURE 10.3. A comparison of the corrosion rates of metallic glasses (×, ●) and crystalline stainless steel (○, △) as a function of HCl concentration at 30°C. No weight changes of the metallic glasses of $Fe{-}Cr_{10}P_{13}C_7$ were detected by a microbalance after immersion for 200 hr.

as an indication of PC. Again the stainless steel (304, 136, 316) all showed significant pitting whereas metallic glasses showed no detectable weight loss after 200 hr. Not all metallic glasses are resistant to corrosion and much more work is needed to understand these differences.

10.3.6. Layer Corrosion

Like GBC, LC is caused by the dissolution of one element in an alloy and the formation of leaflike scale-exfoliation. Some cast irons and brasses show flakelike corrosion products. The corrosion is due to microcells between varying compositions of an alloy.

10.3.7. Stress Corrosion Cracking

This is normally found only in alloys such as stainless steel and in specific environments. This type of corrosion is a result of the combined effects of mechanical, electrochemical, and metallurgical properties of the system.

The residual stress in a metal, or more commonly an alloy, will, in certain corrosive environments, result in mechanical failure by cracking. It first became apparent at the end of the 19th century in brass (but not copper) condenser tubing used in the electric power generating industry. It was then called *season cracking*. It is usually prevalent in cold drawn or cold rolled alloys which have residual stress. Heat treatments to relieve this stress were developed to solve the problem. It was soon realized that there were three important elements of the phenomenon: the mechanical, electrochemical, and metallurgical aspects.

The mechanical aspect is concerned with the tensile stress of the metal alloy. The mechanism of crack formation includes an induction period followed by a propagation period which ends in fracture. The kinetics of crack formation and propagation has been studied for high-strength alloys and the overall process can be resolved into 2 or 3 stages depending on the alloy. The velocity of cracking is usually very slow, and rates of about 10^{-11} m/sec have been measured. Activation energies for stages I and II are usually of about 100 and 15 kJ/mol, respectively. Stainless steel piping in nuclear reactors (BWR) often suffer such SCC and must be replaced before they leak. Zircaloy tubes used to contain uranium fuel in nuclear reactors are also subject to SCC.

An essential feature is the presence of tensile stress which may be introduced by loads (compression), cold work, or heat treatment. The first stage involves the initiation of the crack from a pit which forms after the passive oxide film is broken by Cl^- ions, the anodic dissolution reaction of metal produces oxide corrosion products with high levels of H^+ ions. Hydrogen evolution during the second stage contributes to the propagation of the crack. Stainless steel pipes used in nuclear power plants for cooling often suffer from SCC. This can be reduced by removing oxygen and chloride from the water, by using high purity components, and careful annealing with a minimum of weld joints.

The electrochemical aspect of the process is associated with anodic dissolution, accounting for high cracking velocities. The crack tip is free of the oxide protective coating in the alloy and crack propagation proceeds as the alloy dissolves. Chloride ions present in solution tend to destroy this passivity in the crevice, which is depleted in oxygen. In stainless steels the dissolution of chromium in the crevice occurs by the reactions:

$$Cr \rightarrow Cr^{3+} + 3e^- \quad (10.12)$$

$$Cr^{3+} + 3H_2O \rightarrow Cr(OH)_3 + 3H^+ \quad (10.13)$$

and accounts for the major cause of the autocatalytic process whereby the increased acidity in the crevice increases the rate of corrosion. Titanium is resistant to CC because its passive layer is not attached by chloride ions. This explains the specificity of the corrosive environment for a particular alloy since the reformation of the protective surface layer would stop the crack from propagating further.

The metallurgical aspect is exemplified by the effect of grain size—reducing grain size reduces SCC. SCC is increased by cold working and reduced by heat treatment annealing. Other metallurgical properties of an alloy can contribute to its susceptibility to SCC. Solutions to the problem include heat treatment, the use of corrosion-resistant cladding and in the case of nuclear power plants—the use of a nuclear grade stainless steel.

10.3.8. Cavitation Corrosion

Cavitation is due to ultrasonics or hydrodynamic flow and is associated with the formation of micro bubbles which collapse adiabatically to form thermal shocks and localized *hot spots* sufficient to decompose water and form hydrogen peroxide and nitric acid (from dissolved air). The resulting corrosion is thus due to a mechanical and chemical effect and can be reduced by cathodic protection or by the use of chemically resistant alloys.

Cavitation is normally associated with motion of metal through water which forms low pressure bubbles. These micro bubbles, upon collapsing adiabatically, heat the entrapped oxygen, nitrogen, and water to above decomposition temperatures with the resulting formation of a variety of compounds such as NO_x, HNO_3, H_2O_2, and at times O_3. Cavitation is thus produced in the turbulence formed by propeller blades of ships, water pumps and mixers, and in the steady vibrations of engines. Cavitation also has the effect of disrupting the protective surface coating on metals, and when pieces of the metal are actually removed by the flow of bubbles, the process is called *cavitation erosion* (CE).

Figure 10.4A shows the cylinder casing of a diesel engine which was water cooled. Vibrations caused cavitation resulting in pitting which penetrated the casing. The lower Fig. 10.4B shows the blades of the water pump in the diesel which had also corroded for the same reasons.

Cavitation corrosion can be reduced by the proper design and vibration damping of systems. It has also been shown that the addition of drag reducers (see Appendix B) to the water reduces CE and transient noise. High Reynolds number ($Re = 124{,}000$) can be achieved without cavitation. It would seem advantageous to add water-soluble

FIGURE 10.4 (A) Cavitation corrosion of a water-cooled cylinder casing of a diesel engine. Corrosion holes have penetrated the wall. (B) The water pump propeller in the same diesel engine corroded by cavitation.

drag reducers such as polyethyleneoxide to recirculating water cooling systems to reduce CC.

10.3.9. Hydrogen Embrittlement

The migration of hydrogen dissolved in a metal lattice usually occurs along grain boundaries where cracks occur during stress. The embrittlement of steels is due to hydrogen atoms which diffuse along grain boundaries. They then recombine to form H_2 and produce enormous pressures which result in cracking. The H-atoms are formed during the corrosion of the metal or a baser metal in contact with the steel.

10.4. ATMOSPHERIC CORROSION

The major cause of corrosion of metals in the air is due to oxygen and moisture. In the absence of moisture the oxidation of a metal occurs at high temperatures with activation energies E_a, ranging from 100 to 250 kJ/mol which is determined by the work function ϕ, where

$$E_a(\text{kJ/mol}) = \phi - 289 \quad (10.14)$$

At ambient temperature, however, all metals except gold have a thin microscopic layer of oxide.

An example of a noncorroding steel structure is the Delhi Iron Pillar (India) which dates from about 400 A.D. It is a solid cylinder of wrought iron 40 cm in diameter, 7.2 m high. The iron contains 0.15% C and 0.25% P and has resisted extensive corrosion because of the dry and relatively unpolluted climate.

The industrial corrosive effluents could include NO_x, SO_x, and H_2S whereas natural occurring corrosive substances are H_2O, CO_2 and, in coastal areas, NaCl from sea sprays. These two sources of corrosive substances were enough to corrode the Statue of Liberty in New York Bay. (The statue, which is 46 m high, was a gift from France in 1886, erected on a pedestal 46 m above ground level to commemorate the centenary of the American Revolution.) It was constructed of 300 shaped copper panels (32 tonnes), 2.4 mm thick, riveted together and held in place by 1800 steel armatures which slipped through 1500 copper saddles. Thus though the iron touched the copper, there was no direct bonding of the two metals. This did not stop the electrochemical corrosion when rain water and ocean spray penetrated the structure. More than one third of the 12,000 rivets had popped by 1975. To commemorate the 2nd centennial of the USA, the rebuilt Statue of Liberty was unveiled after renovations costing about $60 million. The copper panels are now sealed on the inside of the structure by silicone sealant to prevent water from entering the statue. The iron armatures were replaced by stainless steel with a Teflon-coated tape to separate the two metals. Though the copper skin is expected to last over 1000 years, the durability of the wrought iron structure is much shorter and it will corrode quickly if not protected from the elements. This normally involves lead-based paints or silicone rubber sealants which are used for bridges.

10.5. CORROSION IN SOIL

The resistivity of soil is an important characteristic which often determines the rate of corrosion — low resistivity is usually associated with high rates of corrosion. This is shown in Table 10.4. Soluble salts and high moisture content account for low resistivity–high conductivity. The density and particle size can control the moisture level and permeability of the soil to water and oxygen.

The ground water level determines the depth of *dry* soil. Oxygen depletion by decaying organic substances or living organisms tend to inhibit corrosion. Oxygen transport from air into soil is facilitated by water and leads to higher corrosion rates above ground water than below.

A low pH of soil (pH 3.5–4.5) — high acid level — contributes to the corrosion rate. Soil of pH > 5 is much less corrosive. Alkaline soil, pH > 7 can be corrosive to aluminum and if ammonia is formed by bacterial activity then even copper will be

TABLE 10.4
Relationship Between the Resistivity of Soil Corrosion Activity and Estimated Lifetime of Buried Steel Pipe

Resistivities (Ω-cm)	Corrosion	Normal duration (years)
<800	Severe	<10
800–500	Moderate	15
5000–10,000	Mild	20
>10,000	Unlikely	>25

attached. The weak organic acids present in humic acid can solubilize surface oxides and lead to corrosion of metals by complexation processes. Anaerobic bacterial action in soil can lead to H_2S (and CH_4 plus CO_2) which, though a weak acid, will form insoluble metallic sulfides, reducing the free metal ions in the soil and shifting the equilibrium toward metal dissolution.

10.6. AQUEOUS CORROSION

As indicated previously, the corrosion of metals in aqueous environments is determined by the Nernst equation in terms of the electrode potential and pH—called a *Pourbaix* diagram. This is shown in Fig. 10.5 for iron where the vertical axis is the redox potential of the corroding system and the pH scale is the horizontal axis. The

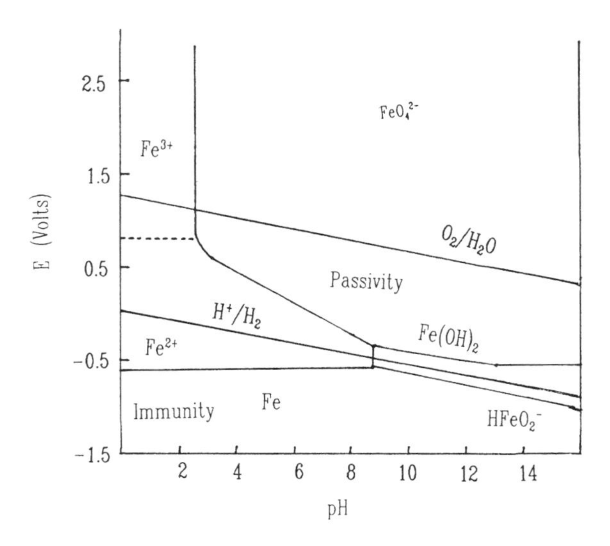

FIGURE 10.5. A Pourbaix diagram for iron showing the general conditions under which the metal is passive, corrosive, or stable, (immunity).

dashed lines show the H^+/H_2 and O_2/H_2O redox reactions which have slopes of 0.059 V/pH. Water is stable between the two lines. Sloping lines indicate that the redox potential is pH dependent. Horizontal lines reflect redox potential which are not pH dependent whereas vertical lines refer to changes which do not involve a change in oxidation state. Above the O_2/H_2O line oxygen is evolved, while below the H^+/H_2 line hydrogen is liberated.

When a solid insoluble product is formed it may protect the metal from further reaction. This corresponds to the passive region and assumes low concentration for metal ions in solution (e.g., 10^{-6} M). Under condition where the metal is stable a state of immunity exists and corrosion cannot occur. Iron corrodes, forming Fe^{2+} at low pH, but at high pH the $Fe(OH)_2$ dissolves to form $HFeO_2^-$. This region is referred to as caustic cracking of steel (pH > 12) analogous to stress corrosion cracking.

Iron will corrode in acids except H_2CrO_4, conc. HNO_3, $H_2SO_4 > 70\%$, and HF > 90%. Pourbaix diagrams are available for most metals and help define the corrosion free conditions.

10.7. CORROSION PROTECTION AND INHIBITION

The Royal Navy's first submarine, *Holland I*, sank in 1913 off the coast of England and for 70 years lay in 63 m of seawater. The wreck was recently located and raised. She was in remarkable condition considering that the hull contained a mass of dissimilar metals, steel, cast and malleable tin, brass, bronze, and lead. The doors opened, springs sprang, the engine turned, rivets were tight, and a battery when cleaned, refilled, and recharged, delivered its specified 30 amps. The explanation for the absence of the corrosion expected is due to the protection given to the surface by the rapid colonization of a coldwater coral and the deposition of a 3–4 mm layer of calcium carbonate. This prevented the diffusion of oxygen and electrolyte from reaching the metal surface. Coatings thus represent a simple and at times effective method of reducing corrosion.

Corrosion can be eliminated by covering metals with more noble ones by plating or cladding. This is impractical because of the expense involved. Protective metal coatings of chromium are familiar, being decorative as well as preventing corrosion. Plated jewelry with silver, gold, and rhodium are common. Steel coated with zinc is protected both in air and water. The standard potentials are: $\mathscr{E}^o_{Zn^{+2}/Zn} = -0.763$ V and $\mathscr{E}^o_{Fe^{+2}/Fe} = -0.440$ V means that any break in the zinc coating on iron will make Zn anodic and iron cathodic (see Fig. 10.6). In hot water exposed iron can be protectively coated with $CaCO_3$ if the water is hard (i.e., contains $CaHCO_3$). The zinc is usually applied by hot-dipping and produces a continuous coat of 80–125 μm thick. Other coating processes include spray plating, electroplating, and for small items, tumbling. A galvanized surface can be repaired by painting with a zinc rich paint consisting of metallic zinc powder bound in an epoxide or resin base.

Ordinary paints may be permeable to oxygen and water vapor (see Chapter 13) and though they may slow the rate of corrosion they cannot prevent it completely. Hence, special paints with chromates or red lead (Pb_3O_4) have been used for many years as a protective coating for steel. Polymeric resins though more expensive than the linseed oil-based paints, last longer and thus are more effective.

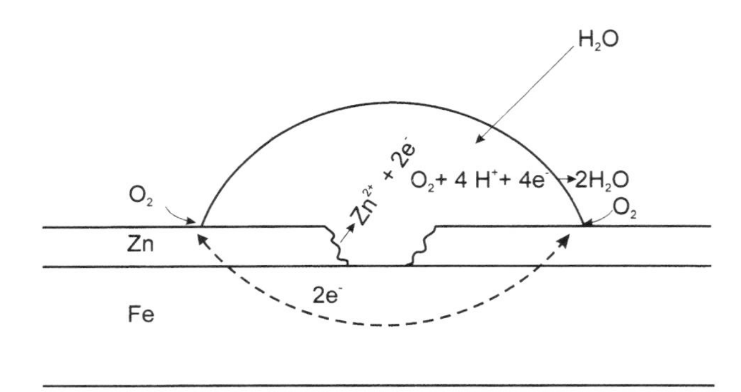

FIGURE 10.6. The Fe/Zn system. A break in the zinc coating on iron (galvanized iron) will, to a limited extent, continue to protect the iron (cathodic) as the zinc (anodic) dissolves.

10.8. CORROSION IN BOILER STEAM AND CONDENSATE

Steam lines with air and CO_2 entrained can be very corrosive. To reduce corrosion oxygen can be removed by the addition of hydrazine (N_2H_4).

$$N_2H_4 + O_2 \rightarrow N_2 + 2H_2O \tag{10.15}$$

or Na_2SO_3

$$2Na_2SO_3 + O_2 \rightarrow 2Na_2SO_4 \tag{10.16}$$

Other additives which are commonly added are basic amines which neutralize the acids (H_2CO_3) present in the water. One important property of the amine besides the pH of its solution is the distribution ratio (DR) which is the ratio of amine in steam to that dissolved in the condensate. A high DR value means that the metal is readily coated with a thin protective film of the amine. Some amines commonly used, and their pH and DR are given in Table 10.5. The amine is slowly lost and it must be replaced continuously. Steam lines invariably have these amines and the use of brass, bronze, or copper results in the corrosive removal of copper.

Stored metallic equipment or parts are subject to corrosion. Sodium nitrite is an inhibitor which is often included in the enclosure or packaging. However, vapor phase corrosion inhibitors (VCI) such as dicyclohexylammonium nitrite and ammonium benzoate are superior corrosion inhibitors because of the film formed on the metal surfaces.

10.9. CATHODIC PROTECTION

It is possible to prevent the corrosion of a metal by connecting it to a more active metal. This active metal becomes anodic and tends to corrode whereas the cathodic metal is preserved. Iron pipes in soil or water will not corrode if they are connected to a sacrificial anode such as aluminum, zinc, or magnesium. Steel pipes for water and gas

TABLE 10.5
Some Characteristics of Selected Amines Used in Steam Systems as Corrosion Inhibitors

	pH	DR
Morpholine	8–10	0.3–0.8
Cyclohexylamine	10–11	6–8
Diethylaminoethanol	11–12	2–4
Benzylamine	8–9	3–4

Values are dependent on concentration of the amine.

are usually protected in this manner. Galvanized iron pipes for hot water lines have a limited life which can be extended by introducing a magnesium rod to act as a sacrificial anode.

The potential needed to protect iron in seawater is -0.62 V with respect to the SHE or -0.86 V relative to SCE. Aluminum can provide this potential, -0.95 V relative to the SCE and its use has been extended to offshore oil platforms, ship's hulls, ballast tanks, jetty piles with life expectancies ranging from 3 to 10 years depending on the mass of aluminum employed.

An alternate approach is to apply a potential onto the steel, making it cathodic relative to an inert anode such as Pb, C, or Ni. A potential of -0.86 V is suitable for the protection of iron.

Though more negative potentials, such as -1.0 V, can be used, it should be avoided in order to prevent hydrogen evolution and hydrogen embrittlement.

EXERCISES

1. Show how different oxygen concentrations in a cell for a single metal can result in corrosion.
2. What are the cathodic reactions which usually accompany the corrosive dissolution of a metal?
3. Explain why the standard reduction potential, $\mathscr{E}^\circ$

$$Hg_2Cl_2 + 2e^- \rightarrow 2Hg + 2Cl^- \ (\mathscr{E}^\circ = 0.2680\ \text{V})$$

   (Table 9.4) is different from that for the SCE ($\mathscr{E}^\circ = 0.2415$ V).
4. Explain why drag reducers may decrease cavitation corrosion (see Appendix B).
5. Why is the corrosion rate of a metallic glass orders of magnitudes lower than the crystalline metal?
6. When Ni and Cd are in contact which metal will corrode?
7. Describe 6 types of corrosion and explain how the corrosion rates can be reduced.
8. How can a metal be made passive—give 3 examples?
9. When two dissimilar metals are joined together a potential is set up due to the Seebeck effect. This is the basis of the thermocouple and is due to differences in work function of the two metals. Explain how this applies to corrosion.

10. The tarnishing of silver by H_2S is a type of corrosion which requires the presence of O_2.

$$2Ag + H_2S + \frac{1}{2}O_2 \rightarrow Ag_2S + H_2O$$

Explain this in terms of a corrosion mechanism.

11. How does polarization affect the rate of corrosion?
12. Why is chloride ion (Cl^-) more corrosive to iron than nitrate (NO_3^-)?
13. Estimate the activation energy for the oxidation of the following metals in dry air. The values of the respective work functions are given in eV units. Cd (4.22), Cr (4.5), Fe (4.5), Mo (4.6), Ni (5.15), Ti (4.33), Zr (4.05).

FURTHER READING

R. W. Revie, *Uhlig's Corrosion Handbook*, 2nd Ed., Wiley, New York (2000).
P. R. Roberge, *Corrosion Engineering Handbook*, McGraw-Hill, New York (1999).
J. R. Becker, *Corrosion and Scale Handbook*, Penn Well Books, Tulsa, Oklahoma (1998).
P. A. Schweitzer, Editor, *Corrosion Engineering Handbook*, M. Dekker, New York (1996).
B. D. Craig and H. D. Anderson, *Handbook of Corrosion Data*, 2nd Ed., ASM International Materials Park, Ohio (1995).
P. Marcus and J. Oudar, Editors, *Corrosion Mechanisms in Theory and Practice*, M. Dekker, New York (1995).
W. F. Bogaerts and K. S. Agena, *Active Library on Corrosion CD-ROM and ALC Network*, Elsevier, Amsterdam (1994).
E. W. Flick, *Corrosion Inhibitors: An Industrial Guide*, 2nd Ed., Noyes, Westwood, New Jersey (1993).
S. Bradford, *Corrosion Control*, Chapman and Hall, New York (1992).
J. C. Scully, *The Fundamentals of Corrosion*, 3rd Ed., Pergamon Press, New York (1990).
K. R. Trethewey and J. Chamberlain, *Corrosion for Students of Science and Engineering*, Longman Scientific, New York (1988).
G. Wrangler, *An Introduction to Corrosion and Protection of Metals*, Chapman and Hall, London (1985).
J. M. West, *Basic Corrosion and Oxidation*. Ellis Harwood Ltd., Chichester, UK (1980).
J. P. Chilton, *Principles of Metallic Corrosion*, 2nd Ed., The Chemical Society, London (1973).
U. R. Evans, *The Rusting of Iron: Causes and Control*, Edward Arnold, London (1972).
Corrosion notes, http://www.cp.umist.ac.uk/
National Association of Corrosion Engineers, http://www.nace.org/index.asp
Corrosion Research Centre, http://www.cems.umn.edu/research/crc
Protective Coatings, http://www.corrosion.com/
Corrosion sources, http://www.corrosionsource.com/
Corrosion, http://www.intercorr.com/
Corrosion experts past and present, http://www.corrosion-doctors.org
Corrosion Prevention Association,
http://www.corrosionprevention.org.uk.cpu/default.htm
Corrosion solution, . http://www.importexporthelp.com/electroshield.htm
Correct corrosion, http://www.repairrustedoutcars.com/
http://www.metalogic.be/projects

11

Polymers and Plastics

11.1. INTRODUCTION

A *polymer* is a large chain molecule of high molecular weight which is composed of a single molecule (monomer) that is repeated many times in the chain. In contrast, a macromolecule is a large molecule composed of many small molecules bound together with chemical bonds, e.g., a protein or DNA. An oligomer is a small polymer of only several monomer units.

Plastics are prepared by the melting, molding, extruding or the compression of polymers. The word polymer implies a molecule consisting of a long chain of units of smaller molecules or monomers. Thus, the polymer is also called a macromolecule. Such large molecules exist in nature and common examples of these are cellulose, rubber, cotton, silk, wool, starch and keratin.

The annual world production of polymers has increased from 11.5 Mtonnes in 1940 to about 27 Mtonnes in 1960, after which time production almost doubled every decade to more than 150 Mtonnes in 1990. Fiber production at about 36 Mtonnes is almost equally divided into natural and synthetic. The production of elastomers (flexible plastics) represent about 1/10 of the total polymers, with production of synthetic elastomers being about twice that of natural rubber.

11.2. MOLECULAR WEIGHT

Normally the number of monomers in a polymer molecule varies considerably but the interesting range for the fabricator of plastics is generally between 10^3 and 10^6 units.

Since the precise number cannot be controlled, the molecular weight (MW) of a polymer is not a unique value and the distribution can vary as a result of the method of preparation. There are two important average molecular weights of a polymer, the number average MW, $\bar{M}_n$, and the weight average MW, $\bar{M}_w$.

If we let w represent the total mass of a sample of polymer and w_i the weight of the ith species of MW M_i, then

$$n_i = \frac{w_i}{M_i} \quad \text{where } n_i \text{ is the number of moles of } ith \text{ species} \tag{11.1}$$

$$\sum_{i=1}^{\infty} n_i = n_T \quad \text{the total number of moles in the sample} \tag{11.2}$$

$$\text{the total weight,} \quad w = \sum_{i=1}^{\infty} w_i = \sum_{i=1}^{\infty} n_i M_i \tag{11.3}$$

The number average MW $\bar{M}_n$ is given by

$$\bar{M}_n = \frac{w}{\sum_{i=1}^{\infty} n_i} = \frac{\sum_{i=1}^{\infty} w_i}{\sum_{i=1}^{\infty} n_i} = \frac{\sum_{i=1}^{\infty} n_i M_i}{\sum_{i=1}^{\infty} n_i} \tag{11.4}$$

The weight average MW, $\bar{M}_w$, is given by

$$\bar{M}_w = \frac{\sum_{i=1}^{\infty} w_i M_i}{\sum_{i=1}^{\infty} w_i} = \frac{\sum_{i=1}^{\infty} n_i M_i^2}{\sum_{i=1}^{\infty} n_i M_i} \tag{11.5}$$

A typical distribution of MW of a polymer is shown in Fig. 11.1.

The MW of a polymer is the single most important physical characteristic of the plastic since it determines its mechanical properties and even solubility among other properties.

Another related concept is the degree of polymerization (DP) which represents the number of monomer units in the polymer chain. Since the value of DP differs from one polymer chain to another, the value of the degree of polymerization is usually an average, and is related to the MW by the relation

$$M_w = M(\text{DP}) \qquad \text{and} \qquad \bar{M}_w = M(\overline{\text{DP}}) \tag{11.6}$$

where M is the MW of the monomer.

The MW of a polymer can be determined by a variety of methods.

The colligative properties of polymers in solution give rise to the number average MW, $\bar{M}_n$. Thus boiling point elevation and osmotic pressure measurement are commonly used though the latter method is much more sensitive, though restricted by the choice of suitable membranes. The weight average MW, $\bar{M}_w$, of a polymer in solution can be obtained by light scattering measurements.

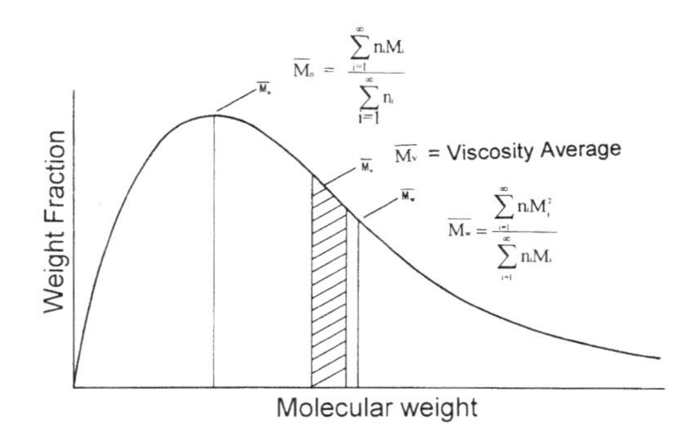

FIGURE 11.1. Fraction of weight having an average MW.

The simplest and most commonly used method of measuring the MW of a polymer are by viscosity measurements of its solution. The relationship is

$$[\eta] = K\bar{M}_v^{\alpha} \tag{11.7}$$

where K and α are empirical constants dependent on the polymer, solvent, and temperature.

$\bar{M}_v$ is the average viscosity MW and $[\eta]$ is the intrinsic viscosity defined as

$$[\eta] = \frac{\text{Limit}}{c \to 0} \frac{\eta sp}{c} \tag{11.8}$$

c is the concentration usually expressed as grams of polymer/100 g solvent and ηsp is the specific viscosity determined from the measurement of the viscosity of the pure solvent η_0, and viscosity of the solution η where $\eta/(\eta_0)$ is usually referred to as the viscosity ratio η_r and

$$\eta sp = \frac{\eta}{\eta_0} - 1 \tag{11.9}$$

Values of α and K are available from handbooks on polymers and range from 0.5 to 1 for α and 0.5 to 0.5×10^{-4} for K. The value of $\bar{M}_v$ is usually about 10 to 20% below the value of $\bar{M}_w$ (see Appendix B).

It is now well established that all important mechanical properties, such as tensile strength, elongation to break, impact strength, and reversible elasticity of polymers, depend on DP. When DP is relatively low, the polymer has little or no strength. As DP increases, the mechanical properties improve and tend toward a constant value. This is illustrated in Fig. 11.2 which shows the typical shape of the curve. The critical value DPc below which the polymer is essentially friable is different for each polymer, as is the bendover point β. However, plastics have little strength when DP < 30 and approach limiting strength at DP > 600.

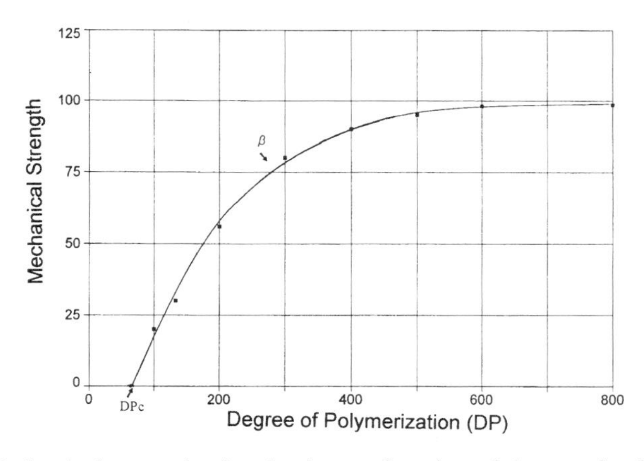

FIGURE 11.2. Mechanical strength of a plastic as a function of degree of polymerization (DP).

11.3. COPOLYMERS

When a polymer is formed from two or more monomers then the polymer is said to be a copolymer. The relative positions of the two monomers can be random or regular or in chunks. Figure 11.3 shows the different possible arrangements.

Blends of copolymers can be used to obtain specific properties of a plastic. Thus polyethylene is brittle at temperatures below 0°C. However, when copolymers are formed with vinyl acetate (15 mol%) the resulting plastic is more flexible down to −40°C.

Another example of a copolymer is vinyl chloride with about 5% propylene. Polyvinyl chloride (PVC) is a hard brittle plastic which is made soft and flexible by dissolving a plasticizer into the PVC. Up to 30% by weight of plasticizers such as dioctylphthalate is used to make plastic tubing. The propylene copolymer is soft without the plasticizer, or less plasticizer is required at lower concentrations in the propylene/PVC copolymer.

The loss of plasticizer from vinyl upholstery is the cause of cracking commonly observed in automobile seats and furniture.

11.4. CLASSIFICATION OF POLYMERS

Many polymers occur naturally, e.g., cotton, wool, silk, gelatine, rubber, leather. Some are even inorganic such as sulfur, glass, silicones. The thermal property of polymers is another important characterization. Thermoplastic polymers become soft and, without crosslinking, can be molded and shaped into various forms which are retained on cooling. The process is reversible and the plastics can be reformed into other shapes when heated. Examples include polyethylene, PVC, nylon, polystyrene. Thermosetting polymers crosslink on setting and once formed cannot be reshaped. Heating decomposes the plastic. Examples include Bakelite, melamine, phenol formaldehyde, and epoxy resins.

The manner in which polymers are formed is also a distinguishing feature. Two common methods are described.

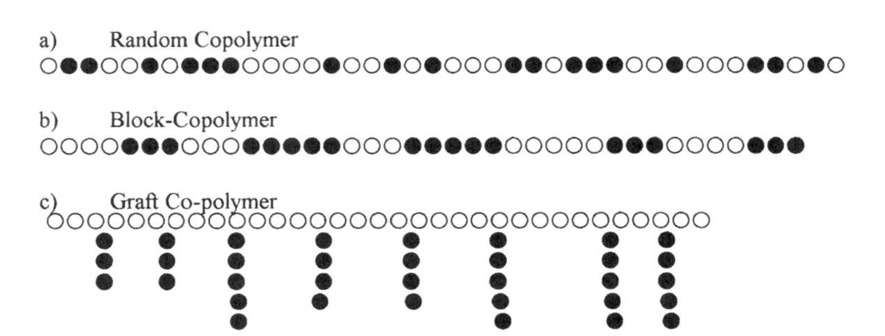

FIGURE 11.3. Schematic arrangements of copolymers made from 2 monomers ○ and ●.

11.4.1. Addition Polymers

The addition process where the monomer is converted into a free radical* which adds to another monomer. The process continues until the two growing chains combine, or one combines with a free radical. The process is as follows:

$$\text{Initiation} \quad \begin{cases} \text{Catalyst} \rightarrow R^{\bullet} \\ M + R^{\bullet} \rightarrow R\text{–}M^{\bullet} \end{cases} \tag{11.10}$$

$$\text{Propagation} \quad \begin{cases} R\text{–}M^{\bullet} + M \rightarrow R\text{–}M\text{–}M^{\bullet} \\ RM_n^{\bullet} + M \rightarrow RM_{n+1}^{\bullet} \end{cases} \tag{11.11}$$

$$\begin{array}{ll} \text{Termination} & RM_x^{\bullet} + R^{\bullet} \rightarrow RM_xR \\ \quad \text{recombination} & RM_x^{\bullet} + R^{\bullet}M_y^{\bullet} \rightarrow RM_xM_yR \\ \quad \text{disproportionation} & RM_x^{\bullet} + RM_y^{\bullet} \rightarrow RM_x\text{–}H + RM_yH \end{array} \tag{11.12}$$

where RM_x–H is $RM_x^{\bullet}$ which has lost a H-atom forming a C=C double bond.

The initiation process is usually by the thermal generation of free radicals from a peroxide such as benzoyl peroxide

$$C_6H_5\overset{\displaystyle O}{\overset{\|}{C}}\text{—O—O—}\overset{\displaystyle O}{\overset{\|}{C}}\text{—}C_6H_5 \rightarrow C_6H_5^{\bullet} + C_6H_5\text{—}\overset{\displaystyle O}{\overset{\|}{C}}\text{—}O^{\bullet} + CO_2 \tag{11.13}$$

The peroxide or other azo initiators can also be decomposed when exposed to UV light. Such processes are used for the setting of polymers which function as fillings of tooth cavities.

11.4.2. Condensation Polymers

Condensation polymers are formed from the reaction of two different bifunctional monomers A and B which form AB by the reaction

$$\begin{aligned} A + B &\rightarrow A - B \\ AB + A &\rightarrow ABA \\ ABA + B &\rightarrow BABA \end{aligned} \tag{11.14}$$

and so on.

Thus, the polymer grows at both ends by condensing and stops when at least one of the reagents is fully consumed. Nylon is a condensation polymer between a 6-carbon diamine and 6-carbon dicarboxylic acid.

$$NH_2\text{–}(CH_2)_6\text{–}NH_2 + HOOC\text{–}(CH_2)_4COOH$$

$$\rightarrow NH_2\text{–}(CH_2)_6\text{–}NHOC(CH_2)_4\overset{\displaystyle O}{\overset{\|}{C}}OOH + H_2O \tag{11.15}$$

resulting in nylon 6,6 when the chain has grown sufficiently.

*A free radical is a molecule or fragment which has one or more unpaired electrons.

11.5. VINYL POLYMERS

The vinyl radical is $CH_2{=}CH^{\bullet}$ and is the basis of a wide variety of monomers having the general formula $CH_2{=}CHX$. For example, when $X = H$ the molecule is ethylene and the polymer is polyethylene. The major vinyl polymers are listed in Table 11.1.

11.5.1. Polyethylene

Also referred to as polythene, *polyethylene* is similar to polymethylene $(-CH_2)-_x$ which was prepared about 100 years ago by the decomposition of diazomethane (CH_2N_2), an explosive gas.

TABLE 11.1
Addition Polymers for Vinyl Polymers $CH_2{=}CHX$

X	Monomer	Polymer	Uses
H	$CH_2{=}CH_2$ ethylene	$-CH_2-CH_2-$	Bottles, plastic tubing
CH_3	$CH_2{=}CHCH_3$ propylene	$-CH_2-CH(CH_3)-$	Carpeting, textiles, ropes
Cl	$CH_2{=}CHCl$ vinylchloride	$-CH_2-CH(Cl)-$	Pipes, floor tiles, tubing
C_6H_5 (benzene ring)	$CH_2{=}CH(C_6H_5)$ styrene	$-CH_2-CH(C_6H_5)-$	Clear film, foam insulation, cups
CN	$CH_2{=}CHCN$ acrylonitrile	$-CH_2-CH(CN)-$	Orlon, ABS, carpet
$-C({=}O)-OCH_3$	Methylmethacrylate	$-CH_2-C(CH_3)(C({=}O)OCH_3)-$	Windows, outdoor signs lighting
$-O-C({=}O)-CH_3$	Vinylacetate	$-CH_2-CH(O-C({=}O)-CH_3)-$	Paints, adhesives
$CF_2{=}CF_2$	Teflon	$-CF_2-CF_2-$	Electrical insulation, heat resistant, lubricant

Polyethylene ($\cdot CH_2\text{–}CH_2\text{–})_x$ was first produced commercially in 1939. The early process was under high pressure (1000–3000 atm) and at temperatures from 80 to 300°C. The polymerization mechanism is via free radical initiators such as benzoyl peroxide

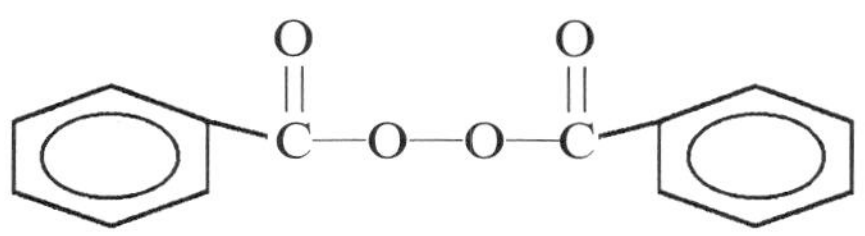

$$C_6H_5\text{—}\overset{O}{\overset{\|}{C}}\text{—O—O—}\overset{O}{\overset{\|}{C}}\text{—}C_6H_5$$

which are added to the reaction mixture.

This process results in low-density polyethylene (0.915–0.94 g/mL).

High density polyethylene is prepared at low pressure at about 70°C in the presence of a special catalyst (usually a titanium complex). The density is approximately 0.95 g/mL because of the higher degree of crystallinity and order in the polymer.

The original high-pressure process gave some branched polymers; polyethylene formed at low pressure has a higher melting point, higher density, and higher tensile strength. It is a linear crystalline polymer which costs approx. 1.5 times that of the high pressure-low density material.

Polyethylene films are commonly used as vapor barriers in housing insulation. For greenhouse covering or window material it is transparent enough, but will slowly disintegrate due to the presence of residual carbon–carbon double bonds (C=C) which are split by ozone. Ultraviolet light will also degrade plastics unless a UV stabilizer is added which converts the absorbed UV light into heat. To make a plastic biodegradable, a substance is added which absorbs UV light from the sun and forms free radicals which attack the polymer chain.

11.5.2. Polypropylene

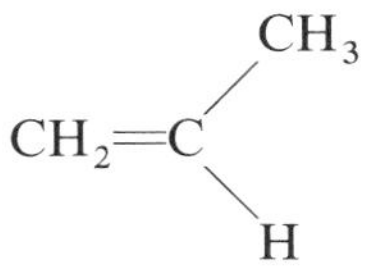

$$CH_2\text{=}C(CH_3)(H)$$

Polypropylene was first produced commercially in 1957. Early attempts resulted in very low MW polymers having poor plastic properties. The titanium complex used to prepare high density polyethylene was found to be effective in polymerizing propylene.

Because of the asymmetry of the propylene molecule three different types of stereochemical arrangements can occur in the polymer chain.

1 Isotactic

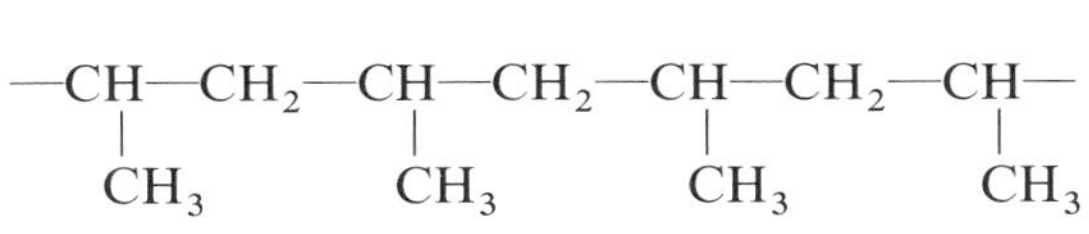

$$\text{—}\underset{CH_3}{\underset{|}{CH}}\text{—}CH_2\text{—}\underset{CH_3}{\underset{|}{CH}}\text{—}CH_2\text{—}\underset{CH_3}{\underset{|}{CH}}\text{—}CH_2\text{—}\underset{CH_3}{\underset{|}{CH}}\text{—}$$

(all methyl groups on one side)

2 Syndiotactic

```
      CH3                  CH3
       |                    |
—CH—CH2—CH—CH2—CH—CH2—CH—CH2—
 |                 |
CH3               CH3
```

methyl groups alternate

3 Atactic

```
CH3        CH3                   CH3
 |          |                     |
CH—CH2—CH—CH2—CH—CH2—CH—CH2—
                    |
                   CH3
```

random orientation of CH_3 groups

Atactic polypropylene is completely amorphous, somewhat rubbery, and of little value. The isotactic and syndiotactic polymers are stiff, crystalline, and have a high melting point. Increasing the degree of crystallinity increases the tensile strength, modulus and hardness.

Polypropylene is the lightest nonfoamed plastic, with a density of 0.91 g/mL. It is more rigid than polyethylene and has exceptional flex life. Polypropylene has found use in a wide variety of products which include refrigerators, radios and TVs as well as monofilaments, ropes and pipes.

11.5.3. Polyvinyl Chloride

Polyvinyl chloride is one of the cheapest plastics in use today. It is prepared by the polymerization of vinyl chloride (PVC) ($CH_2{=}CHCl$, B.P. – 14°C) as a suspension or emulsion in a pressure reactor. The polymer is unstable at high temperatures and liberates HCl at $T > 200°C$. It can be injection molded or formed into a hard and brittle material. It can be readily softened by the addition of plasticizers such as diethylhexylphthalate to the extent of 30%. Plasticized PVC is used as an upholstery substitute for leather. Since the plasticizer is volatile to a small extent it slowly leaves the vinyl which eventually becomes hard, brittle, and then cracks. This can be restored by replacing the plasticizer by repeated conditioning of the vinyl surface.

11.5.4. Polyvinylidene Chloride

Polyvinylidene chloride (PVDC) is prepared by free radical polymerization of vinylidene chloride ($CH_2{=}CCl_2$). This polymer, unlike PVC, is insoluble in most solvents. It forms copolymers with fiber forming polymers. Its films, known as Saran, have a very low permeability for O_2 and CO_2 and thus is used in food packaging. When heated to high temperatures, in the absence of oxygen, it liberates HCl leaving a very active carbon with pores of about 1.6 nm. This "Saran Carbon" has been used to double the storage capacity of CH_4 in cylinders. This is presently being considered for use for CH_4 fueled vehicles.

11.5.5. Polystyrene

Polystyrene (PS) is prepared by the polymerization of styrene (C_6H_5–CH=CH_2), also known as vinylbenzene. Commercial PS is mostly of the atactic variety and is therefore amorphous. The polymer, on decomposition, unzips and forms the monomer with some benzene and toluene. Its major defects are poor stability to weather exposure, turning yellow and crazing in sunlight. In spite of these drawbacks and its brittleness it has found wide use as molded containers, lids, bottles, electronic cabinets. As a foamed plastic it is used in packaging and insulation. The thermal conductivity of the expanded PS foam is about 0.03 Wm^{-1} K^{-1}. The foam can absorb aromatic hydrocarbons usually found in the exhaust of automobiles and buses, causing the foam to disintegrate after long periods of normal exposure to a polluted environment.

The copolymerization of a small amount of divinylbenzene results in a cross-linked polymer which is less soluble and stronger. Crosslinking can sometimes be accomplished by γ-radiation which breaks some C–H and C–C bonds and on rearrangement form larger branched molecules. This is the case for polyethylene which, after crosslinking, will allow baby bottles to withstand steam sterilization.

11.5.6. Polyacrylonitrile

Polyacrylonitrile (PAN) is formed by the peroxide initiated free radical polymerization of acrylonitrile (CH_2=CH–CN). The major application of PAN is as the fiber Orlon. When copolymerized with butadiene it forms Buna N or nitrile rubber, which is resistant to hydrocarbons and oils. As a copolymer with styrene (SAN) it is a transparent plastic with very good impact strength used for machine components and for molding crockery. As a terpolymer of acrylonitrile–butadiene–styrene (ABS) the plastic is known for its toughness and good strength and finds applications in water lines and drains.

Polyacrylonitrile fibers are an excellent source for high strength carbon fibers which are used in the reinforcement of composite (plastic) materials. The process was developed by the British Royal Aircraft Establishment and consists of oxidizing the atactic polymer at about 220°C while preventing it from shrinking. Further heating to 350°C results in the elimination of water and crosslinking of the chains which continues with loss of nitrogen. The fibers are finally heated to 1000°C. The reactions are illustrated in Fig. 11.4. The high tensile strength (3.2 $GN\,m^{-2}$) and Young's modulus (300 $GN\,m^{-2}$) is attributed to the alignment of the polymer chains and their crosslinking.

Carbon fibers have also been made from the pyrolysis of Viscose (cellulose), Rayon, jute, and from pitch. Though these methods produce slightly lower strength carbon fibers as compared to PAN, the lower cost ($\sim\frac{1}{5}$ to $\frac{1}{2}$) makes them excellent reinforcement materials for noncritical items such as golf clubs, tennis rackets, skis, and related sports goods.

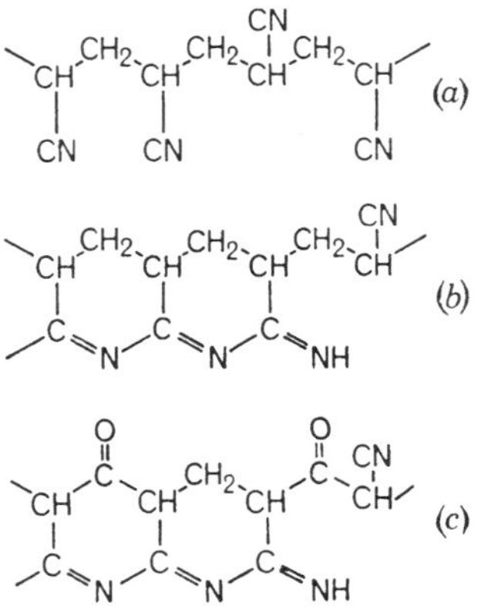

FIGURE 11.4A. Structure of (a) PAN, (b) PAN ladder polymer, (c) oxidized PAN ladder polymer.

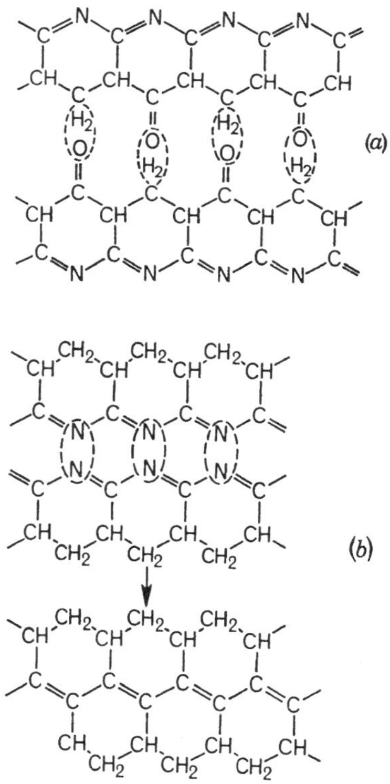

FIGURE 11.4B. (a) Crosslinking of PAN by intermolecular elimination of water, (b) crosslinking of dehydrated PAN by intermolecular elimination of nitrogen.

11.5.7. Polymethyl Methacrylate

Polymethyl methacrylate (PMMA), also called *plexiglass*, *Lucite*, or *Perspex* is a colorless clear transparent plastic with excellent outdoor stability if UV absorbers are added to the polymer—otherwise it yellows on exposure to sunlight. Like styrene it also unzips on heating to reform the monomer. It has poor scratch resistance but was the plastic of choice for early contact lenses.

11.5.8. Polyvinyl Acetate, Polyvinyl Alcohol

Vinyl acetate ($CH_2{=}CH(OCOCH_3)$) is polymerized to polyvinyl acetate (PVAc) which is used in adhesives and lacquers. Its major use, however, is in the preparation of polyvinyl alcohol (PVAl) which cannot be prepared from vinyl alcohol ($CH_2{=}CHOH$) which isomerizes into acetaldehyde (CH_3CHO).

Polyvinyl alcohol is a water soluble polymer which can be crosslinked into a gel by sodium borate ($Na_2B_4O_7$). This is shown in Fig. 11.5. Fibers made from PVAl can be made insoluble in water by crosslinking with formaldehyde, shown in Fig. 11.6. Such fibers are excellent substitutes for cotton because they absorb moisture (sweat) readily.

11.5.9. Polytetrafluoroethylene or Teflon

This polymer was discovered by accident. An old cylinder of gaseous tetrafluoroethylene (C_2F_4 B.P. $-$ 76°C) was found to have no gaseous pressure but still contained the original mass of material. When the cylinder was cut open a white waxy hydrophobic powder was found. The polymerization process is highly exothermic and it must be conducted with caution. The highly crystalline polymer is stable up to 330°C (its melting point), and is inert to strong acids, alkali, and organic solvents. It reacts with sodium leaving a carbon surface and NaF. This surface activation process allows Teflon to be bonded to other surfaces. The reaction of Teflon with hydroxyl free radicals (OH) can make the surface hydrophilic and bondable with ordinary adhesives (see Chapter 12).

$$Na_2B_4O_7 \rightarrow 2Na^+ + B_4O_7^{2-}$$

$$B_4O_7^{2-} + 9H_2O \rightleftharpoons 4B(OH)_4^- + 2H^+$$

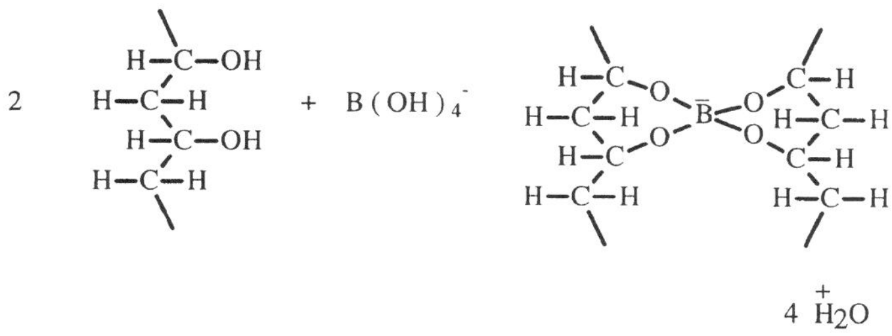

2 H–C–OH / H–C–H / H–C–OH / H–C–H + $B(OH)_4^-$ → H–C–O / H–C–H / H–C–O $\bar{B}$ O–C–H / H–C–H / O–C–H / H–C–H + 4 H_2O

FIGURE 11.5. The crosslinking of polyvinylalcohol by borax.

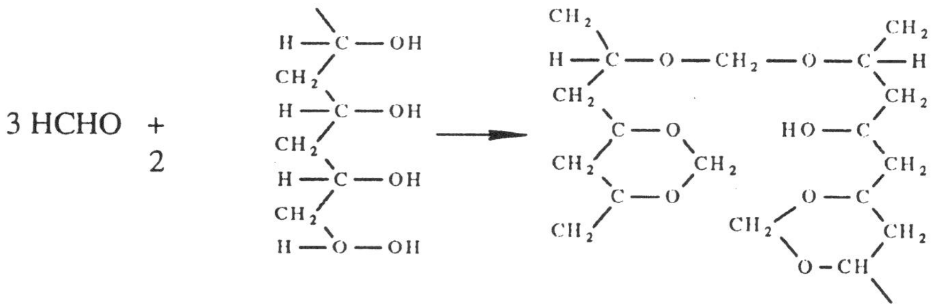

$$3\,HCHO + 2\;\; \begin{matrix} \diagdown \\ H-C-OH \\ CH_2 \\ H-C-OH \\ CH_2 \\ H-C-OH \\ CH_2 \\ H-O-OH \end{matrix} \longrightarrow \begin{matrix} CH_2 & & CH_2 \\ H-C-O-CH_2-O-C-H \\ CH_2 & & CH_2 \\ C-O & & HO-C \\ CH_2 \quad CH_2 & & CH_2 \\ C-O & & O-C \\ CH_2 & CH_2 & CH_2 \\ & & O-CH \end{matrix}$$

FIGURE 11.6. The crosslinking of polyvinylalcohol with formaldehyde.

Teflon tends to flow under pressure and is thus readily distorted. When filled with glass the composite is stabilized and can be machined to precise dimensions.

Teflon cannot be injection molded because of the high viscosity of the melt and must therefore be formed by a compression of its powders. Another fluorinated polymer of comparable properties to Teflon is a blend of PTFE and polyhexafluropropylene (FEP) made by polymerization of perfluoropropylene (C_3F_6). This plastic is not as thermally stable as Teflon (M.P. = 290°C) but it is less opaque than Teflon and can be extruded, injection molded, or blow molded and thus presents some advantage over Teflon in particular applications.

A Teflonlike surface is made when polyethylene bottles are blown with nitrogen containing about 1% F_2. This makes the bottles less permeable to organic solvents and thus increases its usefulness.

11.6. CONDENSATION POLYMERS

Some condensation polymers are listed in Table 11.2.

TABLE 11.2
Some Condensation Polymers

Polymer	Reaction
Nylon 66	$HOOC(CH_2)_4COOH + HN_2-(CH_2)_6-NH_2 \downarrow -[-NH-(CH_2)_6-NH-\underset{\Vert \atop O}{C}-(CH_2)_4-\underset{\Vert \atop O}{C}-]-$
Polyester	$HO-CH_2-CH_2-OH + HOOC-(C_6H_4)-COOH \downarrow HO-CH_2-CH_2-O-[-\underset{\Vert \atop O}{C}-(C_6H_4)-\underset{\Vert \atop O}{C}-O-CH_2-CH_2-O-]-\underset{\Vert \atop O}{C}-$
Polyurethane	$H-[-O-CH_2-CH_2-]-OH + CH_3-(C_6H_3)-(NCO)_2 \downarrow (CH_3)(NCO)(C_6H_3)-NH-C-[-O-CH_2-CH_2-]-O-C-NH-(C_6H_3)(CH_3)NCO$

11.6.1. Nylon

Nylon is classed as a polyamide polymer prepared by the condensation of a dicarboxylic acid [HOOC–$(CH_2)_n$–COOH] and a diamine [H_2N–$(CH_2)_m NH_2$]. The plastic is characterized by the values of n and m, i.e., nylon m, n. Thus nylon 6,6 (M.P. = 250°C) has good tensile strength, elasticity, toughness, abrasion resistance, and has use as a fiber as well as a plastic. The melting temperatures range of nylon is from 250° to 300°C. The aromatic polyamides have very high melting points (>500°C) and unusually high strength/weight ratio, of which the fiber Kevlar is an example.

11.6.2. Polyester

A condensation of a dicarboxylic acid and a diol results in a polyester

$$\mathrm{HOOC{-}(CH_2)_x{-}COOH + HO{-}CH_2{-}(CH_2)_y{-}CH_2{-}OH}$$
$$\rightarrow \mathrm{{-}OCH_2{-}(CH_2)_y{-}CH_2O(OC{-}(CH_2)_x{-}COOCH_2(CH_2)_y{-}CH_2{-}O)_nOC{-}(CH_2)_x{-}}$$
$$+ n\mathrm{H_2O} \qquad \textbf{(11.16)}$$

The aliphatic polyester has a melting point of about 65°C whereas the aromatic substituted dicarboxylic acid has a melting point of 265°C. Thus, the polyester polyethylene terephthalate (PETP) is commercially one of the most popular polymers marketed as Terylene or terene.

11.6.3. Polycarbonates and Epoxides

The condensation of a diphenol, (bis-phenol-A), with dicarboxylic acid

$$\mathrm{HO{-}C_6H_4{-}C(CH_3)_2{-}C_6H_4{-}OH + HOOC{-}(CH_2)_n{-}COOH} \rightarrow$$
$$\mathrm{{-}\overset{O}{\overset{\|}{C}}{-}\left(O{-}C_6H_4{-}C(CH_3)_2{-}C_6H_4{-}O{-}\overset{O}{\overset{\|}{C}}{-}(CH_2)_n{-}\overset{O}{\overset{\|}{C}}\right)_x{-}O{-} + xH_2O} \qquad \textbf{(11.17)}$$

forms the polycarbonate. These polymers melt at about 265°C and have very high impact strength making them useful for helmets, and safety shields.

The reaction of epichlorohydrin with bis-phenol-A forms the diepoxy or diglycidyl ether

$$2\overset{\displaystyle O}{\overbrace{CH_2-CH}}-CH_2-Cl + HO-C_6H_4-\underset{\displaystyle CH_3}{\overset{\displaystyle CH_3}{\underset{|}{\overset{|}{C}}}}-C_6H_4-OH$$

$$\rightarrow \overset{\displaystyle O}{\overbrace{CH_2-CH}}-CH_2-O-C_6H_4-\underset{\displaystyle CH_3}{\overset{\displaystyle CH_3}{\underset{|}{\overset{|}{C}}}}-C_6H_4-O-CH_2-\overset{\displaystyle O}{\overbrace{CH_2-CH_2}} + 2HCl \qquad (11.18)$$

The reaction of the diglycidyl bis-phenyl ether with a polydiol or polydiamine will result in a hardened thermosetting resin. Other thermosetting polymers are discussed in the next section.

11.7. THERMOSETTING POLYMERS

Polymers which form 3-D network solids are thermosetting and decompose when heated and thus cannot be reshaped once they have set. A selection of the common polymers is listed in Table 11.3.

11.7.1. Phenol Formaldehyde (Bakelite)

The first industrial plastic was developed by Baekeland in about 1907 and was called Bakelite. This was prepared by the reaction of phenol and formaldehyde in the presence of catalysts.

$$C_6H_5OH + HCHO \longrightarrow HOC_6H_4CH_2OH \longrightarrow HOCH_2-C_6H_3(OH)-CH_2-C_6H_3(OH)-CH_2OH$$

(11.19)

TABLE 11.3
Some Thermosetting Resins

Phenol formaldehyde

$$
\begin{array}{l}
C_6H_5\text{—}OH + 2HCHO \rightarrow HO\text{—}(C_6H_3)(CH_2OH)_2 \rightarrow HO\text{—}(C_6H_3)\text{—}CH_2\text{—}(C_6H_3)\text{—}OH \\
\qquad\qquad\qquad\qquad\qquad\qquad\qquad\qquad\qquad\quad\; | \qquad\qquad\qquad\; | \\
\qquad\qquad\qquad\qquad\qquad\qquad\qquad\qquad\qquad\quad CH_2 \qquad\qquad\quad CH_2 \\
\qquad\qquad\qquad\qquad\qquad\qquad\qquad\qquad\qquad\quad\; | \qquad\qquad\qquad\; | \\
\qquad\qquad\qquad\qquad\qquad\qquad\qquad\qquad\qquad (C_6H_3)\text{—}CH_2\text{—}(C_6H_3) \\
\qquad\qquad\qquad\qquad\qquad\qquad\qquad\qquad\qquad\quad\; | \qquad\qquad\qquad\; |
\end{array}
$$

Urea-formaldehyde

$$
\begin{array}{l}
O{=}C(NH_2)_2 + HCHO \rightarrow \text{—}NH\text{—}CO\text{—}N\text{—}CH_2\text{—}NH\text{—}CO\text{—}N\text{—}CH_2\text{—}NH\text{—}CO\text{—}N\text{—}CH_2\text{—} \\
\qquad\qquad\qquad\qquad\qquad\qquad\quad | \qquad\qquad\qquad\qquad\quad | \qquad\qquad\qquad\qquad\quad | \\
\qquad\qquad\qquad\qquad\qquad\qquad CH_2 \qquad\qquad\qquad\quad CH_2 \qquad\qquad\quad \text{—}CH\text{—}NH\text{—}CO\text{—} \\
\qquad\qquad\qquad\qquad\qquad\qquad\quad | \qquad\qquad\qquad\qquad\quad | \\
\qquad\qquad\qquad\qquad\qquad\qquad OH \qquad\qquad \text{—}NH\text{—}CO\text{—}N\text{—}CH_2\text{—}NH\text{—}CO\text{—}N\text{—}CH_2\text{—} \\
\qquad\qquad\qquad\qquad\qquad\qquad\qquad\qquad\qquad\qquad\quad | \qquad\qquad\qquad\qquad\quad |
\end{array}
$$

Melamine formaldehyde

$$
\begin{array}{l}
(C_3N_3)(NH_2)_3 + 3HCHO \rightarrow \text{—}NH\text{—}CH_2\text{—}NH\text{—}(C_3N_3)\text{—}NH\text{—}CH_2\text{—}NH\text{—} \\
\qquad\qquad\qquad\qquad\qquad\qquad\qquad\qquad\quad | \\
\qquad\qquad\qquad\qquad\qquad\qquad\qquad\qquad NH\text{—}CH_2\text{—}NH\text{—}(C_3N_3)\text{—}NH\text{—}CH_2\text{—} \\
\qquad\qquad\qquad\qquad\qquad\qquad\qquad\qquad\qquad\qquad\qquad\qquad\quad | \\
\qquad\qquad\qquad\qquad\qquad\qquad\qquad (C_3N_3)\text{—}NH\text{—}CH_2\text{—}NH \\
\qquad\qquad\qquad\qquad\qquad\qquad\qquad\quad |
\end{array}
$$

Epoxy polymer

$$
\begin{array}{l}
\qquad\qquad\quad CH_2\text{——}CH\text{—}CH_2\text{—}Cl + HO\text{—}(C_6H_4)\text{—}C(CH_3)_2\text{—}(C_6H_4)\text{—}OH \\
\qquad\qquad\qquad\;\; \diagdown\;\diagup \\
\qquad\qquad\qquad\quad O \qquad\qquad\qquad\qquad\quad \downarrow \\
CH_2\text{——}CH\text{—}CH_2\text{—}[\text{—}O\text{—}(C_6H_4)\text{—}C(CH_3)_2\text{—}(C_6H_4)\text{—}O\text{—}CH_2\text{—}CH\text{—}CH_2\text{—}]\text{—}O \\
\quad \diagdown\;\diagup \qquad\qquad\qquad\qquad\qquad\qquad\qquad\qquad\qquad\qquad\qquad | \\
\quad\;\; O \qquad\qquad\qquad\qquad\qquad\qquad\qquad\qquad\qquad\qquad\qquad\; OH
\end{array}
$$

When heated in excess formaldehyde, crosslinking occurs and the resin novolac is formed for $P/F \sim 1.25$.

11.7.2. Urea Formaldehyde

Urea

$$
\left(NH_2\text{–}\overset{\displaystyle O}{\overset{\|}{C}}\text{–}NH_2 \right)
$$

reacts with formaldehyde to form a crosslinked resin which is an inexpensive adhesive

$$
NH_2\text{—}\overset{\displaystyle O}{\overset{\|}{C}}\text{—}NH_2 + 2HCHO \rightarrow HOCH_2\text{—}\overset{\displaystyle O}{\overset{\|}{C}}\text{—}NHCH_2OH \qquad (11.20)
$$

On further addition of urea and HCHO, $H(NHCO\text{–}NH\text{-}CH_2)_n\text{–}OH$ is formed. With an acid catalyst it is possible to produce a foam product known as urea formaldehyde foam insulation (UFFI) having a thermal conductivity, K, of about 0.022 $Wm^{-1} K^{-1}$.

In 1977 the Canadian government subsidized the introduction of UFFI in older homes to conserve energy.* The UFFI proved to be unstable in some cases due to improper installation, and as a result formaldehyde levels in some homes exceeded the threshold limit value (TLV) of 0.10 ppm (120 $\mu g/m^3$). Ammonia was able to neutralize the acid and it was also shown that the water soluble polymeric amine, polyethylene imine, could remove the liberated formaldehyde. Nonetheless—the Canadian government then paid the homeowners an estimated $272 million ($5000 to 57,700 homes) to remove the UFFI. The urea formaldehyde resin is commonly used as the adhesive resins in plywood and particle board, and will initially release formaldehyde if not sealed. As more composite wood products find their way into buildings, greater concern about indoor air is warranted.

11.7.3. Polyurethane

This condensation polymer is unique insofar as it can be a coating and varnish, a soft or hard foam, a resilient or rigid elastomer (rubber) as well as an adhesive. It is prepared by the reaction of a diisocyanate (OCNRNCO) with a diol (HOR′OH) where R can be an aromatic radical such as toluene (TDI-2,4, toluenediisocyanate).

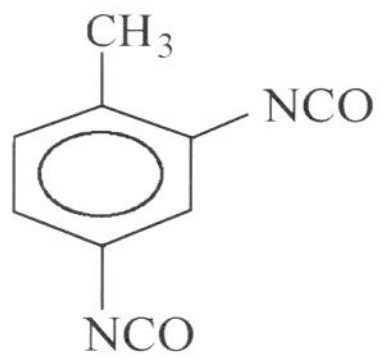

CH_3
NCO
NCO

and R′ is an aliphatic radical $(CH_2)_n$ where the length n determines strength, toughness and elasticity of the plastic. The reaction is

$$\text{OCN–R–NCO} + \text{HO–R}'\text{–OH} \rightarrow \text{–R}'\text{–[OCO–NH–R–NH–CO–OR}'\text{–]}_x\text{O–CO} \quad (11.21)$$

For the preparation of foams, the R component is a polyether or polyester with reactive end groups of hydroxyl and carboxyl. The reaction is

$$\text{–RCOOH} + \text{OCN–R}'\text{–} \rightarrow \text{–R–CO–NH–R}' + CO_2 \quad (11.22)$$

where the liberated CO_2 foams the plastic into an open or closed cell sponge with densities of 25–50 g/dm^3 and which is often used in upholstery. The hard and rigid foams, having a density of 50–300 g/dm^3, are used as insulation and elastomers.

It is possible to replace the air in inflatable tires by *polyurethane* foam. This is feasible for low speed vehicles used in road construction, service equipment, snowplows, street sweepers as well as many other applications. The two components are blended together to produce the resilient foam in the tire which is then not susceptible to flats or punctures, a feature which reduces down time and tire replacement costs. Poly-

*Each of the 100,000 homeowners was given $500 towards the cost of adding UFFI.

urethane foam (closed cell) has a thermal conductivity of 0.022 Wm^{-1} K^{-1} and is usually covered with aluminum foil to reduce the heat loss due to the transmission of radiation.

11.8. GLASS TRANSITION TEMPERATURE

The melting point of a polymer is not a unique value unless it can be formed into a crystalline solid. The amorphous glassy solid is really a supercooled liquid. A polymer which does not have long-range order cannot exist in a crystalline state. As the temperature of an amorphous plastic is increased, the polymer chains begin to achieve segmental mobility. This is called the *glass transition temperature* (T_g), and the material is in a rubbery state. On further heating the polymer chains begin to move and have molecular mobility—the plastic begins to flow. A graph showing the transition in terms of the variation of the specific volume (the reciprocal of density) as a function of temperature is shown in Fig. 11.7. The T_g and melting points of some polymers are listed in Table 11.4.

The value of T_g increases as molecular weight of a polymer increases or as the branching or crosslinking increases. Thus, for PS #3, Table 11.4, $T_g = 100°C$ which is much higher than for PE #5, $T_g = -125°C$. Similarly the difference in T_g between polybutadiene (−102°C) #1 and #2 polyisoprene (−75°C) shows the effect of replacing a H by CH_3 in the side of a chain.

The T_g of a polymer can be reduced by the addition of a plasticizer to the solid plastic. This reduces the van der Waals interaction between the polymer chains and allows the molecules to move. The plasticizer may be considered as an internal lubricant. The plasticizer can also be considered to increase the free volume of the polymer by allowing increased motion of the chain ends, the side chains or even the main chain. Another possible mechanism by which the plasticizer lowers the T_g is in terms of the solvent/solute system that forms when the plasticizer can be considered to solubilize the polymer. The plasticizer is usually a low volatile, low molecular weight organic compound which is compatible with the polymer.

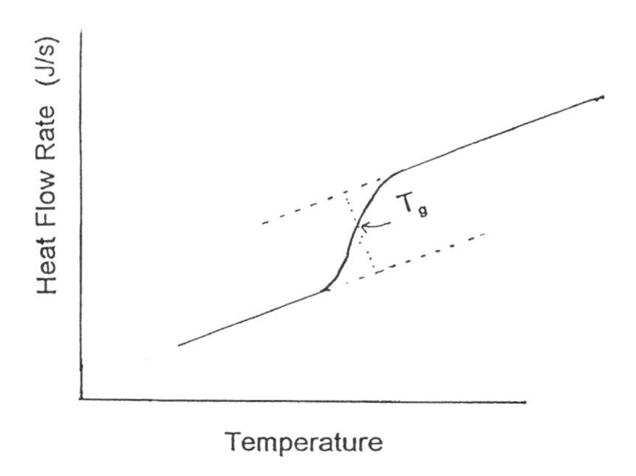

FIGURE 11.7. The glass transition temperature is indicated by a change in heat flow of the material while the temperature increases linearly with time.

TABLE 11.4
The Glass Transition Temperature (T_g) and Melting Temperature (T_m) of Selected Polymers

Polymer	Structural unit	T_g (°C)	T_m (°C)
1. Polybutadiene	$—CH_2—CH{=}CH—CH_2—$	*cis* −102 *trans* −58	6 100
2. Polyisoprene	$—CH_2—C(CH_3){=}CH—CH_2—$	−75	65
3. Polystyrene	$—CH(C_6H_5)—CH_2—$	100	240
4. Nylon 6,6	$—NH—(CH_2)_6—NH—O—CO—(CH_2)_4CO—O—$	50	270
5. Polyethylene	$—CH_2—CH_2—$	High density −125	140
6. Polypropylene	$—CH(CH_3)—CH_2—$	Atactic −13 Isotatic −8	 200
7. Polymethacrylate	$—CH_2—CH(C(=O)—OCH_3)—$	5	
8. Polymethyl-methacrylate	$—CH_2—C(CH_3)(C(=O)—OCH_3)—$	72	
9. Polyvinyl chloride	$—CH_2—CHCl—$	80	310
10. Polyethylene terephthalate	$—O—CH_2—CH_2—O—C(=O)—C_6H_4—C(=O)—$	70	260
11. Polycarbonate bisphenol A	$—O—C_6H_4—C(CH_3)_2—C_6H_4—O—C(=O)—$	150	225

11.9. ELASTOMERS

Flexible plastics composed of polymers with T_g well below room temperature are classed as elastomers or rubbers. Natural rubber was known to the natives of South America for centuries, though it was not until Goodyear's discovery of vulcanization in 1839 that it became a practical product. Prior to this, rubber was used for water

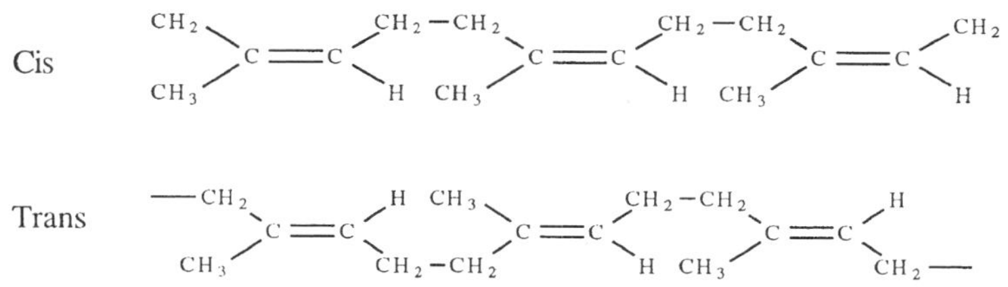

Cis

$$\ce{CH_2 \backslash C=C / CH_2-CH_2 \backslash C=C / CH_2-CH_2 \backslash C=C / CH_2}$$

CH_3 H CH_3 H CH_3 H

Trans

—CH_2 H CH_3 CH_2-CH_2 H

C=C   C=C   C=C

CH_3 CH_2-CH_2 H CH_3 CH_2—

FIGURE 11.8. Structures of natural rubber: cis, natural, herea and trans; gutta, percha, balata.

proofing boots, clothing, and other weather-proofing surfaces. Goodyear showed that sulfur crosslinked the rubber and made it a manageable product. The polymer is based on the monomer isoprene

$$CH_2{=}\underset{\displaystyle CH_3}{\underset{|}{C}}{-}CH{=}CH_2$$

which results in *cis* and *trans* forms (see Fig. 11.8). Of the two forms, the *trans* is less elastic because of the more ordered structure and more close packing of the molecules.

In general, *elastomers* differ from plastics only because the elastomer is in a mobile "liquid" state whereas the plastic is in a glassy state. The transition between these two states occurs at the glass transition temperature T_g when the glassy state changes into the rubbery state. Below this temperature the molecules are frozen into position and held in place by van der Waals forces.

Because of the residual carbon–carbon double bonds (C=C) in natural rubber, it is readily degraded by ozone, which adds to double bonds forming ozonides that eventually decompose, splitting the polymer chain.

Synthetic rubbers are made from chloroprene and butadiene which form neoprene and buna, respectively. The copolymer of acrylonitrile with butadiene, (1,3) is known as nitrile rubber and styrene with butadiene (1,3) is Buna S. The combination of acrylonitrile, butadiene and styrene in various formulations is used to form the thermoplastic ABS.

Some fluorinated polymers which show exceptional thermal stability and chemical inertness are Kel-F elastomers, made of a copolymer of chlorotrifluoroethylene-vinylidene fluoride $ClCF{=}CF_2/CH_2{=}CF_2$ and Viton, a copolymer of hexafluoropropylene and vinylidene fluoride $CF_2{=}CF{-}CF_3/CH_2{=}CF_2$.

Though stable at high temperature these fluorocarbons show limited low temperature flexibility. Silicone rubbers are made from dimethyl dichlorosilane which under controlled hydrolysis form oils, gels, and rubbers.

Silicone rubber is more permeable to oxygen and carbon dioxide than most other polymers. A comparison of the permeability of these gases and water through various plastics is given in Table 11.5.

Saran plastic shows the lowest permeability to O_2, CO_2, and very low for water. This feature makes Saran wrap an excellent packaging material for food in which freshness and flavor are to be preserved.

Silicone rubber, however, shows the highest permeability rates for these gases, and in fact, silicone rubber is used in blood oxygenators required for open heart surgery. It

TABLE 11.5
The Gas Permeability of Various Plastics at 25°C

	$P_r = \frac{\text{mL gas (NTP) cm}}{\text{sec, cm}^2, \Delta P\text{ (cmHg)}} \times 10^9$		
Film	O_2	CO_2	H_2O
1. Polyvinylidene chloride (Saran)	6.3×10^{-4}	4.8×10^{-4}	5.3
2. Monochlorotrifluorethylene (Trithene A)	9.2×10^{-4}	0.010	0.8
3. Polyester (Mylar)	6.7×10^{-3}	0.012	32
4. Cellulose acetate	0.067	0.35	1900
5. Opaque high density polyethylene	0.087	0.218	5.3
6. Polypropylene	0.115	0.400	15
7. Clear high density polyethylene	0.14	0.845	
8. Polystyrene	0.1	0.98	152
9. Low density polyethylene	0.35	1.09	25
10. Tetrafluoroethylene (Teflon)	0.67	1.88	67
11. Ethyl cellulose (Ethocel)	0.98	4.07	1600
12. Natural rubber	2.5	13	
13. Fluorosilicone	1.1	64	
14. Nitrile silicone	8.5	67	
15. Silicone rubber (Silastic 372)	60	325	3600
16. MEM-213 (G.E. silicone block copolymer)	16 ($N_2$7)	97 ($H_2$21)	

is also used in extended wear contact lenses since the transport of O_2 and CO_2 through the lens allows the cornea to respire. Thus cloudiness and rainbows are not generally experienced, even after continuous wear for a month.

The permeability of various gases through silicone rubber is given in Table 11.6 and shows a broad variation. The permeability (P_r) of a gas through a plastic film is usually considered as equal to the product of the solubility (S) of the gas in the plastic and the rate of diffusion (D) of the gas in the plastic (actually the diffusion coefficient). Thus

$$P_r = D\frac{\left(\frac{\text{cm}^2}{\text{sec}}\right) \times 10^6 \times S\left(\frac{\text{mL NTP}}{\text{mL atm}}\right)}{76\text{ cm/atm}} \tag{11.23}$$

Some values of P_r, D and S for O_2 and CO_2 at various temperatures are given in Table 11.7. When comparison is made for other gases and other plastic films it becomes obvious that the solubility of gases in silicone rubber is not much different from other elastomers. Hence the higher permeability of gases such as O_2 and CO_2 in silicone rubber is primarily due to higher diffusion coefficients due to more flexible O–Si–O bonds and to a much lower T_g (T_g, silicone rubber, is -123°C).

The high permeability of oxygen and CO_2 through silicone rubber suggests its possible use as an artificial gill. This is demonstrated in Fig. 11.9 in which a hamster lived in a (0.03 m^2) silicone rubber lined cage (30 L) submersed in air saturated water. When 35 L/min of the air saturated water is pumped around the cage, oxygen can be

TABLE 11.6
The Permeability of Various Gases through Dimethyl–Silicone Rubber at 25°C

$$P_r = \frac{\text{mL gas (NTP) cm}}{\text{sec, cm}^2, (\Delta P \text{ cmHg})} \times 10^9$$

Gas	P_r	Gas	P_r
H_2	65	C_3H_8	410
He	35	n-C_4H_{10}	900
NH_3	590	n-C_5H_{12}	2000
H_2O	3600	n-C_6H_{14}	940
CO	34	n-C_8H_{18}	860
N_2	28	n-$C_{10}H_{22}$	430
NO	60	Freon 11	1500
O_2	60	Freon 12	138
H_2S	850	H_2CO	1110
Ar	60	CH_3OH	1390
CO_2	325	Acetone	1980
N_2O	435	Pyridine	1910
NO_2	760	Benzene	1080
SO_2	1500	Toluene	913
CS_2	9000	$COCl_2$	1500
CH_4	95	Phenol	2100
C_2H_6	250	Freon 22	382
C_2H_4	135	Freon 114	211
C_2H_2	2640	Freon 115	51
CCl_4	5835	Xe	171

TABLE 11.7
Effect of Temperature on P_r, D, and S for O_2 and CO_2 in Silicone Rubber

Temp. (°C)		$P_r \times 10^9$	$D \times 10^6$	S
O_2	28	62	16	0.31
	−40	20	3.9	0.39
	−75	0.74	0.0012	47
CO_2	8	323	11	2.2
	−40	293	2.7	8.2
	−75	22	0.0022	770

Note: Unit for $P_r = \frac{\text{mL gas (NTP) cm}}{\text{sec, cm}^2 (\Delta P \text{ cmHg})} \times 10^9$

$$D = \frac{\text{cm}^2}{\text{sec}} \times 10^6$$

$$S = \frac{\text{mL (NTP)}}{\text{mL atm}}$$

FIGURE 11.9. Hamster in submerged cage fitted with silicone rubber membrane sides.

supplied to the 30 g hamster at a rate of 2.5 mL/min which is enough for its needs. The CO_2 is removed by the water flow and though the experiment could be continued for days, the molding of the food limited the duration of the experiment.

It may be noted in Table 11.6 that the permeability of O_2 is about twice that of nitrogen. Hence it is possible to obtain air enriched in oxygen by collecting the gases which pass through several large membranes. The use of oxygen instead of air is advantageous in many processes such as combustion, steel manufacture, heating, welding and many others. For example, the removal of N_2 from the air used in the burning of natural gas results in a higher temperature and therefore more heat for the same amount of gas burnt. This is because when nitrogen is present, some of the heat of combustion is used to heat up the nitrogen which is both reactant and product.

This effect is even more pronounced if the nitrogen were to be removed from the air used in an internal combustion engine. The result would be a higher temperature of combustion and less work expended in the compression of the gases (see Exercise 11.8).

It is difficult to obtain continuous sheets of silicone rubber having a large area free of holes. It is, however, easier to draw capillary tubes and to stack these together giving very large areas. However, silicone rubber is not a thermoplastic and capillary tubing cannot be extruded. To get around this difficulty, General Electric prepared a block copolymer of silicone rubber and polycarbonate and because the polycarbonate is thermoplastic—the plastic (MEM 213) made from the copolymer can be molded and small bore capillary tubing can be readily fabricated. The polymer has most of the properties of pure silicone rubber with permeability rates of about 60% of the pure material.

11.10. MECHANICAL STRENGTH OF PLASTICS

The mechanical properties of materials are usually studied by means of tensile testing machines or dynamometers. The stress–strain curve obtained characterizes the plastic and determines its usefulness for specific applications.

Some design properties of common plastics are shown in Fig. 11.10. Shown also are relative costs as well as useful temperature range. Many of these properties can be improved by incorporating solid fillers into the plastics. Fillers which increase the mechanical strength are called *active fillers*, and include carbon black, titania, limestone, kaolin, silica, and mica. Their application to rubber and elastomers has been practiced for many years. The T_g of elastomers are usually increased by the addition of fillers. Good wetting of the filler by the polymer is essential for maximum effect. Thus coupling agents are used (see Chapter 12) which bond to the solid and can react with the polymer. The size and shape of the filler particles also has an influence on its effectiveness. Thus mica, which is a layer lattice, is not spherical particles but thin platelets which can be split into thinner particles by ultrasonics. The length to thickness dimensions is called the *aspect ratio*. High aspect ratio (HAR) mica is much superior as a filler to ordinary mica or comparable amounts of silica or other fillers. The role of reinforcing fibers and binders in composite materials is discussed in Chapter 16.

11.11. FIRE RETARDANTS IN PLASTICS

Plastics composed of polymer which have carbon and hydrogen are combustible. During the flammable process both thermal decomposition and combustion occur. A substance is classified as noncombustible if it does not produce flammable vapors when heated to 750°C. Few organic polymers can pass this test. Hence most plastics burn, producing combustion products which can be toxic. Some of these gases are listed in Table 11.8. In the case of hydrogen cyanide (HCN), the amount produced is from 20 to 50% of the nitrogen present in the polymer. The major fire hazard is not the toxic gases but the smoke and lack of oxygen. Thus, smoke and fire retardants are essential ingredients in the formulation of plastics. The relative decrease in light transmission or degree of obscuration for some materials as determined in a specific apparatus is given in Table 11.9. The addition of flame retardants to materials may reduce fire but can at

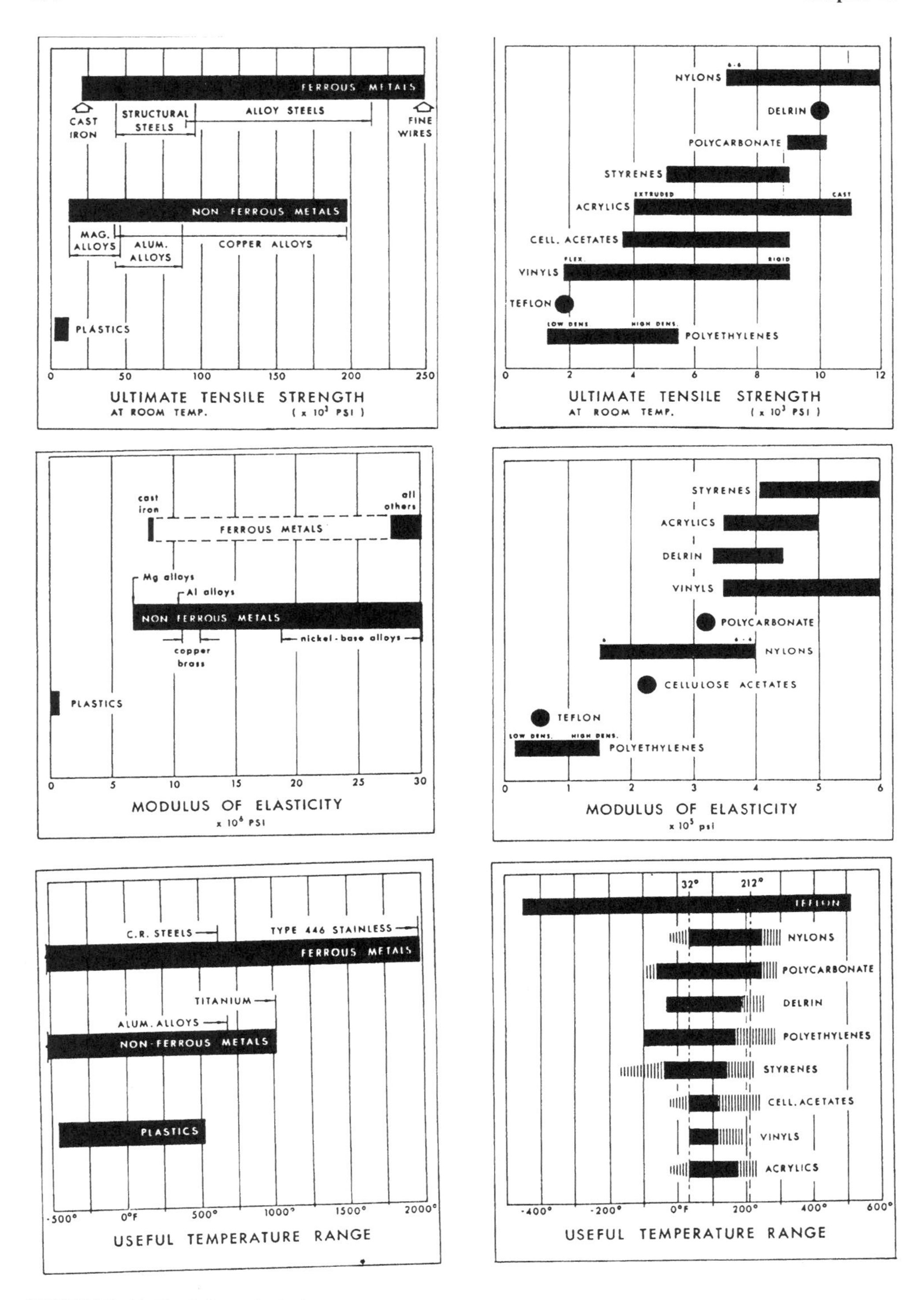

FIGURE 11.10. Selected design properties of some plastics as compared to other common materials.

TABLE 11.8
Toxic Degradation Products from Combustible Materials in Air

CO and CO_2	From all substances containing carbon
NO_x	Polyurethanes
HCN	Wool (150), silk, nylon (100), polyurethanes (40), polyacrylonitrile (200)
SO_2	Rubber
Hydrohalides	PVC, fluorinated plastics, polyvinylidene chloride
Phosgene	Chlorohydrocarbons
NH_3	Melamine, nylon, urea formaldehyde
Benzene	Styrene
Phenol	Phenol formaldehyde
CO	PUF (600), nylon (450), wool (300)

Values in parentheses are average values in mg/g of material.

times increase the formation of smoke. The ease with which a substance will burn is determined by the minimum O_2 concentration (in N_2 as %) which will support combustion. This is called the limiting oxygen index (LOI) and some selected valves are given in Table 11.10. Thus those materials with LOI $\leqslant 21$ are combustible in air and must be treated to increase the LOI values.

Fire retardants are additives to plastics and are usually based on some of the following elements: Al, B, Br, Cl, Mg, N, P, Sb, Sn, Zn. Halogen compounds (RX) produce halogen atoms (X) which act as chain terminators in the reacting vapors. Bromine compounds are often used but the formation of HBr in a fire makes it a corrosive retardant. A more inert fire retardant is alumina trihydrate ($Al_2O_3 \cdot 3H_2O$) which also acts as a smoke suppressant. It absorbs heat while liberating water at 230° to 300°C. Another inorganic fire retardant is zinc borate ($ZnO \cdot B_2O_3 \cdot H_2O$) which liberates water at 245°–380°C and is used to supplement halogen retardants. A list of selected fire retardants is given in Table 11.11. The increase in LOI of 10% (from 18%–28%) represents an effective application of retardants.

TABLE 11.9
Relative Obscurance due to Smoke Formation During the Combustion of Various Materials

Material	Obscuration (%)
Oak	2
Pine	48
Acrylic	2
Polystyrene	100
PVC	100
Polyester	90
Plaster board	1
Flame retarded acrylic	97
Polyester	99
Plywood	15

TABLE 11.10
Limiting Oxygen Index (LOI) Values of Selected Materials

Material	LOI
Polyurethane foam	16.5
Polymethymethacrylate	17.3
Polyethylene	17.4
Polypropylene	17.4
Polystyrene	18.0
Acrylic fiber	18.2
Cotton	18.4
Nylon fiber	20.1
Polyester fiber	20.6
Oak	23
Polycarbonate	25.0
Wool	25.2
Nylon 6,6	28.7
Kevlar	29
Polyvinylidene fluoride	43.7
Polyvinyl chloride	47
Teflon	95
Polyvinylidene chloride	60
Carbon	65

The flame characteristics of a plastic and a measure of its density can often be used to identify the polymer. A short list of tests of a few plastics is given in Table 11.12. In general aromatic substances burn with smokey flames. Chloride can be tested with the plastic coated copper wire which shows a green color in the colorless part of a Bunsen flame.

The growth of the polymers and plastics industries has meant that a major fraction of the workforce in developed countries is either directly or indirectly employed by plastic related jobs. As more stable and less costly plastics are developed, more

TABLE 11.11
Selected Fire Retardants

Al_2O_3—$3H_2O$	Alumina trihydrate
$ZnOB_2O_3$—H_2O	Zincborate
Sb_2O_3	Antimony oxide
NH_2CONH_2	Urea
$(NH_4)_2HPO_4$	Diammonium hydrogen phosphate
$Mg(OH)_2$	Magnesium hydroxide
$Br_3C_6H_4$—$CH{=}CH_2$	Tribromostyrene
C_6Br_5—CH_2—O—CO—$CH{=}CH_2$	Pentabromobenzylacrylate
MoO_3	Molybdenum trioxide
CuC_2O_4	Cupric oxalate

TABLE 11.12
Combustion and Density Tests for Plastics Identification

Polymer	Density* (g/cm^3)	Type	Odor	Color	Character
PE	0.91–0.98	TP	Candle	Yellow	Burns, melts, drips
PP	0.89–0.92	TP	Wax, candle wax	Yellow	Burns, melts, drips
PS	1.05–1.07	TP	Styrene	Smoky	Burns
PMMA	1.24	TP	Methanol	Yellow	Burns slowly
PET	1.39	TP	Aromatic	Smoky	Slightly acidic fumes
PC	1.2	TP	Benzene	Smoky	Neutral vapors, difficult to ignite, SE
PAN	1.2	TP	HCN	Smoky	
PF	1.28	TS	Phenol	Yellow smoky	Basic Fumes, SE, difficult to ignite
PU	1.2	TS	Acid	Yellow	Acidic or basic fumes
PVC	1.39	TP	Acid	Yellow	Acidic fumes, SE
N	1.14	TP	Ammonia	Bluish	Burns, basic fumes, SE
ABS	1.0	TP	Styrene	Yellow	Burns

*Values $\pm$0.1 to 0.2. SE, self extinguishing; TP, thermoplastic; TS, thermosetting; PE, polyethylene; PP, polypropylene; PS, polystyrene; PMMA, polymethylmethacrylate; PET, polyethylene terephthalate; PC, polycarbonate; PAN, polyacrylonitrile; PF, phenolformaldehyde; PU, polyurethane; PVC, polyvinylchloride; N, nylon; ABS, acrylonitrile, butadiene, styrene.

applications are found and growth continues. Because polymers and plastics are based on petroleum and since petroleum is a limited resource, it is essential that continued efforts be made to recycle our plastic wastes—something that is slowly being realized.

EXERCISES

1. From the data given in Section 11.1 calculate the annual production of rubber in 1990.
2. The table below gives the fraction of molecules of a polymer sample having a given average molecular weight. Calculate the number average molecular weight, $\bar{M}_n$ and the weight average MW, $\bar{M}_w$.

MW	10,000	14,000	16,000	19,000	22,000	24,000	28,000	32,000
Fraction	0.12	0.14	0.16	0.22	0.12	0.1	0.08	0.06

3. What conclusion can you reach if both $\bar{M}_w$ and $\bar{M}_n$ are determined to be identical?
4. The MW of a polymer dissolved in a solvent was determined from viscosity measurements at 25°C for various concentrations

C (g/100 mL)	0.25	0.50	1.00	2.00
η/η_0	1.36	1.8	2.8	6.1

The ratio η_{sp}/C is to be plotted against C and extrapolated to zero concentration in order to obtain the intrinsic viscosity $[\eta]$. The constants for Eq. (11.7) are $K = 3.8 \times 10^{-4}$ and $\alpha = 0.92$ for this polymer–solvent–temperature system when the concentration is in g/100 mL. Calculate the MW of the polymer.

5. Polyvinylalcohol (PVA) is soluble in water. What will be the freezing point of a 3% solution of PVA (MW 50,000 g/mol) in water? [*Note:* 1 molal solution depresses the F.P. by 1.86°C.]
6. Write the chemical reaction showing the formation of the following polymers from initial reactants: (a) nylon, (b) lexan, (c) terylene, (d) polyurethane.
7. Floor tiles of plasticized PVC can be made more flexible, more scratch resistant, to have a longer lifespan and have increased color fastness when treated with γ-radiation. Explain.
8. The permeability equation of a gas through a membrane is

$$N = A(P_i X_i - P_0 X_0)P_r/l$$

   where N = mL/s (NTP) gas flow; l = film thickness in cm; P_r = permeability; P_i = pressure of feed gas in cmHg; X_i = mole fraction of diffusing component in feed gas; P_0 = pressure of product gas in cm of Hg; X_0 = mole fraction of component in product gas; A = area in cm^2. For air $P_i = 76$ cm, $X_i = 0.2$.
   (a) If we assume l = 1 miL (0.025 mm), $P_0 = 38$ cm, $X_0 = 0.3$ calculate the area required for an internal combustion engine burning up to 150 mL/min of gasoline.
   (b) If P_i is increased to 3 atm, would the required area be reduced significantly?
9. A man at rest uses 300 mL O_2/min. Seawater contained 4 mL O_2/L. What area of membrane is required to act as an artificial gill if water is initially saturated at 160 mmHg of O_2 and can be depleted to 100 mm? What volume of water must be passed over the membrane each minute to achieve this oxygen flow to support a man under water?
10. It has been suggested that polymeric membranes be used to separate alcohol from water. What advantage would this have over distillation?
11. Distinguish between thermosetting plastics, thermoplastics and elastomers.
12. How is polyvinylalcohol prepared?

FURTHER READING

J. A. Brydson, *Plastic Materials*, 7th Ed., Butterworth-Heinemann, Newton, Massachusetts (1999).
A. B. Strong, *Plastics*, 2nd Ed., Prentice-Hall, Upper Saddle River, New Jersey (1999).
H. G. Elias, *An Introduction to Polymer Science*, VCH, New York (1997).
C. A. Harper, Editor, *Handbook of Plastics, Elastomers, and Composites*, 3rd Ed., McGraw-Hill, New York (1996).
R. R. Luise, *Applications of High Temperature Polymers*, CRC Press, Boca Raton, Florida (1996).
J. E. Mark, Editor, *Physical Properties of Polymers Handbook*, American Institute of Physics, Cincinnati, Ohio (1996).
J. C. Salamone, Editor, *Polymeric Materials Encyclopedia*, CRC Press, Boca Raton, Florida (1996).
J. R. Fried, *Polymer Science and Technology*, Prentice-Hall, Englewood Cliffs, New Jersey (1995).

A. Ravve, *Principles of Polymer Chemistry*, Plenum, New York (1995).
T. A. Osswald and G. Menges, *Materials Science of Polymers for Engineers*, Hanser-Gardner, Cincinnati, Ohio (1995).
M. Santappa, *State of the Art in Polymer Science and Engineering*, International Special Book, Portland, Oregon (1995).
G. E. Zaikov, Editor, *New Approaches to Polymer Materials*, Nova Science, Commack, New York (1995).
G. E. Zaikov, Editor, *Polymers in Medicine*, Nova Science, Commack, New York (1995).
G. E. Zaikov, Editor, *Flammability of Polymeric Materials*, Nova Science, Commack, New York (1995).
G. E. Zaikov Editor, *Kinetic and Thermodynamic Aspects of Polymer Stability*, Nova Science, Commack, New York (1995).
G. E. Zaikosv, Editor, *Degradation and Stabilization of Polymers: Theory and Practice*, Nova Science, Commack, New York (1994).
I. M. Ward and D. W. Hadley, *An Introduction to the Mechanical Properties of Solid Polymers*, Wiley, New York (1993).
H. Domininghaus, *Plastics for Engineers: Materials, Properties, Applications*, Hanser-Gardner, Cincinnati, Ohio (1993).
S. L. Rosen, *Fundamental Principles of Polymeric Materials*, 2nd Ed., Wiley, New York (1993).
H. Ulrich, *Introduction to Industrial Polymers*, Hanser-Gardner, Cincinnati, Ohio (1993).
L. H. Sperling, *Introduction to Physical Polymer Science*, 2nd Ed., Wiley, New York (1992).
W. R. Vieth, *Diffusion in and Through Polymers: Principles and Applications*, Hanser-Gardner, Cincinnati, Ohio (1991).
J. Brandrup and E. H. Immergut, Editors, *Polymer Handbook*, 3rd Ed.,. Wiley, New York (1989).
V. R. Gowariker, N. V. Viswanathan, and J. Sreedkar, *Polymer Science*, Wiling Eastern, New Delhi (1986).
F. W. Billmeyer, *Textbook of Polymer Science*, 3rd Ed., Wiley-Interscience, New York (1984).
A. Rudin, *The Elements of Polymer Science and Engineering*, Academic Press, Toronto (1982).
J. Chem. Educ., Nov. (1981), Full issue devoted to polymer chemistry.
J. A. Brydson, *Plastic Materials*, 2nd Ed., Newnes-Butterworths, London (1980).
E. A. MacGregor and C. T. Greenwood, *Polymers in Nature*, Wiley, New York (1980).
A. Tager, *Physical Chemistry of Polymers*, 2nd Ed. (English) MIR, Moscow (1978).
H. Lee, and K. Neville, *Handbook of Biomedical Plastics*, Pasadena Technology Press, Pasadena, California (1971).
G. R. Palin, *Plastics for Engineers*, Pergamon Press, Oxford (1967).
American Plastic Council, http://www.ameriplas.org/
History of Plastics, http://www.polymerplastics.com
About Silicone Rubber, http://www.chemcases.com/silicon/
Sources of polymers, http://www.polymer-search.com
Chemical Searcher, http://www.chemconnect.com
Polymer processing, http://www.polymerprocessing.com/sites.html
http://www.plastics.com/
http://www.canplastics.com/
Polymers in Industry, http://www.chemindustry.com/

12

Adhesives and Adhesion

12.1. INTRODUCTION

Adhesives play an important part in modern technology. They are usually defined as any substance which holds two solids together—by forming bonds between the solids as distinct from bolting, rivetting or even brazing, welding, or soldering. One of the most common examples of *adhesion* and one not yet understood completely is the barnacles which strongly adhere to ships' hulls and which are supposed to have influenced the battle of Salamis in 480 B.C.

Early glues were materials found in nature and include resins from trees, gums such as gum arabic, pitch and tar, egg, cheese, and fish extracts and flour paste. The introduction of rubber improved the versatility of the glues but it was not until aircrafts were produced for the World Wars (I and II) that the demand for strong lightweight adhesives forced the development of special glues. The Mosquito aircraft of World War II was an example of the application of modern resin adhesives.

Adhesives offer many advantages over normal or conventional methods such as bolting, rivetting, welding, stitching, clamping, or nailing. It offers:

1 The ability to join different materials such as plastic to metal, glass, or rubber.
2 The ability to join thin sheets together effectively.
3 Improved stress distribution in the joint which imparts a greater resistance to fatigue in the bonded components.
4 Increase in the flexibility of design.
5 Convenient and cost effective method in production.

These advantages have led to an ever increasing application of adhesives to a wide variety of industries which include automobiles and aircrafts, garments, furniture, appliances, buildings, and floor coverings.

It is interesting that the theory of adhesion is less developed than its application and it is only as a result of increasing demands by industry that the science of adhesion is developing. One difficulty is that the study of adhesion is a multidisciplinary subject and involves aspects of surface chemistry and physics, organic and inorganic chemistry, polymer chemistry and physics, rheology, stress analysis, and fracture phenomena. Thus, it is easier at present to practise the art of adhesion than the science of adhesion. Nevertheless, some simple principles have been shown to be effective in designing a strong adhesive joint and these will be presented and discussed.

12.2. CLASSIFICATION AND TYPES OF ADHESIVES

Adhesives can be classified in many different ways, e.g., by application or setting, chemical composition, cost, materials to be bonded, and end products or use.

Adhesives can be applied as: (a) a melted solid which sets when cooled, (b) a solution or suspension in a solvent which evaporates, (c) unpolymerized or partly polymerized blend which sets when polymerization is completed.

(a) Hot melt adhesives are usually thermoplastic polymers or waxes which are heated to a temperature sufficiently high to allow the melt to wet the substrate and readily flow in the joint. Examples of such adhesives are asphalt, ethyl cellulose, cellulose acetate butyrate, polyethylene copolymers, and waxes. Safety glass is made by softening polyvinyl butyral by heat and bonding the two glass pieces under pressure.

(b) These adhesives are usually slow setting if the solvent must diffuse through the joint or the material to be bonded. To circumvent this aspect—contact (adhesive) cements have been developed where the solvent is allowed to evaporate from the two surfaces before they are brought together. The cement consists of low molecular weight elastomers, or for aqueous systems suspended latexes are used usually at over 50% solids, but still fluid enough to flow and wet the surfaces. The elastomers used are natural and synthetic rubbers, vinyl resins, and acrylics.

(c) This type of reaction adhesive includes all thermosetting resins, as well as elastomers which can be crosslinked, and some vinyl thermoplastics. The adhesive is usually applied as a low viscosity wetting blend which polymerizes to a strong hard adherent glue. Examples of these types include epoxy adhesives, cyanoacrylate esters (crazy glue), urea–formaldehyde resins and urethanes.

The further classification of adhesives can be made by referring to the nature of the material, organic, inorganic, or hybrid. Thus organic adhesives are either natural (e.g., starch, gelatin, shellac), semi-synthetic (e.g., cellulose nitrate, castor-oil based polyurethanes), or synthetic (e.g., all vinyl polymers, plastics, etc.). The inorganic adhesives include cement, silicates (water glass), and sulfur and ceramic cements, whereas hybrid adhesives include litharge (PbO + glycerol 3:1), $AlPO_4$ in kerosene, and silicone rubbers.

Some of the adhesives can be used in strong acid, at high temperatures, in high vacuum, or other highly specialized applications. A detailed comprehensive treatment of adhesives can devote a full volume to classification. We shall hereafter concern ourselves with the form of the adhesive, e.g., some adhesives have been encapsulated (NCR), others are pressure sensitive (adhesive tapes) and some are flexible (rubbers). The electrical or thermal conductivity of adhesives (usually achieved by the addition of metal powders) are required in special applications. Adhesives are slowly penetrating into the garment industry where their replacement of stitching can speed up production and reduce costs. Similar changes are being made in the automobile where plastic is replacing metal at ever increasing rates.

12.3. THE ADHESIVE JOINT

The Adhesive Joint (AJ) is primarily composed of 5 parts (see Fig. 12.1); the two solid materials bonded (A & A′)—the two interfaces or boundary layers between the materials and the adhesive (B & B′), and the bulk adhesive (C). A strong AJ implies:

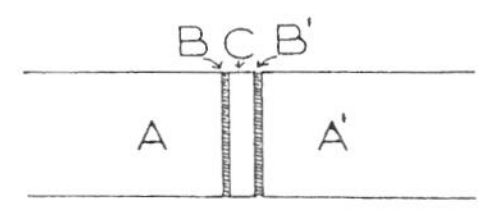

FIGURE 12.1. The adhesive joint: A representation of the 5 component parts on an adhesive joint. The adherend A & A′, the interfaces or boundary layers B & B′, and the adhesive C.

(1) strong boundary layers, (2) strong interfacial bonds, (3) a strong or hard set adhesive. Weak boundary layers have been attributed by Bikerman to be responsible for most weak joints. In the case of elastomers or plastics it is possible to show that even if interfacial or surface bonding of the adhesive is strong, a weak boundary layer results in a weak joint as the materials are separated and the surface molecules are pulled out of the bulk of the material. Thus, though strong bonds may be formed between the materials and the adhesive layers, the surface bound molecules are not held very strongly by the material bulk. This can be corrected by crosslinking the surface and creating one giant surface molecule which is anchored to the bulk by numerous bonds thereby making its withdrawal from the surface energetically too high, resulting in strong boundary layers. This process of crosslinking the surfaces of a plastic or elastomer is called CASING (crosslinking by activated species of inert gases) and was developed in 1966 by Hansen and Schonhorn at the Bell Telephone Laboratories. Casing can be effected by subjecting the surface to hydrogen atom reactions in a vacuum discharge through He–H_2 mixtures.

The surface of a solid is usually contaminated by adsorbed gases and vapors, and as a result the adhesives may form a poor or weak contact with the actual surface. The surface adsorbed vapors can be displaced by a liquid adhesive which "wets" the solid. In some cases a surface active agent may be added to an adhesive to facilitate the wetting process.

An alternate method of increasing the strength of an AJ is to add coupling agents to the solid surface, thereby introducing an intermediate molecule which has reactive functional groups that can bond to the adhesive and thereby strengthen the bond. This is done to the surface of glass fibers used to reinforce radial tires and in fiberglass plastics. Some examples of coupling agents and their use are listed in Table 12.1.

These compounds which have alkoxide groups will hydrolyze on a hydroxylated surface to form an alcohol and the surface-bonded organic residue. For example,

```
   O
  /
Si—OH      + NH2—CH2CH2CH2Si(OC2H5)3
  \
   O      O
  /        \
Si—OH       Si—O
  \        /     \
   O      O       \
  /        \       \
Si—OH →     Si—O—Si—CH2CH2CH2NH2     + 3CH3OH      (12.1)
  \        /       /
   O      O       /
           \     /
            Si—O
           /
```

TABLE 12.1
Some Selected Commercial Silane Coupling Agents

Name	Formula	Application
Vinyltriethoxysilane	$CH_2{=}CHSi(OC_2H_5)_5$	Unsaturated polymers
Vinyl-tris(b-methoxyethoxy)silane	$CH_2{=}CHSi(OCH_2CH_2OCH_3)_3$	Unsaturated polymers
Vinyltriacetoxysilane	$CH_2{=}CHSi(OOCCH_3)_3$	Unsaturated polymers
γ-Methacryloxypropyltrimethoxysilane	$CH_2{=}C(CH_3)COO(CH_2)_3Si(OCH_3)_3$	Unsaturated polymers
γ-Aminopropyltriethoxysilane	$H_2NCH_2CH_2CH_2Si(OC_2H_5)_3$	Epoxies, phenolics, nylon
γ-(2-Aminoethyl)aminopropyltrimethoxysilane	$H_2NCH_2CH_2NH(CH_2)_3Si(OCH_3)_3$	Epoxies, phenolics, nylon
γ-Glycidoxypropyltrimethoxysilane	$\overset{O}{\overbrace{CH_2CH}}CH_2O(CH_2)_3Si(OCH_3)_3$	Almost all resins
γ-Mercaptopropyltrimethoxysilane	$HSCH_2CH_2CH_2Si(OCH_3)_3$	Almost all resins
β-(3,4-Epoxycyclohexyl)ethyl-trimethoxysilane	$OC_6H_9{-}CH_2CH_2Si(OCH_3)_3$	Epoxies
γ-Chloropropyltrimethoxysilane	$ClCH_2CH_2CH_2Si(OCH_3)_3$	Epoxies

The organic amino group can react further with the adhesive and contribute to the bond strength of the joint.

Another method used to improve the strength of a bond is to introduce the chemical reactive functional groups directly on the surface of the material. For example, the bonding of Teflon or polyethylene can be improved by introducing OH groups onto the plastic surface by sparking moist air next to the surface to be joined. This is best done by means of a Tesla coil or a Corona discharge, but a vacuum high voltage ac discharge through water vapor is most effective. Such surfaces become wet with respect to water and can form strong bonds with common glues such as epoxy adhesive.

More examples of the AJ will be given after a discussion of the theory of the adhesive bond.

12.4. THE THEORY OF THE ADHESIVE BOND

The theory of the adhesive bond (AB) is basically an attempt to generalize the mechanistic account of adhesion into a comprehensive explanation. One of the major difficulties in realizing a uniform consistent theory is due to the many variable factors which determine the strength of an AJ and the methods used to evaluate them, e.g., geometrical factors, loading factors, rheological energy losses, and interfacial interactions, to mention only a few. There are four principal mechanisms used to account for adhesion. These are:

i Mechanical interlocking (E_M).
ii Diffusion theory (E_D).
iii Electrostatic theory (E_E).
iv Adsorption theory (E_A).

12.4.1. Mechanical Interlocking

This early theory was successfully applied to joints of porous or rough adherends such as wood. The liquid adhesive penetrates the porous and irregular surface, and when hardened adds strength to the adhesive joint. Thus,

i the greater the surface irregularity and porosity the greater is the strength of the joint;
ii the joint strength will be proportional to the film strength of the adhesive when the adherend is stronger than the adhesive.

Some examples in which the mechanical mechanism is important is in the adhesion of polymers (elastomers and rubbers) to textiles. Another example, though somewhat contentious, is the metal plating of a plastic which usually requires a pretreatment to modify the surface topography of the polymer. Usually the increase in adhesion is also attributed to an improved surface force component due to the increased rugosity.

The bonding of maplewood samples with urea–formaldehyde resins at 5 psi gluing pressure was tested in shear as a result of surface treatment.

	Increasing roughness →			
Surface treatment	Planed	Sanded	Sawn	Combed
Shear strength (psi)	3120	2360	2690	2400

This indicates that the smoother surface forms the stronger joint. This is due to the wood fibers becoming damaged by treatment and are easily removed from the bulk when the joint is stressed and separated. The trapping of air within the pores prevents the anchoring of the glue. Thus mechanical interlocking has limited application and usually strength is associated with other forces.

12.4.2. Diffusion Theory

This theory, put forth by Voyutskii, accounts for the autoadhesion of polymers by the interdiffusion of polymers across the interface and their mutual solubility. Direct radiometric and luminescence experiments have confirmed that the diffusion boundary may be as deep as 10 μm. Interdiffusion is also important in solvent-welding of plastics where the solvent essentially plasticizes the surfaces and promotes diffusion between the two materials. Thus, though diffusion does occur in special cases the theory has very little general application because many adhesives show no solubility in the adherends, e.g., glass, metal and wood.

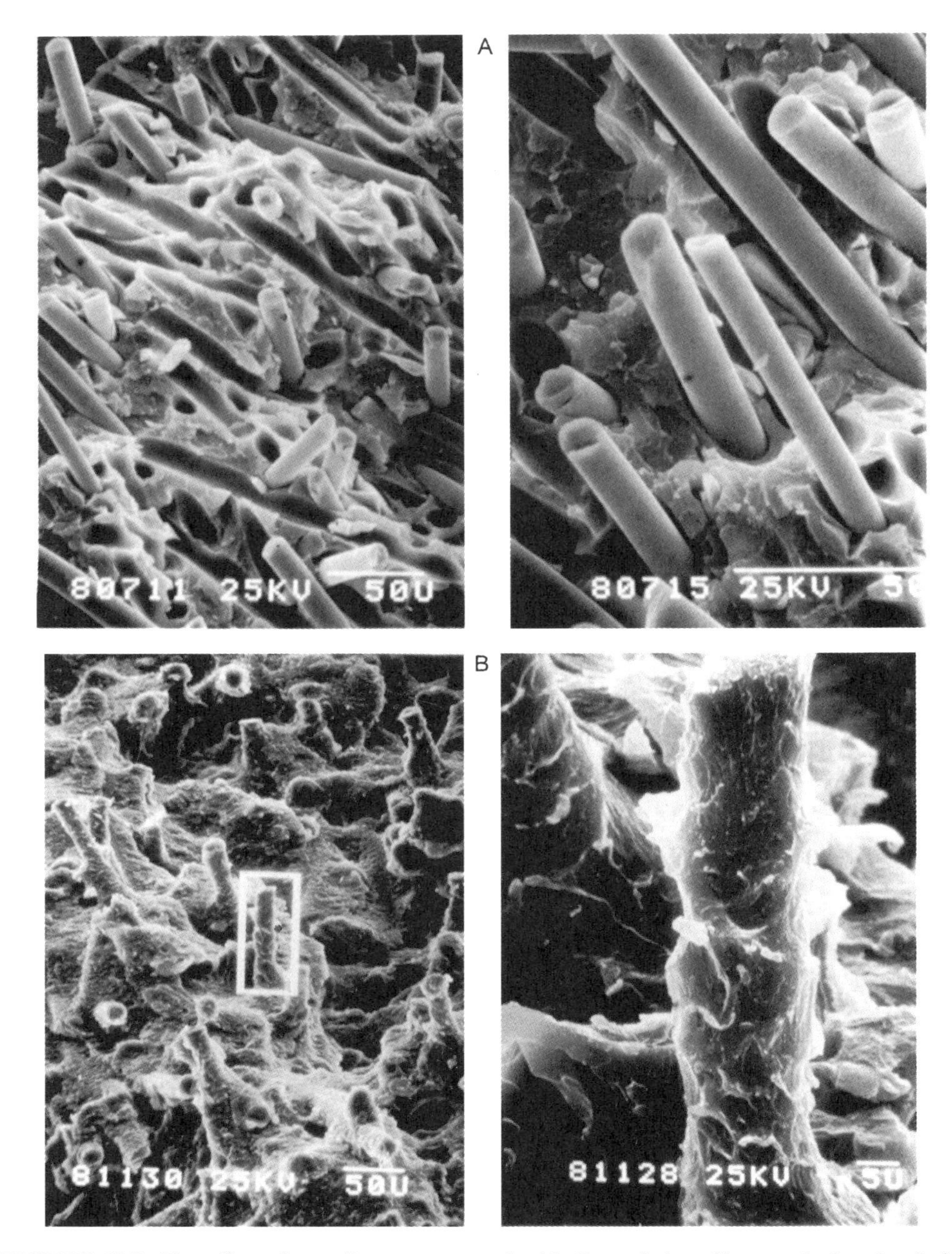

FIGURE 12.2. The effect of coupling agents on the binding of glass fibers and glass beads in adhesives and plastics. The example is that of a reaction injection molded polyurethane composite which has been fractured to expose the reinforcing glass fibers (A and B) and glass beads (C and D). (A) Untreated glass fibers: The photomicrograph shows holes where the poorly bonded fibers have been clearly removed. The extreme closeup shows a typical glass fiber which is not coated with polymer. (B) Glass fibers treated with coupling agent: The chemically coupled system shows fewer fibers pulled out in comparison with (A). The extreme closeup shows

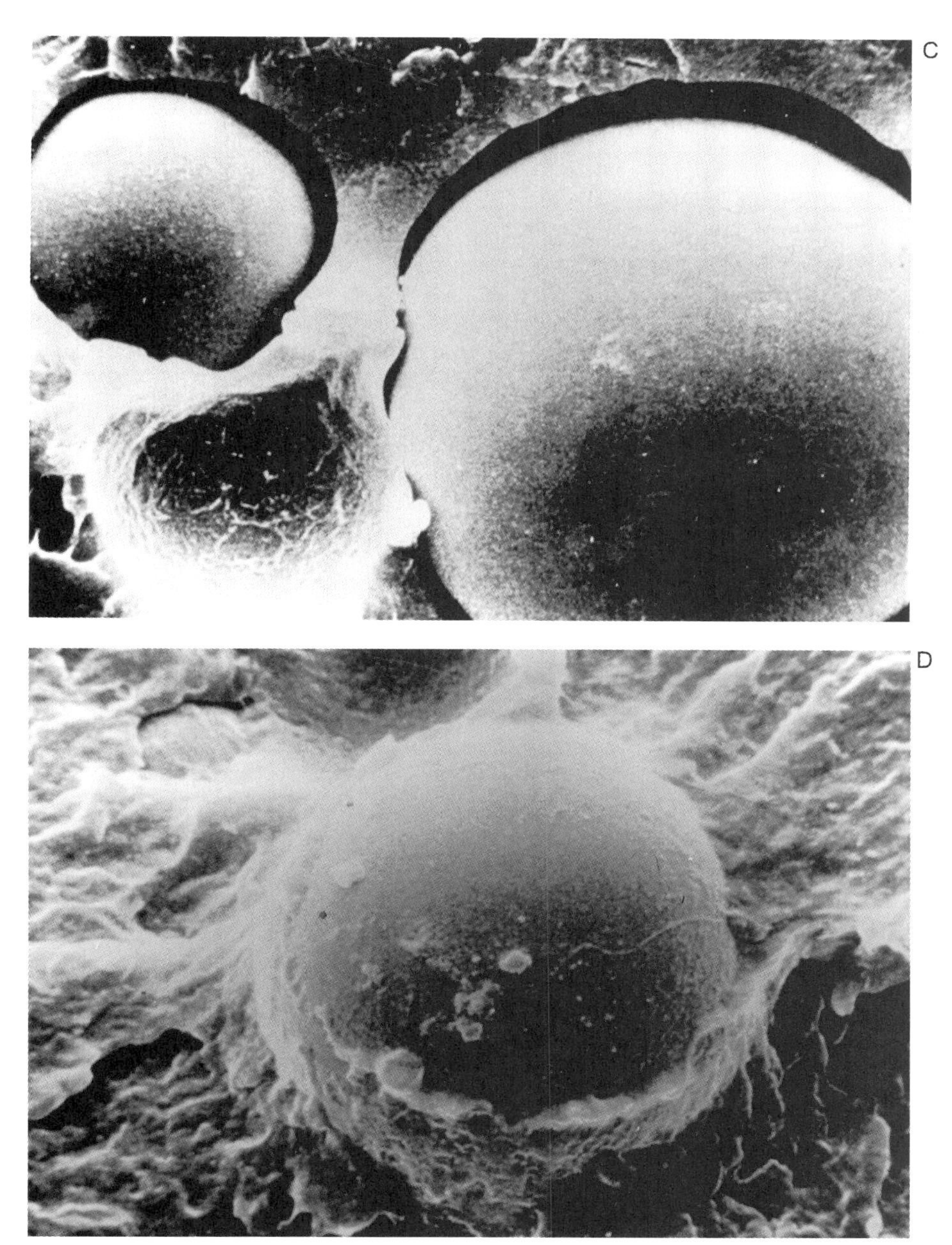

FIGURE 12.2. Continued.
excellent coating and adhesion of polymer attainable when silanes produce a chemical bond between the organic matrix and the inorganic reinforcement. (C) Untreated glass beads: The filler is clean and free of adhering polymer. (D) Treated glass beads: The glass shows strongly adhering polymer.

12.4.3. Electrostatic Theory

This theory, proposed by Deryaguin, arose from the crackling noises and flashes of light observed when adhesive films are peeled off rapidly in the dark. Likewise film strengths (of the order of 1–100 mJ/cm^2) depend on the speed at which it is peeled. The Russians proposed that an electrical double layer is formed at the junction of two different materials due to the difference in the energy bands of the materials. The separation of the adhesive–adherend interface is then analogous to the plate separation of a capacitor. This should depend on the dielectric but such peel strengths show no difference when measured in air or vacuum. Calculated values of the work of adhesion for gold, copper, and silver on glass show that the electrostatic contribution is 5, 80, and 115 mJ/m^2, respectively, whereas the van der Waals contribution is 950, 400, and 800 mJ/m^2, respectively. Measured values for the three metals were 1400, 800, and 1000 mJ/m^2, respectively, with errors of about ± 200 mJ/m^2. Thus, though electrostatic forces contribute to the strength of an adhesive joint, the major contributing factor is the ever present van der Waals force.

12.4.4. Adsorption Theory

This theory relies on the orientation of surface forces resulting in van der Waals interaction. There are three general types of van der Waals forces: (a) Keesom, (b) Debye, (c) London.

(a) Keesom forces are due to the dipole–dipole interaction. Since the dipoles are assumed to be aligned this is also called the *orientation effect*, and is given by the potential energy of attraction V_K.

$$V_K = \frac{-2\mu^4}{3kTr^6} \quad \text{or} \quad V_K = \frac{-2\mu_1^2\mu_2^2}{3kTr^6} \tag{12.2}$$

for molecules with different dipole moments μ_1 and μ_2, and where k is the Boltzmann constant, T is the absolute temperature and r is the distance of separation.

(b) Debye forces are due to an induced dipole effect or polarization. The attractive potential energy V_D of two different molecules is:

$$V_D = -\frac{\alpha_1\mu_2^2 + \alpha_2\mu_1^2}{r^6} \tag{12.3}$$

where α is the polarizability and μ is dipole moment. This force is temperature independent and therefore important at high temperatures.

(c) London forces are due to induced dipole–induced dipole attraction. Thus a symmetrical molecule will exhibit oscillations inducing dipoles in neighboring molecules. The London attractive potential energy V_L is:

$$V_L = \frac{-3hv_1v_2}{2(v_1 + v_2)} \cdot \frac{\alpha_1\alpha_2}{r^6} \tag{12.4}$$

where α_1 and α_2 are the polarizabilities, h is Planck's constant, v_1 and v_2 are the characteristic frequencies of the molecules.

Since $hv = I$, the ionization energy of the molecule, then

$$V_L = -\frac{3}{2}\frac{I_1 I_2}{(I_1 + I_2)} \cdot \frac{\alpha_1 \alpha_2}{r^6} \tag{12.5}$$

The London forces are also called *dispersive forces* and contribute the major proportion of the bonding for nonpolar or weakly polar molecules.

The adsorption theory requires intimate molecular contact between surfaces for a strong adhesive bond. Van der Waals forces when operative are considered as secondary bonds whereas ionic, covalent, and metallic forces contribute to primary adhesive bonds.

The hydrogen bond is intermediate in strength between primary and secondary forces. The results of corona discharge on polymers leads to the formation of hydroxyl groups on the surface which improves the adhesive strength by hydrogen bonds or by covalent bonds (as in the case of coupling agents).

Intimate interfacial contact is promoted by the wetting of surfaces by a fluid adhesive which also displaces air from the surface if it has low viscosity prior to setting. This is the major characteristic of crazy glue that makes it so effective.

Wetting alone may not always give rise to a strong adhesive joint. A weak boundary layer can cause failure at what appears to be the interfacial boundary of adhesive–substrate but which is claimed to be at the surface layer of the substrate. CASING is supposed to strengthen this layer and thereby improve the strength of the joint. An alternate explanation for the observed effects of glow discharge treatment is that the surface free energy of the polymer substrate may be increased or chemically active species may be formed on the surface—both of which could account for the observed increase in adhesion.

Insofar as it is possible to cite examples which could fit each of the above mechanism it is best to describe the overall adhesive strength to result from a combination of each such that

$$E_{\text{total}} = \alpha E_M + \beta E_D + \gamma E_E + \delta E_A \tag{12.6}$$

where the coefficients vary depending on the adhesives and adherends. It is not possible, at present, to assign values to these coefficients except in special cases. This is of little consequence—especially if the adhesive does the job and the bonded joint holds together.

12.5. CHEMISTRY OF SELECTED ADHESIVES

One of the most common and versatile adhesives is the epoxy resin, usually used as a two part formulation which sets into a hard adherent mass in from 5 min to 24 hr depending on the length of the prepolymer and the choice of the hardener.

The epoxy resins were discovered in 1938 during the search for a denture resin. The epoxy group,

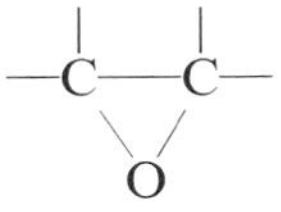

$$-\overset{|}{C}\underset{\diagdown O \diagup}{-}\overset{|}{C}-$$

is highly reactive and the ring is readily opened by acidic or basic catalysts. Thus, an amine (primary or secondary) will react with an epoxide to form an amine alcohol

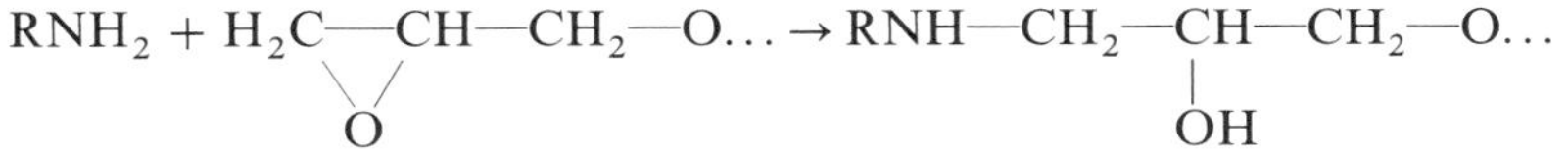

$$RNH_2 + H_2\underset{\diagdown O \diagup}{C-CH}-CH_2-O\ldots \rightarrow RNH-CH_2-\underset{OH}{\underset{|}{CH}}-CH_2-O\ldots \quad (12.7)$$

Other catalysts which react with the epoxide ring are anhydrides (which form ester alcohols).

The epoxide intermediates (short chain nonviscous polymers) commonly used are made by the reaction of epichlorohydrin

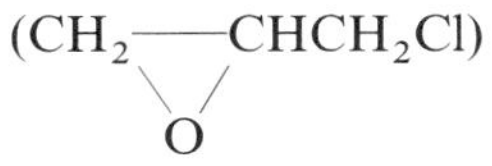

$$(\underset{\diagdown O \diagup}{CH_2-CH}CH_2Cl)$$

with bisphenol A

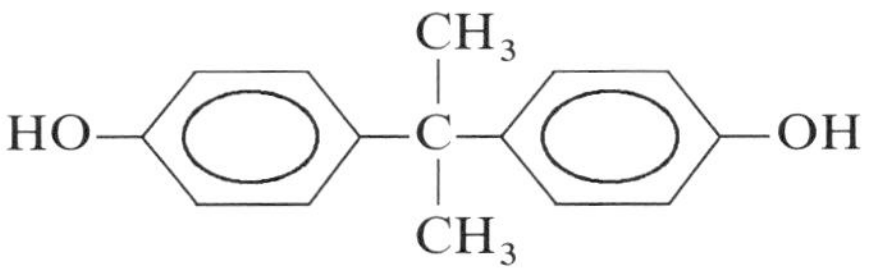

$$HO-C_6H_4-\underset{CH_3}{\overset{CH_3}{C}}-C_6H_4-OH$$

e.g.,

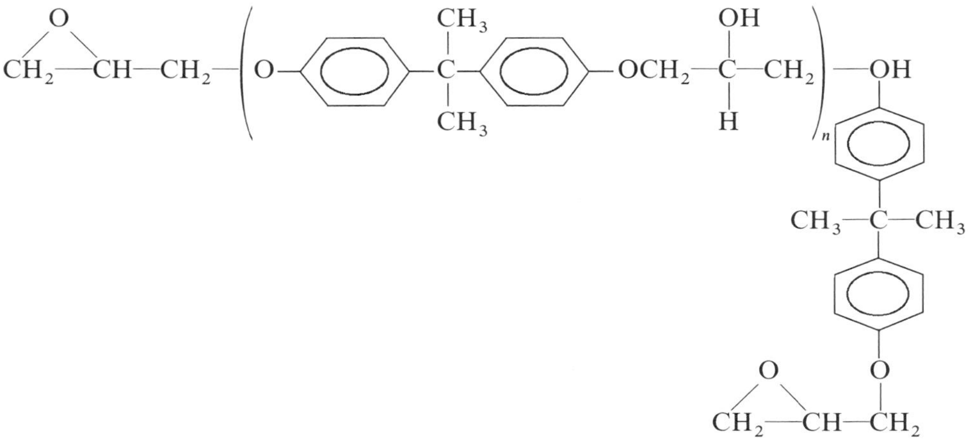

$$\underset{\diagdown O \diagup}{CH_2-CH}-CH_2-\left(O-C_6H_4-\underset{CH_3}{\overset{CH_3}{C}}-C_6H_4-OCH_2-\underset{H}{\overset{OH}{C}}-CH_2\right)_n-O-C_6H_4-\underset{CH_3}{\overset{CH_3}{C}}-C_6H_4-O-CH_2-\underset{\diagdown O \diagup}{CH-CH_2} \quad (12.8)$$

where n varies from 0 to 20 or the molecular mass is between 900 and 5000. When n is greater than 2, the resin is a solid with a melting point which increases as n increases. Since it is desirable for adhesives to be liquid, and thus flow in the joint, the prepolymer is small, i.e., $n \leqslant 2$. However, each polymer molecular has n hydroxyl groups (—OH) which increase the curing rate of the adhesive and because of its polarity increases the adhesion of the polymer to polar surfaces such as metal and glass. Thus a balance between viscosity and strength results in a choice of n which is a compromise.

Catalysts are called reactive hardeners or curing agents and become bound to the polymer by opening the epoxide linkage. When the hardener is an amine the reaction is

$$RN_2 + \overset{\overset{O}{\diagup\;\diagdown}}{CH_2\text{—}CH}\text{—}X \rightarrow R\text{—}NH\text{—}CH_2\text{—}\overset{\overset{OH}{|}}{CH}\text{—}X \tag{12.9}$$

followed by

$$R\text{—}NH\text{—}CH_2\text{—}\overset{\overset{OH}{|}}{CH}\text{—}X + \overset{\overset{O}{\diagup\;\diagdown}}{H_2C\text{——}CH}\text{—}X \rightarrow R\text{—}\underset{\underset{\underset{\underset{OH}{|}}{CH_2\text{—}CH\text{—}X}}{|}}{N}\text{—}CH_2\text{—}\overset{\overset{OH}{|}}{CH}\text{—}X \tag{12.10}$$

resulting in crosslinking which can be extended if the hardener is a diamine such as ethylene diamine $NH_2\text{—}CH_2CH_2\text{—}NH_2$. The concentration of the hardener is 3 to 100 phr (parts per hundred resin) depending on the system and curing time. Fillers commonly used include TiO_2, PbO, Fe_2O_3, Al_2O_3, SiO_2 which are added to reduce cost, decrease shrinkage on setting, lower the coefficient of thermal expansion of the joint, and to improve the heat resistance of the joint.

The high strength of epoxy adhesives has led to their application over a wide variety of conditions which include the fabrication of helicopter rotor blades, fiberglass lined boats, aircraft, and in the building trades.

Another class of adhesives are those based on isocyanate (—N=C=O) of which the most common is the urethanes. The isocyanate group reacts with active hydrogen such as an alcohol (ROH) or an amine (R_2NH) as follows:

$$ROH + \phi NCO \rightarrow R\text{—}O\text{—}\underset{\underset{O}{\|}}{C}\text{—}\overset{\overset{H}{|}}{N}\text{—}\phi \tag{12.11}$$

$$R_2NH + \phi NCO \rightarrow R_2\text{—}N\text{—}\underset{\underset{O}{\|}}{C}\text{—}\overset{\overset{H}{|}}{N}\text{—}\phi \tag{12.12}$$

where $\phi = C_6H_5$. Thus, by reacting a diisocyanate with a polymeric diol, a condensation reaction occurs resulting in bonding to a hydroxylated surface such as glass or metal. The diisocyanates commonly used are toluene diisocyanate TDI, or hexamethylene diisocyanate HDI.

$$\text{2,4 TDI} \qquad \text{2,6 TDI} \qquad OCN{-}(CH_2)_6{-}NCO \ \text{(HDI)} \tag{12.13}$$

The high reactivity of the NCO group make the diisocyanates toxic and great care must be taken in using these adhesives. The isocyanates react with water to form CO_2 and an amine which can also react with the NCO group,

$$R{-}NCO + H_2O \rightarrow RNH_2 + CO_2 \tag{12.4}$$

$$RNH_2 + R'{-}NCO \rightarrow RNH{-}\underset{\underset{O}{\|}}{C}{-}\overset{\overset{H}{|}}{N}{-}R' \tag{12.15}$$

The evolved CO_2 can lead to a highly crosslinked foam which is soft or hard depending on the prepolymers used. Similarly the hardness of the set adhesive is determined by the hydrocarbon chain length in the polyol or polyamine used to react with the diisocyanate. Crosslinking by short chain triols leads to a hard adhesive whereas long diols results in elastomeric material, often used in textiles or where a flexible joint is required.

A new urethane based adhesive has recently been developed which combines the properties of "hot melts" and chemical reactivity. Usually a hot melt adhesive is applied at 150–200°C. The heat weakens the van der Waals forces and allows the polymer to flow. When the material has cooled the bonds reform and sets within minutes. Chemically reactive adhesives rely on the reaction of polymerization or crosslinking to set a monomeric or prepolymeric adhesive and setting may take hours during which the joint must be held together with clamps or in a jig. The new adhesive is a reactive polymer of short chains held together by weak forces which break upon heating to only 100°C and sets again within minutes upon cooling. However, unlike previous material the bond continues to strengthen with time because of further reaction with water vapor in the air which crosslinks the polymer.

Another class of thermosetting resins used as adhesives involves the polymerization of formaldehyde with urea or phenols. These materials are very cheap and find extensive use in binding wood, e.g., plywood, chipboard, and particle-board. Foamed urea–formaldehyde has also been used as insulation (UFFI) in homes because of its ability to be injected into the walls of older homes. Unfortunately, the foam is not very stable and shrinks and hydrolyses to liberate formaldehyde and possibly other toxic vapors. Recent work has, however, shown that the addition (and subsequent removal) of ammonia (NH_3) to the foam can reduce the level of formaldehyde released by the polymer. This release of formaldehyde also occurs in plywood and particle-board and these materials must be sealed to avoid the formation of toxic levels (TLV = 0.10 ppm or 120 $\mu g/m^3$) of formaldehyde.

The urea–formaldehyde polymer is formed as follows:

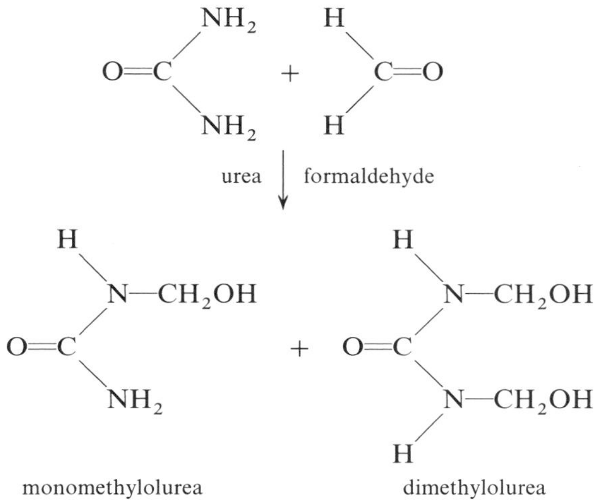

$$
\begin{array}{ccccc}
 & \mathrm{NH_2} & & \mathrm{H} & \\
 & / & & \backslash & \\
\mathrm{O{=}C} & & + & & \mathrm{C{=}O} \\
 & \backslash & & / & \\
 & \mathrm{NH_2} & & \mathrm{H} & \\
 & \text{urea} & \Big\downarrow & \text{formaldehyde} &
\end{array}
$$

$$
\begin{array}{cccccccc}
\mathrm{H} & & & & & \mathrm{H} & & \\
 & \backslash & & & & & \backslash & \\
 & & \mathrm{N{-}CH_2OH} & & & & \mathrm{N{-}CH_2OH} & \\
 & / & & & & / & & \\
\mathrm{O{=}C} & & & + & \mathrm{O{=}C} & & & \quad (12.16) \\
 & \backslash & & & & \backslash & & \\
 & & \mathrm{NH_2} & & & & \mathrm{N{-}CH_2OH} & \\
 & & & & & / & & \\
 & & & & & \mathrm{H} & & \\
 & \text{monomethylolurea} & & & & \text{dimethylolurea} & &
\end{array}
$$

When the methylols eliminate water a methylene urea is formed which can condense to form a crosslinked polymer (under acid conditions)

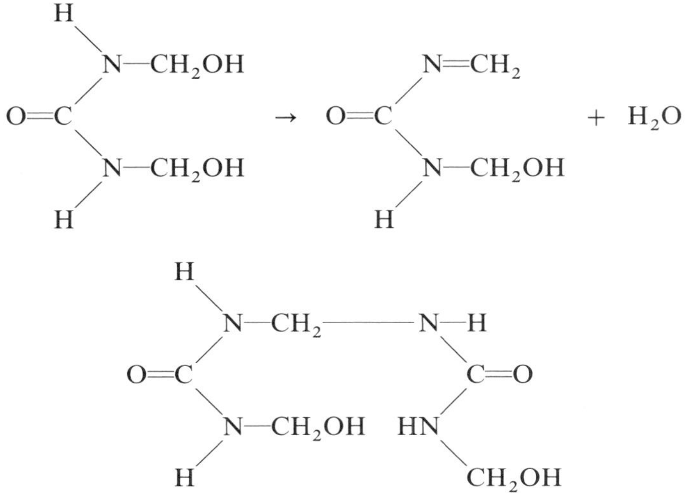

$$
\begin{array}{ccccccc}
\mathrm{H} & & & & & & \\
 & \backslash & & & & & \\
 & & \mathrm{N{-}CH_2OH} & & & \mathrm{N{=}CH_2} & \\
 & / & & & / & & \\
\mathrm{O{=}C} & & & \rightarrow \quad \mathrm{O{=}C} & & & + \; \mathrm{H_2O} \\
 & \backslash & & & \backslash & & \\
 & & \mathrm{N{-}CH_2OH} & & & \mathrm{N{-}CH_2OH} & \\
 & / & & & / & & \\
\mathrm{H} & & & \mathrm{H} & & &
\end{array}
$$

$$
\begin{array}{cccccc}
\mathrm{H} & & & & & \\
 & \backslash & & & & \\
 & & \mathrm{N{-}CH_2{-}\!\!-\!\!-\!\!-\!\!-} & \mathrm{N{-}H} & & \\
 & / & & & \backslash & \\
\mathrm{O{=}C} & & & & \mathrm{C{=}O} & \quad (12.17) \\
 & \backslash & & & / & \\
 & & \mathrm{N{-}CH_2OH} & \mathrm{HN} & & \\
 & / & & & \backslash & \\
\mathrm{H} & & & & \mathrm{CH_2OH} &
\end{array}
$$

Excessive residual acid in the resin is partly responsible for the subsequent release of formaldehyde from the set polymer.

The reaction of formaldehyde (H_2CO) with phenol,

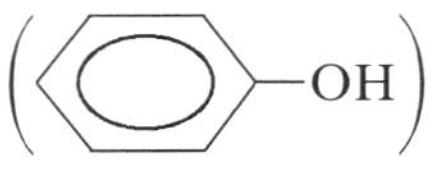

(C$_6$H$_5$–OH)

has been studied for over 100 years. In 1909 Baekeland patented the resin which has become known as *Bakelite*. The prepolymers of phenol–formaldehyde are of two types: (1) resitols in which $\phi OH/H_2CO < 1$, and (2) novolacs in which $\phi OH/H_2CO > 1$.

(1) Resitols are formed (under alkaline catalysis) by the reaction

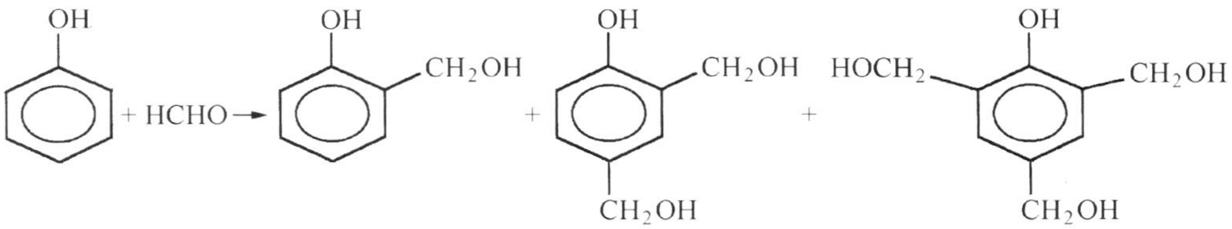

(12.18)

The phenol alcohols condense to form resitols

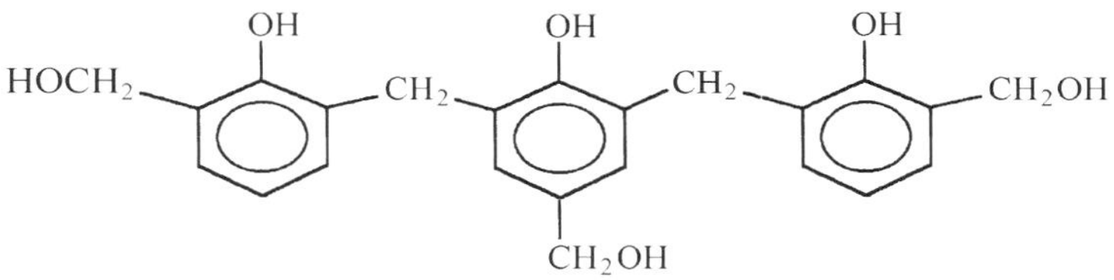

which can further condense to form ether linkages

$$2R—CH_2OH \rightarrow R—CH_2—O—CH_2R + H_2O$$

The crosslinked resin does not melt when heated, but carbonizes. This characteristic is used in efficient heat shields where the resins are used in the nose cone of missiles and space vehicles.

(2) Novolacs are formed (under acid catalysed with $\phi OH/H_2CO > 1$):

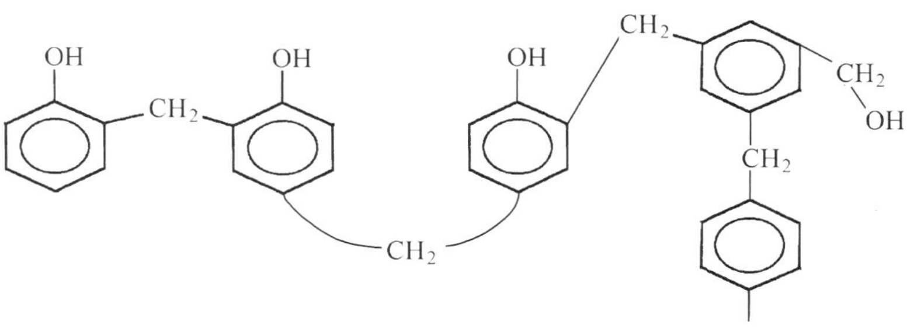

(12.19)

The novolac resins are usually applied as solid adhesives incorporating some phenol and formaldehyde (from hexamethylene tetramine) to crosslink the prepolymer when heat is applied. Resitol resins are applied as a dispersion in dilute aqueous NaOH with fillers, surfactants, and extenders.

Phenol formaldehyde resins have recently been applied to the fabrication of reconstituted lumber called *Parallam*. Small pieces of wood, approx. 0.3 cm × 20 cm are layed end to end and staggered side by side, are impregnated with about 5% of the resin and cured by microwave energy to give a continuous length of lumber with a width and thickness which can be readily controlled and varied. The strength of this lumber is superior to the equivalent natural lumber which it soon will be displacing. Other applications of phenol–formaldehyde resins include bonding abrasive grits in grinding wheels, wood and veneers, paper laminates, and many others.

A popular and highly advertised adhesive is based on cyanoacrylate—the most common being methyl-2-cyanoacrylate

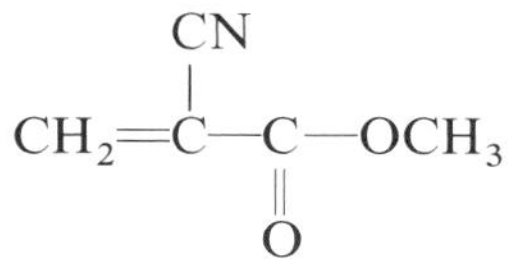

$$CH_2{=}\overset{\displaystyle CN}{\overset{|}{C}}{-}\underset{\displaystyle O}{\underset{\|}{C}}{-}OCH_3$$

also known as *crazy glue* (Eastman 910). This was first introduced commercially in 1959 and because of its unique properties has gained widespread use in industry, medicine, and by hobbyists. The monomer has a low viscosity and so can readily penetrate surfaces. Its rapid polymerization is initiated by water, alcohol, or other weak bases. Hence, when the monomer is placed on a dry grease-free joint, the air usually has sufficient moisture to initiate the polymerization while the two surfaces are held together. Though very expensive, its use without solvent, heat cures, or excessive pressure, makes it an ideal adhesive. One important disadvantage is that it tends to glue things indiscriminately, e.g., hand, fingers, eyelids, etc. Since the body is seldom free of surface oils it has been possible to loosen a crazy glue joint (of an eyelid) by applying mineral oil (Nujol) to dissolve the underlying oil film. In some cases surgery is required.

Special adhesives are required for high temperature use. Silicone rubber is an excellent flexible adhesive with a temperature limit of 250°C. For higher operating temperatures one must resort to inorganic adhesives and cements. These are usually based on silicates (water glass), phosphates (phosphoric acid), and Portland cement.

Silicate adhesives using water soluble sodium silicate (Na_2SiO_3, water glass) have fillers of silica (SiO_2), salt (NaCl), clay, talc, metals, asbestos, or limestone ($CaCO_3$) depending on its ultimate use.

These silicate adhesives are fire resistant and can withstand temperatures up to 1000°C if the adhesives cures to a ceramic (homogeneous phase). Since water soluble silicates are basic it is essential that metals such as aluminum and zinc which form amphoteric oxides (the metal dissolves in basic solutions) be avoided since they would dissolve in the joint and form hydrogen.

Phosphate adhesives are made by reacting zinc oxide, (ZnO), with phosphoric acid (H_3PO_4), forming $ZnHPO_43H_2O$ which sets to a crystalline material and is used as a dental cement. More elaborate dental materials are constantly being introduced and tested as we all verify on our trips to the dentist's office. However, progress is slow because the demands and requirements are high.

An old and still useful cement for joining glass to metal is litharge cement made by mixing litharge (PbO) with glycerol ($CH_2OHCHOHCH_2OH$) in the ratio PbO: glycerol: water of 6:2:1. The cement takes a day to set and will resist the action of dilute acid, ammonia and hydrocarbons.

There are many different adhesives but to all the common requirements of a good and lasting joint, is the surface preparation. The following factors must be observed:

1 Dust, scale, grease, oil, plasticizer, and other surface contaminants must be removed.
2 The surfaces should be etched or roughened to give a larger surface area for bonding.
3 The surfaces may require planing or smoothing to allow intimate contact.
4 A priming coat may be applied to wet the surface and to displace air from the crevices.
5 The surfaces can be sealed to prevent the adhesive from penetrating the porous surface.
6 The surface can be treated chemically (e.g., CASING or coupling agents) to increase the surface energy and facilitate bonding.
7 The surface should be dried and any adsorbed liquids must be removed.
8 A metal may be electroplated with another metal which forms stronger bonds with the adhesive, (e.g., Ni on Au).
9 Use ultrasonics where appropriate to clean the surface to the true surface interface.
10 Remove residual cleaning substances from the surface.

The main problem in the selection of a proper adhesive is to match the adhesive to the adherend. Only experience and the ability to test the strength of trial sample joints can solve the problem. Since adhesives are here to stay, it is always worthwhile to follow the developments in new and interesting adhesive products.

EXERCISES

1. Airplanes today can have as much as a tonne of adhesives. Give some reason for this.
2. Classify adhesives according to their application and setting, giving examples of each type.
3. In what way does the chemical composition of a surface affect the strength of an adhesive joint?
4. What are coupling agents, how do they work and give two examples?
5. Give a short account of the four different theories of the adhesive bond.
6. What is an epoxy group and how do epoxy adhesives work? (Write chemical reactions.)
7. Which hardener will result in a more crosslinked adhesive:

   ethylene diamine ($H_2N—CH_2—CH_2—NH_2$)

   or

   diethylene triamine ($H_2N—CH_2—CH_2—NH—CH_2—CH_2—NH_2$)?

8. Of what use are fillers in an adhesive joint? List 5 fillers.
9. (a) What is a urethane adhesive? (b) Why must care be taken when using them?
10. What surface feature of glass and metals makes urethane adhesive ideal for these materials?
11. How do the formaldehyde adhesives work? (Write chemical reactions.)
12. What is crazy glue and why is it an exceptional adhesive?
13. Write a short note on inorganic adhesives. What conditions are they best suited to?
14. List the primary surface treatments for a good adhesive joint.
15. (a) What is CASING? (b) To what type of surfaces can it be applied?
16. Draw typical strength time curves for: (a) Hot melt adhesive; (b) Reactive adhesion; (c) Reactive melt adhesion.

FURTHER READING

D. S. Rimai et al., Editors, *Fundamentals of Adhesion and Interfaces*, Coronet Books, Philadelphia, Pennsylvannia (1995).
T. Young, *The Crafter Guide to Glues*, Chilton, Radnor, Pennsylvannia (1995).
A. Pizzi and K. Mittal, Editors, *Handbook of Adhesive Technology*, Dekker, New York (1994).
J. Hodd, *Epoxy Resins*, Pergamon Press, New York (1991).
J. D. Minford, Editor, *Treatise on Adhesion and Adhesives*, vol. 7, Dekker, New York (1991).
L. H. Lee, Editor, *Fundamentals of Adhesion*, Plenum, New York (1991).
I. Skeist, *Handbook of Adhesives*, Chapman and Hall, New York (1989).
A. J. Kinloch, *Adhesion and Adhesives: Science and Technology*, Chapman and Hall, New York (1987).
J. J. Bikerman, *The Science of Adhesive Joints*, Academic Press, New York (1961).
The Adhesion Society, http://www.adhesionsociety.org/
Glossary of Terms, http://www.pprc.org/pprc/
Sources and Newsletter, http://www.adhesivesandsealants.com/
Epoxy adhesives, http://www.airproducts.com/
Theory of Adhesive Bonding, http://joinplastics.eng.ohio-state.edu/kkwan/main.html
http://www.umist.ac.uk/lecturenotes/JDS

13

Paint and Coatings

13.1. INTRODUCTION

The purpose of *paints and coating* is primarily to prevent corrosion and wear. The aesthetic aspect is usually of secondary importance. The nature of the protective coating is usually determined by the surface, and the duration of the desired protection. Needless to say, the cost of the coating is another factor which can vary considerably from place to place, as well as with time. The concern for the environment is another factor which must be taken into consideration since the drying process may involve the release of volatile organic compounds (VOC), which can contribute to smog or other undesirable atmospheric conditions.

Special paints are available for ships' hulls, hot mufflers, luminous surfaces, water proofing, etc.

13.2. CONSTITUENTS OF PAINT AND COATINGS

There are four normal constituents of a paint or coating. These are:

1 *Binder:* Designed to hold the film together and provides the adhesive forces required to bond it to the surface.
2 *Pigment:* A fine powdered material that provides the color as well as the hiding characteristics and weathering resistance of the film. Pigment without binder, as in whitewash, has little permanence. Binder, such as linseed oil, without pigment can be an excellent sealer and preservative for wood. Varnish is resin and solvent without pigment.
3 *Solvent:* A volatile liquid which dissolves the binder and acts as a thinner to dilute the coating, allowing it to spread easily on the surface.
4 *Additives:* Small amounts of wetting agents, flattening agents, driers, plasticizers, emulsifiers, stabilizers, crosslinking agents.

13.3. BINDER

The *binder* is also the paint vehicle and for oil-based paints, consists of a drying oil which contains double bonds. A list of various oils and their average compositions in terms of the fatty acids is given in Table 13.1. These oils are called triglycerides

TABLE 13.1
Composition and Characteristics of Vegetable Oils

	% In seed	Saturated (%)	Oleic (%)	Ricinoleic (%)	Linoleic (%)	Linolenic (%)	Eleostearic (%)	Licanoic (%)	I_2 (No.)	Saponification (No.)
No. of double bonds		0	1	1	2	3	3	3		
Linseed	32–40	10	20		20	50			180	190
Perilla	30–50	8	14		14	64			200	
Tung	40–50	4	7		8		81		165	
Oiticica	52–62	12	7					81		
Soyabean	16–22	12	28		54	6			132	
Tall		7	45							
Safflower	24–36	10	14		76					
Castor		3	7	88	2				110	190
Olive		12	83		5				85	
Coconut		90	8						10	
I_2 no.		0	89.9		181	273.5				

Stearic acid ($C_{18}H_{36}O_2$), $CH_3(CH_3(CH_2)_{16}$—COOH.
Oleic acid ($C_{18}H_{34}O_2$), CH_3—$(CH_2)_7$—CH=CH—$(CH_2)_7$—COOH.
Ricinoleic ($C_{18}H_{34}O_3$), $CH_3(CH_2)_5$—CH(OH)—CH_2—CH=CH$(CH_2)_7$—COOH.
Linoleic ($C_{18}H_{32}O_2$), $CH_3(CH_2)_4$CH=CH—CH_2—CH=CH$(CH_2)_7$COOH.
Linolenic ($C_{18}H_{30}O_2$), CH_3—CH_2—CH=CH—CH_2—CH=CH—CH_2—CH=CH—$(CH_2)_7$—COOH.
Eleostearic ($C_{18}H_{32}O_2$), $CH_3(CH_2)_3$CH=CH—CH=CH—CH=CH$(CH_2)_7$COOH.
Licanoic ($C_{18}H_{24}O_3$), $CH_3(CH_2)_3$CH=CH—CH=CH—CH=CH—$(CH_2)_4$—C(=O)—$(CH_2)_2$—COOH.

because they are fatty acid esters of glycerol.

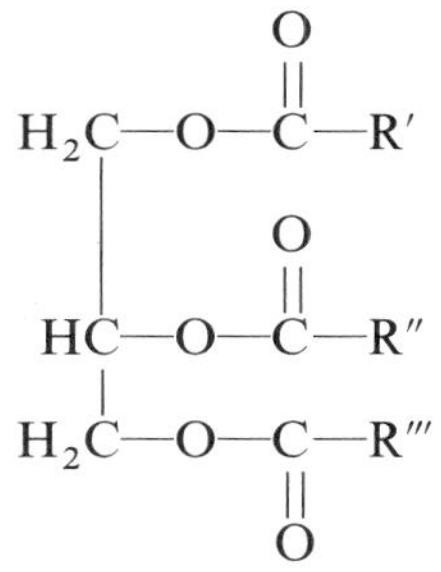

$$
\begin{array}{l}
H_2C-O-\overset{\displaystyle O}{\overset{\|}{C}}-R' \\
\ \ | \\
HC-O-\overset{\displaystyle O}{\overset{\|}{C}}-R'' \\
\ \ | \\
H_2C-O-\underset{\displaystyle O}{\underset{\|}{C}}-R'''
\end{array}
$$

where R′, R″, and R‴ are the fatty acid chains which may or may not be identical radicals.

The oils with fatty acids having 3 double bonds are classed as fast drying oils. The rate of drying can be increased by boiling the oil in the presence of driers—(metal oxides or salts of organic acids) such as those containing cobalt, manganese, etc. Lead as a drier or pigment is being phased out because of its toxic nature.

Boiled oils essentially contain the catalyst necessary to initiate the drying process. A raw oil which takes 4–6 days to dry can be made to dry in 12–15 hr when "boiled."

Blowing air through hot oil can form peroxides which can also activate the drying process.

The iodine number is the number of centigrams of I_2 absorbed by 1 g of oil. The reaction is

$$
-CH{=}CH- + I_2 \rightarrow -CHI-CHI- \tag{13.1}
$$

Thus the iodine number is a measure of the degree of unsaturation of the oil and an indication of its drying rate.

The saponification value is the mass in mg of KOH needed to completely neutralize 1 g of oil.

$$
RC\begin{matrix} {\nearrow\!\!\!\!\nearrow} O \\ \searrow OR' \end{matrix} + KOH \rightarrow RC\begin{matrix} {\nearrow\!\!\!\!\nearrow} O \\ \searrow OK \end{matrix} \quad R'OH \tag{13.2}
$$

Usually an excess of base is used and the excess is back-titrated with standard acid.

Other characteristics of an oil are the melting point, the peroxide value, the free fatty acid content, and color.

Linseed oil is the most commonly used fast drying oil for paint. It has one major disadvantage because it tends to yellow when used indoors. When exposed to sunlight the yellowing tendency is inhibited or reversed. Hence linseed oil-based, paints are generally restricted to outdoor use or dark colored indoor paints.

The natural vegetable oils can be improved by crosslinking the glycerides. This is done by first converting the triglycerides into monoglycerides by the reaction

$$\begin{array}{l} CH_2-OH \\ \quad | \\ 2CH-OH \\ \quad | \\ CH_2-OH \end{array} + \begin{array}{l} H_2C-C-\overset{\overset{\displaystyle O}{\|}}{C}-OR' \\ \quad | \\ \quad | \qquad \overset{\displaystyle O}{\|} \\ HC-C-OR \\ \quad | \\ H_2C-\underset{\underset{\displaystyle O}{\|}}{C}-OR \end{array} \rightarrow 3\left[\begin{array}{l} CH_2OH \\ \quad | \\ CHOH \\ \quad | \\ H_2C-\underset{\underset{\displaystyle O}{\|}}{C}-OR \end{array}\right] \tag{13.3}$$

It is possible to control the degree of crosslinking by the use of trifunctional acids and the molecular weight of the resin. The alkyl resin, called *alkyd paint*, dries faster than the corresponding oil which is one of its major advantages.

13.4. DRIERS

Driers are additives which accelerate the drying process of a paint film. This is effected by: (a) catalyzing the uptake of oxygen, (b) sensitizing the decomposition of the peroxides forming free radicals which add to the double bonds, causing polymerization to occur. The driers are generally metallic salts of the naphthenic acids. The mechanism by which they catalyze the decomposition of peroxides is shown for cobalt

$$Co^{2+} + ROOH \rightarrow Co^{3+} + RO + OH^- \tag{13.4}$$

$$Co^{3+} + ROOH \rightarrow Co^{2+} + RO_{2'} + H^+ \tag{13.5}$$

$$Co^{3+} + OH^- \rightarrow Co^{2+} + OH \tag{13.6}$$

The overall reaction—the sum of reaction (13.4) and (13.5) is

$$2ROOH \rightarrow RO_2 + RO + H_2O \tag{13.7}$$

The effect of various amounts of cobalt in a paint film on the time required to reach tack free dryness is shown in Fig. 13.1.

Lead has been a common drier in the past and its use is limited to less than 0.5% in the dry film. This quantity may still be too large for use where children may ingest the paint.

13.5. PIGMENT

The *pigment* in a paint is usually fine powder of 0.01 μm in diameter and is meant to cover the surface, provide color, improve the strength of the film, improve the adhesion of the paint to the surface, improve the abrasion and weathering of the film, to reduce gloss, and to control the flow and application of the paint. An extender is cheaper than pigment and serves the same purpose except it does not provide the

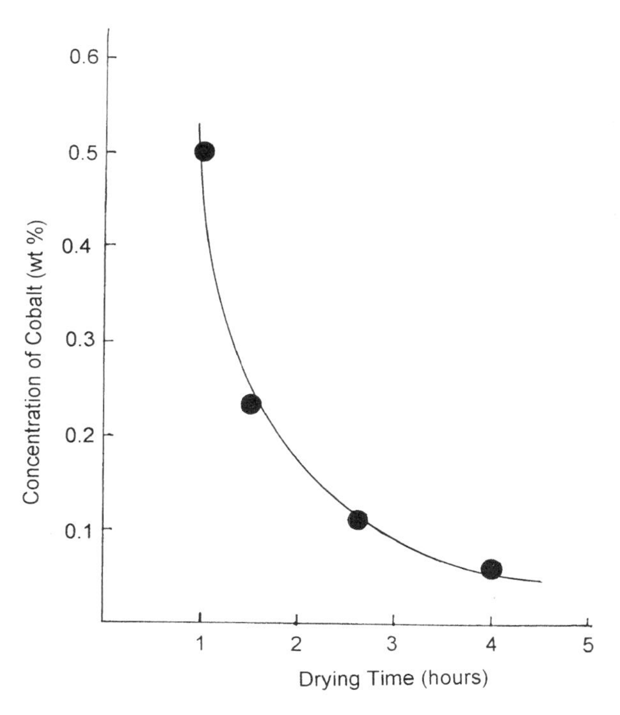

FIGURE 13.1. Effect of increasing concentration of cobalt drier on the drying time of a coating film.

required color to the film. The hiding power of a pigment is determined by four properties of the interaction of light with a solid: absorption, transmission, reflection, and diffraction. Particle size, particle density, and refractive index of the pigment contribute to the hiding power. The optimum particle size (diameter in μm) of a pigment has been shown empirically to be given by

$$d = \frac{\lambda(\mu m)}{1.414\pi n_b M} \tag{13.8}$$

where M is the measure of scattering given by the Lorentz–Lorenz equation

$$M = \frac{\left(\frac{n_p}{n_b}\right)^2 - 1}{\left(\frac{n_p}{n_b}\right)^2 + 2} \tag{13.9}$$

Some of the properties of the major pigments used today are given in Table 13.2. The hiding power of a pigment is a function of the difference in refractive index of the pigment (n_p) and that of the binder (n_b). It should be noted that extenders have a very low hiding power whereas pigments such as TiO_2 are very effective, especially when used at the optimum particle diameter where scattering is at a maximum.

TABLE 13.2
Physical Characteristics and Hiding Power of Paint Components

	Refractive index* $(Np)_D$	Hiding power m^2/kg	Tinting[†] strength	Optimum particle diameter[‡] (μm)
Pigments				
TiO_2 rutile	2.76	32	1850	0.19
TiO_2 anatase	2.55	24	1750	0.22
ZnS	2.37	12	640	0.29
Sb_2O_3	2.09	4.5	300	
ZnO	2.02	4.0	210	0.4
$Pb(OH)_2PbCO_3$	2.00	3.7	160	0.25
$Pb(OH)_2PbSO_4$	1.93	2.8	120	
Extenders				
$BaSO_4$	1.64			1.65
$CaSO_4$	1.59	Very low		
$CaCO_3$	1.57			1.75
SiO_2	1.55			
Binders				
Tung oil	1.52			
Linseed oil	1.45			
Soyabean oil	1.48			
Alkyl resins	1.5–1.6			

*Sodium D line (589 nm).
[†]Tinting strength is the relative ability of a pigment to impart color to a white base mixture.
[‡]At 560 nm.

Organic pigments are used for colored paints. Carbon blacks are classed as organic and come in a variety of properties which depend on the method of production. Metallic powders and flakes are often used especially for hot surfaces. Aluminum, copper and bronze powders also serve as decorative coatings.

13.6. SOLVENTS—THINNERS

Oil- and resin-based paints require a *solvent* to disperse the paint, reduce the viscosity and allow for easy application. The essential requirement of the paint *thinner* is that it not be highly volatile, i.e., have a boiling point within the range of 120–175°C. The earliest paint thinners were the terpenes, cyclic hydrocarbons ($C_{10}H_{16}$) which come from wood and can range in boiling points of 150–180°C.

Some of these are shown in Fig. 13.2 with a few of the oxygenated compounds which are important flavor and perfume additives. These terpenes are formed by the dimerization of isoprene

$$\left(CH_2{=}\overset{\overset{\displaystyle CH_3}{|}}{C}{-}CH{=}CH_2 \right)$$

Other solvents commonly used include the alkyl benzenes and xylenes.

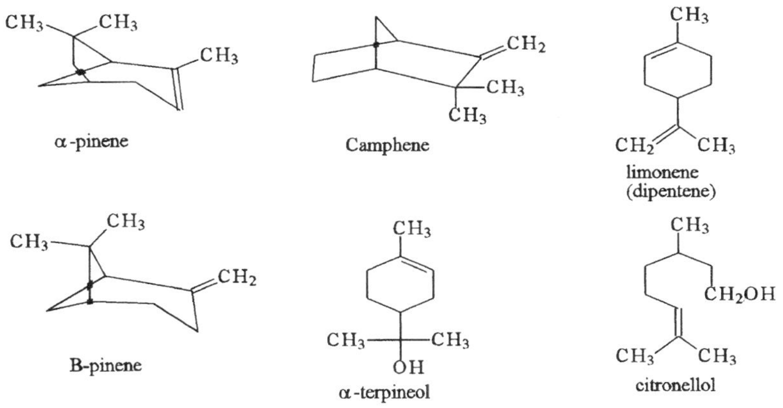

FIGURE 13.2. Structure and name of selected terpenes and oxygenated derivatives.

The trend to reduce VOC in our environment has resulted in the rapid development of water-based latex paints which have little or no volatile solvent.

13.7. WATER-BASED PAINTS

Water-based paints make use of an emulsion of a polymer composed of vinyl acetate, vinyl chloride, acrylics, acrylonitrile, ethylene, styrene, butadiene, and isoprene. The paint consists of an emulsion of polymer, known as latex because their appearance is similar to rubber latex or sap from the rubber tree. The oil-in-water emulsion has plasticizers to smooth out the drying surface and to lower the glass transition temperature of the polymer to below room temperature. As drying occurs and water is lost, the polymer particles coalesce forming a continuous impermeable film.

Emulsifiers are usually bifunctional compounds, like soaps, in which one end is water soluble and the other end is oil or polymer miscible. Other additives include surface active agents, thickening agents, preservatives, and coalescing aids. Other additives such as freeze–thaw stabilizers and wet–edge extenders can be satisfied by the addition of high boiling glycols which evaporate slowly from the surface and allow the latex particles to merge in the film.

13.8. PROTECTIVE COATINGS

The protective nature of coatings with regard to the corrosion of steel is best illustrated by the reduced transport of water and oxygen to the steel coating interface. We have shown in Chapter 10 that the overall corrosion reaction is

$$2Fe + 3H_2O + 1.5O_2 \rightarrow 2Fe(OH)_3 \quad (13.10)$$

or for an intermediate step

$$Fe + H_2O + 1/2O_2 \rightarrow Fe(OH)_2 \quad (13.11)$$

TABLE 13.3
The Permeability of Coating Films to Water Vapor from Air at 95%

Binder	Pigment	Permeability H_2O ($mg/cm^2/yr/100\,\mu m$)
Asphalt	—	190
Pheenolic resin	—	718
Phenolic resin	Al leaf powder	191
Alkyd	—	825
Alkyd	Al leaf powder	200
Polystyrene	Lithopone 30% ZnS, 70% $BaSO_4$	485
Linseed oil	Zn	1125

When ordinary unpainted steel is exposed to an industrial atmosphere the corrosion rate is approximately 70 mg/cm^2/year. Based on Eq. (13.10), the water requirements are 34 mg/cm^2/year and the oxygen needed is 30 mg/cm^2/year. Thus any protective coating must have low permeability for H_2O and O_2 if corrosion is to be reduced.

Some permeability values for H_2O and O_2 by various coatings are given in Tables 13.3 and 13.4 and show that some coatings are not suitable as rust inhibitors.

A Mössbauer spectroscopic study on rust formed under coated steel panels exposed to a marine environment for 18 months showed the presence of γ-Fe_2O_3, αFeOOH, and γFeOOH next to the paint surface. The rust at the steel interface consisted primarily of γFeOOH. The coatings used were:

- a three coats of red mud zinc chromate alkyd-based primer and two coats of alkyd-based olive green paint;
- b two coats of zinc chromate epoxy-based primer and two coats of epoxy grey paint;
- c two coats of red oxide zinc phosphate alkyd-based primer and two coats of chlorinated rubber-based grey paint;
- d two coats of manganese phosphate barium chromate alkyd-based primer and two coats of vinyl–alkyd-based paint;

TABLE 13.4
Permeability of Coating Films to O_2 from Air

Binder	Pemeability ($mg\ O_2/cm^2/yr/100\,\mu m$)
Cellulose acetate	4–9
Polystyrene	13
Poly (vinyl butyral)	27
Asphalt + talc	39
Ethyl cellulose	51

e two coats of basic lead silica chromate alkyl-based primer and two coats of chlorinated rubber grey paint.

In all cases there was sufficient permeability of water and oxygen to corrode the steel resulting in a delamination of the paint film which detaches the corrosion products from the metal surface.

13.9. SURFACE PREPARATION

The inhibition of rusting can be achieved by passivating the iron. This can be accomplished by including chromates, nitrites, or red lead in the paint. Proper surface preparation is an important aspect of effective protection against corrosion.

Metal surface treatment is usually divided into 3 classes: (1) the metal surface is not altered, (2) the metal surface is etched either chemically or mechanically, (3) metallic salts are produced at the metal surface.

(1) Solvent degreasing and oil removal treatment of a metallic surface is a common requirement. This can be done by dipping, hand wiping, or ultrasonic degreasing. Basic alkali solutions for surface cleaning are often used, though etching of the metal surface will occur unless inhibitors are present in the cleaner. Removal of the alkali requires a second rinse–wash if the last trace of a residue film of oil is to be removed.

Vapor degreasing is another form of surface cleaning which is superior to solvent dipping because the tendency to leave an oil film on the surface is greatly reduced.

(2) Sand blasting is an effective cleaning process since it can remove a thin layer of material which would include surface contaminants. Weathering, wire brushing, and flame cleaning are also used to clean surfaces. The most general surface treatment is, however, acid pickling where sulfuric acid at 1–10% by weight and at temperatures of 20–80°C is used to clean the metal by removing the oxide layers. Some etching, which usually occurs, can aid in the adhesion of the new paint.

(3) When phosphoric acid is used for surface cleaning it can leave an inert metallic phosphate coat on the surface which can then act as a corrosion inhibitor. Chromic acid either by itself or with phosphoric acid will form insoluble metallic chromates which remain on the metal surface and inhibit corrosion while binding the coat film. This treatment is used for Al, Mg, Zn, Sn, and alloys of these metals.

The treatment of nonmetallic surfaces is generally determined by the porosity of the material, its moisture content, and roughness. Such surfaces are often treated with a primer or undercoat to seal the surface and act as an intermediate bond between substrate and coating.

13.10. SPECIALIZED COATINGS

13.10.1. Formaldehyde Resins

Formaldehyde polymerizes by condensation with phenol and urea to form phenol– and urea–formaldehyde resins which are popular and inexpensive adhesives. They are often used in coatings when combined with alkyds, epoxides, polyesters, or acrylic to give strong flexible films.

Epoxy coatings are two part systems which, similar to the adhesive, polymerize by condensation to form a thermosetting film. There are basically three types of epoxy resins used in coatings.

(a) Epoxy novolac resins

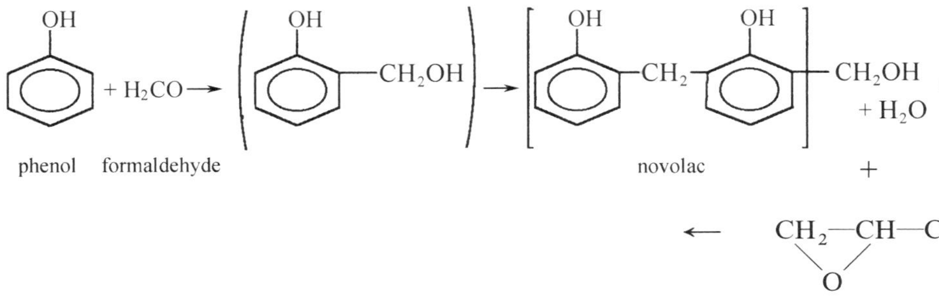

phenol + formaldehyde (H_2CO) → ($C_6H_4(OH)CH_2OH$) → novolac [$-C_6H_3(OH)-CH_2-C_6H_3(OH)-$]$CH_2OH$ + H_2O **(13.12)**

novolac + $CH_2-CH-CH_2Cl$ (epoxide ring, O) →

3HCl + glycidyl ether novolac: $CH_2-CH-CH_2-O-$ (epoxide ring) on each phenyl ring, rings linked by $-CH_2-$; terminal $O-CH_2-CH_2-CH_2$ (epoxide ring) **(13.13)**

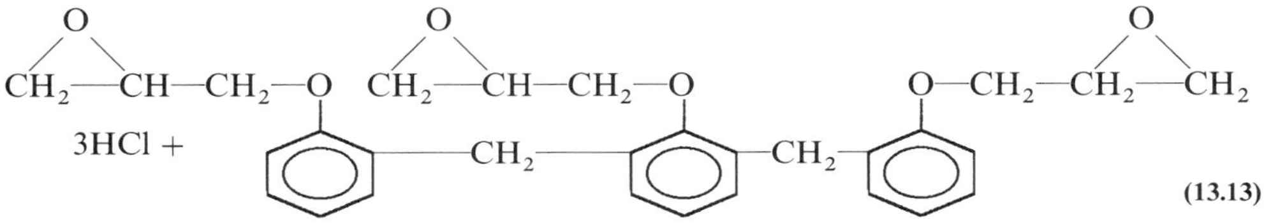

(b) Cycloaliphatic epoxy resins where the epoxy group is part of the cyclic ring.

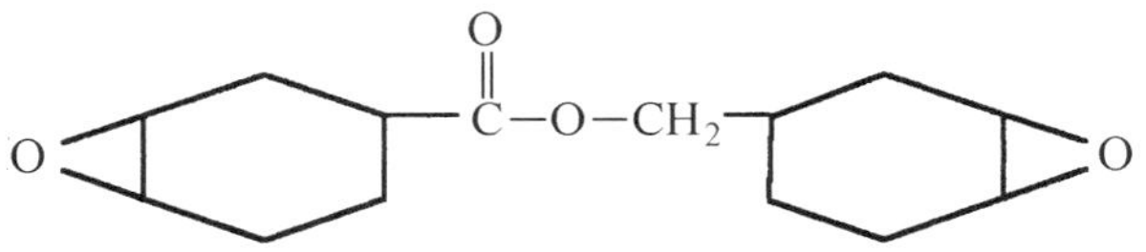

epoxycyclohexyl $-C(=O)-O-CH_2-$ epoxycyclohexyl **(13.14)**

(c) Acrylic epoxide resins where the epichlorhydrin is reacted with methacrylic acid to form glycidyl methacrylate.

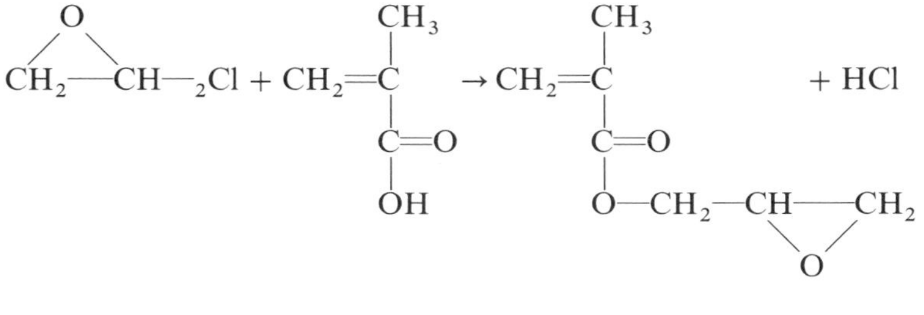

$$CH_2-CH-{}_2Cl\ (\text{epoxide, O}) + CH_2{=}C(CH_3)-C(=O)OH \rightarrow CH_2{=}C(CH_3)-C(=O)-O-CH_2-CH-CH_2\ (\text{epoxide, O}) + HCl \quad \textbf{(13.15)}$$

The resin is prepared by mixing the epoxide with a diol to form the polymer. Some triols are usually added to effect crosslinking. The curing agent used is normally an amine which is also a reactive hardener and each N—H group can react with the epoxy group by

$$\rangle NH + CH_2\text{—}\underset{\diagdown O \diagup}{}CH\text{—} \rightarrow \rangle N\text{—}CH_2\text{—}\overset{OH}{\overset{|}{C}}H\text{—} \qquad (13.16)$$

Hence each primary amine (R—NH_2) can crosslink the epoxy polymer. Silicone resins make excellent durable coating which, except for cost, provide very satisfactory protection in corrosive environments. However, silicone rubbers and elastomeric coatings are highly permeable to oxygen and water, and thus the need to include corrosion inhibitors when used on steel surfaces. Their high temperatures stability (250°C) make them ideal for heat resistant coatings. They are nonyellowing nonchalking, and ozone and UV resistant.

Polyurethane coatings are based on the reaction of a monoglyceride with a polyol and a diisocyanate to form a urethane oil or urethane alkyd. All the isocyanates react, leaving no toxic groups. These resins are alkali and water resistant, and superior to ordinary alkyds as a weather resistant coating and corrosion inhibiting surface coating.

Though water-based paints meet most of the requirements of reducing the release of volatile organic compounds (VOC), other recent coatings are being developed as more restrictive measures are introduced. Three solvent-free systems based on thermoplastic powders applied to preheated surfaces are:

1 Dipping heated substrate into a heated fluidized bed of the plastic powders.
2 Electrostatic spraying.
3 Flame spraying of the powders which as molten particles coat the heated substrate. This method can be applied in the field. In the first two methods the coating must be baked on to the base metal to allow the polymer to melt, crosslink and flow to create a uniform continuous coating.

The ability of such coatings to prevent or reduce corrosion of the metal substrate can be evaluated by the electrochemical methods such as the electrochemical noise method (ENM) and the electrochemical impedance spectroscopy (EIS). These methods rely on the permeation of electrolyte (usually a mixture of NaCl and $(NH_4)_2SO_4$) through the coating at elevated temperatures changing the resistance noise or decreasing the impedance modulus when the T_g of the coating has been exceeded. This has been interpreted to indicate that electrolyte has penetrated the coating.

The common method of testing the durability of a paint or coating and its effectiveness in inhibiting or reducing corrosion of the metal substrate is to expose coupons or samples to an accelerated weathering conditions: that is, by exposure to harsh climate conditions that includes intense sunlight and rain or in a weatherometer which is an instrument that cycles these conditions rapidly and thus attempts to predict long-term exposure effects in very short time intervals.

13.11. FIRE RETARDANT PAINTS

Paint can have an important role to play in the control of fire. If a coating releases a nonflammable gas or vapor it can prevent the oxygen from reaching the surface.

Effective additives include urea, ammonium phosphate, magnesium carbonate, antimony oxide, alumina hydrate, silicates, and borates. Though brominated and chlorinated organic compounds are fire retardants, they decompose at high temperatures to produce HBr and HCl which, as corrosive gases, can often do more harm than good.

Another mechanism of flame retardance is the intumescent coating which forms an expanded carbon layer at the burning surface and so reduces heat transfer to the coated material. The additives include mixtures of paraformaldehyde, melamine, ammonium phosphate, and chlorinated hydrocarbons. Most intumescent paints are water sensitive and thus have limited applications.

13.12. ANTIFOULING PAINTS

The problem of finding a suitable coating for ships' hulls is 2-fold—first, the need to prevent corrosion of the steel; second, the need to keep the hull from fouling due to the growth of organisms such as barnacles, mussels, tube worms, and algal spores on the surface. The fouling results in a surface roughness which increases friction and therefore fuel consumption. A typical 10,000 tonne cruiser after 6 months at sea in temperate waters required almost 50% more fuel to maintain a speed of 20 knots.

Antifouling paints usually incorporate substances such as mercury or butyl tin which are toxic to the organisms but which must slowly leach out of the coatings. However, the prevalent use of such material has contaminated the coastal waters and such materials are being replaced by less dangerous toxins such as Cu_2O. The rate at which the copper is leached from the surface must be more than 10 mg $Cu/cm^2/day$ in order to be effective. Hence, the binder pigment and other paint components must be controlled to permit such regular release of the copper.

The best type of surface coating for a ship's hull that will give the least resistance for the movement of the ship through water, i.e., the fastest speed has at times been proposed to be hydrophobic (Teflonlike surface) or hydrophillic (glasslike).

Teflon sprays are available as well as thin Teflon sheeting which have adhesive backing so it can be laminated to a ships hull.

Hydrophillic paints have been claimed to be superior to hydrophobic surfaces for ships hulls if speed is desired. One such coating is the following:

Paint

Hydroxyethyl methacrylate	10%
Ethyleneglycol monomethyl ether	40%
Ethanol	30%
Water	20%

Catalyst Ammonium dichromate 1.75% in water. The ratio is about 10:1 for paint:catalyst by volume.

If the polymer coating is slightly soluble in water then it may be able to act as a drag reducer, reduce turbulence and so increase the speed of the boat.

The continuous development in polymers and coatings make the optimum choice difficult for any specific application. However, adequate protection is usually readily available.

EXERCISES

1. Name the constituents of a paint and lacquer.
2. What type of coating would you use for: (a) steel, (b) concrete, (c) diesel exhaust pipes?
3. What are the components of a water-based paint?
4. What are the characteristic components of a fire-retardant paint?
5. What are the active components of an antifouling paint?
6. What are the characteristics of the ideal coating?
7. Calculate the iodine number of: (a) ricinoleic acid, (b) eleostearic acid, (c) licanic acid.
8. A ship's hull can be treated with a coating which can be hydrophobic or hydrophillic. Which type would you recommend and why?
9. (a) Calculate the thickness of the corrosion layer of iron corresponding to the 70 $mg/cm^2/year$. (b) Calculate the O_2 and H_2O requirements for iron rusting according to Eq. (13.11).
10. A white pigment is needed for a coating. Explain why TiO_2 is a good choice.
11. What precoating surface treatment would you consider for: (a) a nonmetal surface, (b) a metallic surface?

FURTHER READING

R. Lambourne and T. R. Strivens, *Paint and Surface Coatings: Theory and Practice*, 2nd Ed., Woodhead, New York (1999).
M. Ash and I. Ash, *Handbook of Paint and Coating Raw Materials*, vols. 1 and 2, Gower (UK), Ashgate Brookfield, Vermont (1996).
H. Innes, *Environmental Acceptable Coatings*, BBC, New York (1995).
S. LeSota, Editor, *Coatings Encyclopedic Dictionary*, Fed. Soc. Coat. Tech., Blue Bell, Pennsylvania (1995).
F. L. Bouquet, *Paint and Surface Coatings for Space*, 2nd Ed., Systems Co., Carlborg, Washington (1994).
G. Boxbaum, *Industrial Inorganic Pigments*, VCH, New York (1993).
R. Lambourne, *Paint and Surface Coatings*, Prentice Hall, Paramus, New Jersey (1993).
D. Stage, Editor, *Paints, Coating and Solvents*, VCH, New York (1993).
B. S. Skerry and D. A. Eden, 1991, Characterisation of coatings performance using electrochemical noise analysis, *Prog. Organ. Coat.* **19**, 379.
R. Woodbridge, Editor, *Principles of Paint Formulation*, Chapman and Hall, New York (1991).
S. Labana, *Advanced Coatings Technology*, ESD, The Engineering Society, Southfield, Michigan (1991).
T. Doorgeest et al., *Elsevier's Paint Dictionary*, Elsevier, New York (1990).
W. M. Morgan, *Outline of Paint Technology*, 3rd Ed., Wiley, New York (1990).
G. Schuerman and R. Bruzan, 1989, Chemistry of paint, *J. Chem. Educ.* **66**, 327.

D. J. DeRenzo, Editor, *Handbook of Corrosion Resistant Coatings*, Noyes, Westwood, New Jersey (1987).
B. S. Skerry and D. A. Eden, 1987, Electrochemical testing to assess corrosion protective coatings, *Prog. Organ. Coat.* **15**, 269.
G. E. Weesmetal, Editor, *Paint Handbook*, McGraw-Hill, New York (1981).
G. P. A. Turner, *Introduction to Paint Chemistry and Principles of Paint Technology*, 2nd Ed., Chapman and Hall, London (1980).
Surface Coatings—Oil and Color Chemists' Association of Australia 2 vol., Chapman and Hall (1974).
N. A. Toropov, *Heat Resistant Coatings*, Consultants Bureau, Transl. from Russian, New York (1967).
D. H. Parker, *Principles of Surface Coating Technology*, Interscience, New York (1966).
P. Nylin and E. Sunderland, *Modern Surface Coatings*, Interscience, New York (1965).
C. J. A. Taylor and S. Marks, Editors, *Paint Technology Manuals*: I, Nonconvertible Coatings. II, Solvents, Oils, Resin and Driers. III, Convertible Coatings, Oil and Colour Chemists Association, Chapman and Hall, London (1962).
N. A. Toropov, *Heat Resistant Coatings* Consultants Bureau, Transl. from Russian (1967).
National Paint and Coating Association, http://www.paint.org/index.htm
International Centre for Coating Technology, PRA, . http://www.pra.org.uk
Paint and Coating Mfg. Resources, http://www.pprc.org/pprc/sbap/painting.html
Paint and Coating Industry, http://www.pomag.com
Canadian Paint and Coating Association, http://www.cdnpaint.org
Finishing Info., http://www.paintcoatings.net/
Resource center, PCRC, http://www.paintcenter.org/
History of coatings, http://www.paint.org/history.htm
Papers on coatings, http://www.corrosion.com/techpage.html
Marine coatings, http://www.boatingreview.com

14

Explosives

14.1. INTRODUCTION

The first recorded account of an *explosive* was the description of a crude form of gunpowder by Cheng Yin in China about 850. He cautioned about the dangers of burning the experimenter and the house. These warnings still apply today. The first European to refer to gunpowder was Roger Bacon who concealed the formula in a code which was not revealed for another century. By 1346 gunpowder (or black powder) was used to fire a cannon in battle. In the 17th century it was used in mines as a blasting agent. The reaction still used today is

$$4KNO_3 + 2S + 7C \rightarrow K_2CO_3 + K_2S_2 + 2N_2 + 3CO_2 + 3CO \tag{14.1}$$

During the Napoleonic wars, England produced over 20,000 barrels of gunpowder a year with each barrel weighing 100 lb. The modern era of explosives began in 1846 when Soburo discovered nitroglycerine.

An explosive substance is one which undergoes a very rapid chemical reaction that is highly exothermic and which is accompanied by high pressures at the reaction site and the evolution of a large quantity of gaseous products. When noise is also produced due to a shock wave, the process is called a *detonation* and the substance is called a *high explosive.* The distinguishing characteristics of an explosive relative to fuels and propellants is given in Table 14.1, where it can be seen that the heat of reaction is not important. The most significant property of an explosive is its high rate of reaction, or the power generated, and the pressure produced. When the linear speed of reaction of a substance (the propagation of the flame front) is in the range of meters/sec the substance is classed as a low explosive and the process is called a *deflagration.*

Explosives are further divided into primary and secondary explosives. Primary explosives are detonated by heat, spark, flame, or mechanical impact, whereas secondary explosives can only be detonated by an externally applied shock wave such as commonly produced by a primary explosive.

Secondary explosives when ignited by a flame will normally burn without detonating. However, even low explosives can be made to detonate under suitable conditions depending on the material.

TABLE 14.1
The Distinguishing Characteristics of Burning Fuel, Propellants, and Explosives

Property	Burning fuel	Propellant	Explosive
Typical material	Coal, oil, gas	Hydrazine	TNT
Linear reaction rate (m/s)	10^{-6}	10^{-2}	10^{3}
$t^{1/2}$ (sec)	10^{-1}	10^{-3}	10^{-6}
Factors controlling reaction rate	Heat transfer	Heat transfer	Shock transfer
Energy output (J/g)	10^{4}	10^{3}	10^{3}
Power output (W/cm^2)	10	10^{3}	10^{9}
Common initiation mode	Heat	Hot gases and particulate	High temp. and high pressure shock waves
Pressure developed (atm)	1–10	10–1000	10^{4}–10^{6}

14.2. PRIMARY EXPLOSIVES

Some selected *primary explosives* and their properties are given in Table 14.2. Most of these initiators are toxic as well as unstable. They are usually combined with other initiators or other substances such as $KClO_3$, KNO_3, $Ba(NO_3)_2$, PbO_2, and Sb_2S_3.

The azides are usually unstable and are formed by the reaction of hydrazoic acid (HN_3) with the metal oxide. Sodium azide (NaN_3) is used to explosively fill air bags in car crashes. Silver azide (AgN_3) is more expensive than the lead azide, $Pb(N_3)_2$, which is the common detonator. Care must be taken during its production since large crystals readily explode. Contact with copper must be avoided since copper azide (CuN_3) is extremely sensitive. The high thermal stability of $Pb(N_3)_2$ makes it suitable for long term storage.

The amounts of primary initiator needed to detonate a secondary explosive varies from 10 to 400 mg and depends on both components. The major requirements of a good initiator are that it must be sufficiently stable for safe manufacturing, compatible with metal casing, easily loaded into detonators, and not too expensive. Its storage under adverse conditions must not alter its properties or stability.

The initiators are sensitive to friction and shock (blow). The percussion sensitivity is measured by the drop hammer method where a 2 kg steel ball is dropped from increasing heights until detonation occurs. The amount of electrical energy needed to initiate the detonation depends on the initiator.

Diazodinitrophenol, though sensitive to impact, is not as sensitive to friction or electrostatic energy but somewhat less stable to heat than lead azide.

Lead styphnate ($C_6HN_3O_8Pb$) or lead trinitroresorcinate is very sensitive to electrostatic discharge and often used to sensitize lead azide.

Tetrazene ($C_2H_8N_{10}O$) is readily decomposed in boiling water and is usually used as an ignition agent for lead azide. It is very sensitive to percussion and friction.

Mercury fulminate ($Hg(ONC)_2$) was the first initiating explosive which has found extensive use. However, in comparison with other primary explosives, it is relatively weak and does not store well under adverse conditions, and hence is not used much at present.

TABLE 14.2
Some Primary Explosives and Their Properties

	Mercury fulminate	Lead azide	Silver azide	Lead styphnate	Tetrazene[a]	Diazodinitro-phenol
Formula	$Hg(ONC)_2$	$Pb(N_3)_2$	AgN_3	$C_6H(NO_2)_3$	O_2PbH_2O	$C_6H_2N_4O_5$
Formula wt (g/mol)	285	291	150	468	188	210
Density (g/mL)	4.43	4.71	5.1	3.08	1.7	1.63
Color	Grey	White	White	Tan	Yellow	Yellow
Melting point (°C)	100 exp.	Dec.	252	206 explodes	Dec.	150 ± 10 exp.
Ignition temp. (°C)	215	330	300	280	130	170
Minimum value for igniting TNT (mg)	360	90	70			
Heat of formation (kJ/g)	−0.941	−1.48	−2.07	1.83	1.13	−1.59
Heat of explosion (kJ/g) to $(H_2O)g$	1.78	1.53	1.90	1.91	2.75	3.43
Specific gas volume ml (STP)/g	316	310		368	1190	865
Activation energy (kJ/g)	30	170	150	260		230
Detonation	4.5	4.5	6.8	5.0		6.9
Velocity (km/s) at density (g/mL)	3.3	3.8	5.1	2.7		1.6
Sensitivity to impact (kgm)	0.18	0.41		0.15	5	16
Sensitivity to friction (kgf)	0.43	0.12		0.08		
Drop hammer, 2 kg height (cm)	4	10		35		
Sensitivity to static electricval energy max. for non-ignition (J)	0.07	0.01	0.007	0.001	0.036	0.25

[a] Tetrazole ring (N—N, N—N, N—H) —C—N=N—NH—NH—C(NH$_2$)=NH·H_2O

14.3. SECONDARY EXPLOSIVES

The substances which have high detonation velocities but are relatively insensitive to shock can act as *secondary explosives.* Some common secondary explosions are listed in Table 14.3. There are about 120 different chemical compounds which have been used or can be used as explosives. Many are too unstable to be used industrially or for military applications. A practical explosive is one that is not too difficult to produce on a large scale, stable when stored under various conditions of temperature, humidity and vibrations, and is not too expensive.

TABLE 14.3
Properties of Some Selected Secondary Explosives of Importance

Name	Composition	Density (g/mL)	Ignition (°C)	Det. velocity (km/sec)	Gap test related to TNT	ΔH^0 expl. (kJ/g)	Spec. gas vol. (mL (STP)/g)	Sensitivity to friction (kg f)	Height of drop hammer 2 kg/(cm)	Comments
Black powder (BP)	KNO_3, 15 C, 10 S	1.2	310	0.4		2.8	280		60	Cheap, smoky, sensitive
Nitrocellulose (NC)	Nitrated cotton	1.5	180	6.3		4.3	765	10–36	20	Smokeless, grain size controls rate
Nitroglycerin (NG)	$C_3H_5(ONO_2)_3$	1.6	200	7.8		6.2	715	0–36	4	Volatile, very high sensitivity
Blasting gelatine	92/8: NG/NC (BG)	1.5	190	7.5		6.4	712		12	Strongest high brisance
Gelatine/dynamite	BG + Dope (20–90%)*	1.5	190	6.3		5.3	600		17	Jelly; waterproof, powerful
Trinitrotoluene	TNT, $CH_3C_6H_2(NO_2)_3$	1.57	225	6.9	100	4.0	690	29.5	60	Easily melted, powerful
Tetryl	$C_7H_5N_5O_8$	1.7	195	7.2	50	4.6	710	27–36	40	Cast. waterproof
Amatol	50/50: NH_4NO_3 + TNT1:1	1.5	220	6.5	45	4.0	815		55	Low sensitivity (like TNT) hygroscopic
	80/20: NH_4NO_3 + TNT4:1	1.4	220	5.4	150	3.9	860		60	High thermal stability, 1.5 times stronger than TNT
Hexogen, RDX cyclonite	$C_3H_6N_6O_6$	1.6	260	7.5	40	4.3	910	1.5	20	Plastic easily molded (vaseline)
Plastic explosive	RDX + elastomer	1.65		8.0		5.0				

*Dope is oxygen balanced explosive of either $NaNO_3$, NH_4NO_3 or explosive oils such as ethylene glycol dinitrate.

Explosives are generally characterized by five features: (1) chemical stability, (2) sensitivity to ignition, (3) sensitivity to detonation, (4) velocity of detonation, (5) explosive strength. These characteristics can have quantitative values with statistical variations which must meet stringent requirements, especially for military use. However, the occurrence of an unexpected explosion or dud may be statistically predictable when thousands or millions of events have been tried. Perfect reliability is something which can seldom be achieved.

14.3.1. Chemical Stability

The *chemical stability* of an explosive is determined by its ability to maintain its reactive characteristics and to remain chemically unchanged while stored or aged under specific conditions.

Thus, extreme temperatures of $-30^\circ C$ or $+45^\circ C$ often encountered during military operations can destabilize the materials of an explosive. Chemical instability may thus be due to the nonexplosive degradation or decomposition of the explosive substance with the result that the reliability and strength is decreased.

14.3.2. Sensitivity to Ignition

Explosive solids can be detonated by heat, mechanical impact, friction, or electrical spark or discharge. The sensitivity of an explosive is the effect of the stimulus on its spontaneous detonation.

(a) The thermal stability is determined by immersing a 5 mg sample in a thin walled metal container in a heated bath. The temperature, ranging from 100 to 300°C, at which the sample detonates after 5 sec, is the ignition temperature. The thermal stability is also determined by increasing the temperature of the sample by 5°C/min until detonation occurs.

(b) Sensitivity to percussion and impact is illustrated by the percussion cap in ammunition and the impact fuses in bombs and shells. The test involves a given mass of steel which is dropped from increasing heights onto a given mass of sample of particle size <1 mm and spread over a given area. The height from which the sample detonates 50% of the time is an indication of the hazards involved in handling the explosive.

(c) The frictional force (kg f) required to detonate the sample 50% of the time is determined by the sliding torpedo test, or the pendulum scoring test. This value determines the ease of handling the material and precautions which must be taken.

(d) The sensitivity of an explosive to a spark or electric discharge is determined by the minimum energy required to detonate the sample. The electrostatic charge on clothing has often accidentally detonated explosives while being handled. Nonsparking tools (some made of beryllium–copper alloys) are normally used in factories where explosives are handled. The maximum static energy which will not detonate a secondary explosive is usually much higher than for a primary explosive.

14.3.3. Sensitivity to Detonation

The sensitivity to detonation of an explosive is determined by the ease with which the explosive can be ignited by the primary detonation or another explosive in the vicinity which generates a shock wave. A maximum gap between the initiated detonator and the test explosive determines the sensitivity of the explosive to detonation. This gap test is an important characteristic which is usually determined relative to trinitrotoluene (TNT).

14.3.4. Velocity of Detonation

The *velocity of detonation* (VOD) of an explosive is the rate at which the detonation wave passes through the explosive. The greater the VOD the larger is the power of the explosive. The VOD is determined by several methods; an optical, electrical, and a comparative method.

In the optical method, a high speed streak camera is used to follow the flame front at about 10 million frames/sec, with exposures of as little as 0.01 μsec.

In the electrical method, a resistance element is embedded along the explosive axis, and as the detonation proceeds the resistance changes and can be followed as a function of time on an oscilloscope.

A comparative method uses a standard detonating fuse with known VOD. A fixed known length of the standard fuse is placed in a parallel loop with the test explosive. The two ends of the standard fuse are in contact with the test explosive and when detonated will ignite both standard and test explosive. The larger path of the standard fuse means that the start and end parts both ignite, causing the two flame fronts to meet at a point which is identified by an indentation on a lead plate adjacent to the standard fuse. The VOD of the sample can be calculated from the distances and the VOD of the standard explosive (see Exercise 14.10).

14.3.5. Explosive Strength

The strength of an explosive is its most important characteristic and is a measure of the conversion of its exothermic energy of combustion into mechanical disruption or power. The strength is usually determined on the basis of unit weight but the more practical basis is in terms of unit volume. Weight strength of an explosive is readily determined by comparison with a standard explosive such as Blasting Gelatine which is classified at 100%. For comparison, dynamite with 40% nitroglycerine has a weight strength of 40. The shattering power or brisance (B) of a explosive is usually compared with TNT which is set at 100. The strength values of some explosives are listed in Table 14.4. Most of the explosives are about the same strength as TNT. The B is difficult to determine experimentally and a proposed calculated value is

$$\mathrm{B} = d(\mathrm{VOD})^2 \tag{14.2}$$

where d is the density. Other expressions used to calculate B include a force factor, but

TABLE 14.4
Properties of Some Common Explosives

Explosive	Formula	Mol.wt. (g/mol)	Blasting strength	Oxygen balance g/atom/kg
Trinitrophenol	$C_6H_3N_3O_7$	229.1	106	−28.4
Trinitrotoluene	$C_7H_5N_2O_6$	227.1	100	−46.2
Tetryl	$C_7H_5N_5O_8$	287.2	126	−29.6
RDX (hexogene)	$C_3H_6N_6O_6$	222.1	150	−13.5
PETN	$C_3H_8N_4O_{12}$	316.2	146	−6.3
Nitroglycerine	$C_3H_5N_3O_9$	227.1	140	+2.2
Ammonium nitrate	NH_4NO_3	80.1		+12.5

experimental values are still the most reliable. The experiment called the *sand-bomb* test consists of 80 g of Ottawa sand −20 + 30 mesh placed in a heavy walled cylinder onto which is placed 0.400 g of the test explosive. An additional 120 g of sand is placed on top and the explosive is detonated. The mass of sand that passes through the +30 mesh screen after the detonation is the B.

14.4. OXYGEN BALANCE

The detonation of an explosive is an oxidation reaction in which it may be assumed that all the carbon forms CO_2, all the hydrogen forms H_2O, and all the nitrogen forms N_2. On this basis an explosive with the composition $C_xH_yO_zN_p$ will have an *oxygen balance* (OB) of

$$OB = z - 2x - y/2 \tag{14.3}$$

A positive value indicates a surplus of oxygen within the explosive whereas a negative value indicates that oxygen must be supplied, usually from the surrounding air. The values are expressed in terms of moles of oxygen (O) per kg of explosive. Explosives are often blended to give an oxygen balance of zero or slightly positive value. Hence, Table 14.3 shows an amatol 80/20 with 4:1 NH_4NO_3:TNT which would give an OB of $-46.2 + (4 \times 12.5)$ or a total value of +3.8. The detonation of mixture of explosives is, however, not only dependent on the OB but also on the combustion reactions as well as other physical properties.

14.5. MODERN EXPLOSIVES

The secondary explosives listed in Tables 14.3 and 14.4 represent the more common compounds. Some comments on a few of these will illustrate their relative characteristics.

14.5.1. Nitroglycerine

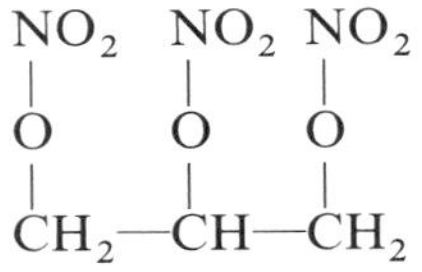

$$\begin{array}{ccc} NO_2 & NO_2 & NO_2 \\ | & | & | \\ O & O & O \\ | & | & | \\ CH_2 - & CH - & CH_2 \end{array}$$

This colorless oily liquid has a density of about 1.6 g/mL, a freezing point of 13.2°C, and decomposes before it boils at 1 atm. It is slightly soluble in water but much more soluble in acetone, ether, benzene, and chloroform. It is very sensitive to shock, especially when the liquid contains air bubbles. It is made by reacting glycerol with a mixture of HNO_3 and H_2SO_4 at -20°C.

When *nitroglycerine* is adsorbed onto kieselguhr (25% diatomaceous earth, SiO_2) a plastic cheesy mass is formed. Present stabilizers use combustible material such as sawdust, flour, starch, or cereal products. When mixed with about 8% collodion cotton, it forms blasting gelatin. Plasticizers such as ethyleneglycol dinitrate, $(CH_2ONO_2)_2$ are added to reduce the freezing point of the dynamite and to increase its OB.

14.5.2. Trinitrotoluene (TNT)

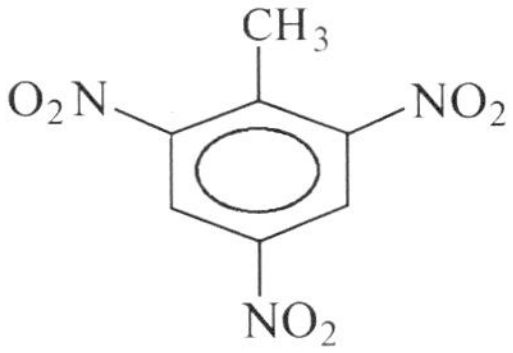

CH_3

O_2N NO_2

NO_2

Trinitrotoluene is easily made by the reaction of HNO_3 and H_2SO_4 with toluene. The order in which the reagents are mixed influences the output and safety of the process. It is thermally stable for over 40 h at 150°C and can be stored for 20 years at ambient temperatures. It is insensitive to shock and friction. It is slightly toxic at concentrations greater than 1.5 mg/m^3. Its use is extensive in the military as bombs, grenades, shells, torpedoes, and depth charges.

14.5.3. Tetryl

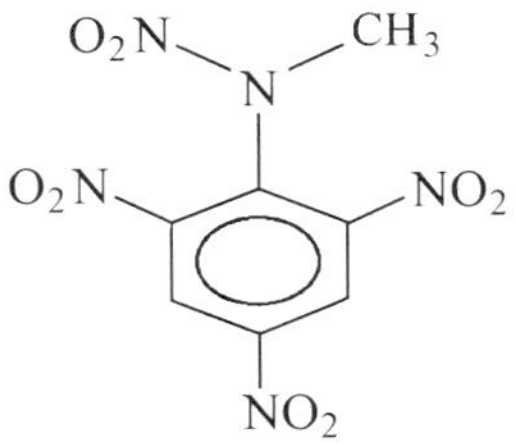

O_2N CH_3

N

O_2N NO_2

NO_2

Tetryl, also called *tetranitromethylaniline*, has a melting point of about 130°C with some decomposition. It is nonhygroscopic and practically insoluble in water but highly soluble in acetone and benzene. It can be stored for over 20 years at ambient temperatures with no noticeable change in properties.

Tetryl is prepared by the nitration of dimethyl aniline followed by the oxidation of one of the *n*-methyl groups to CO_2 and more nitration.

Tetryl is used as a booster where it is mixed with TNT (called Tetrytol) and some graphite to help pressing the mixture.

14.5.4. Ammonium Nitrate, NH_4NO_3

Ammonium nitrate is easily made by reacting nitric acid with ammonium hydroxide

$$HNO_3 + NH_4OH \rightarrow NH_4NO_3 + H_2O \tag{14.4}$$

The reaction is exothermic and can be carried out in a borehole prior to its detonation. The thermal decomposition of NH_4NO_3 yields nitrous oxide, N_2O.

$$NH_4NO_3 \xrightarrow{220^\circ C} N_2O + 2H_2O \tag{14.5}$$

The ignition temperature of NH_4NO_3 is about 465°C with heat of combustion of 2.62 kJ/g, an oxygen balance of + 12.5 g/atom/kg. It is difficult to detonate, but accidental explosions of stored tonnes of the material on land or in ships have caused thousands of deaths and injuries during the last 100 years.

Its VOD is between 1.1 and 2.7 km/s. When fuel oil is added to the salt it is known as ammonium nitrate fuel oil (ANFO). It is usually added to other explosives to increase the overall oxygen balance of the mixture.

14.5.5. Hexogen

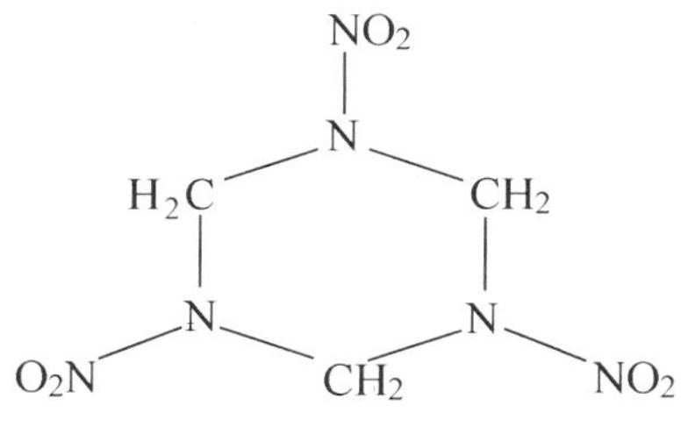

NO2
N
H2C CH2
N N
O2N CH2 NO2

Cyclonite or *Hexogen* (RDX) was first prepared in 1899 and was initially recommended for medical use. Later applications (1916) included its use as a smokeless propellant. During World War II, Germany manufactured 7000 tonnes of RDX per month, whereas the USA production was more than twice this rate.

This substance was the explosive of choice in the last world war. It is prepared by the reaction of ammonia with formaldehyde to form hexamethylene tetramine that when nitrated forms RDX with tetramethylene tetranitramine (HMX) as a byproduct.

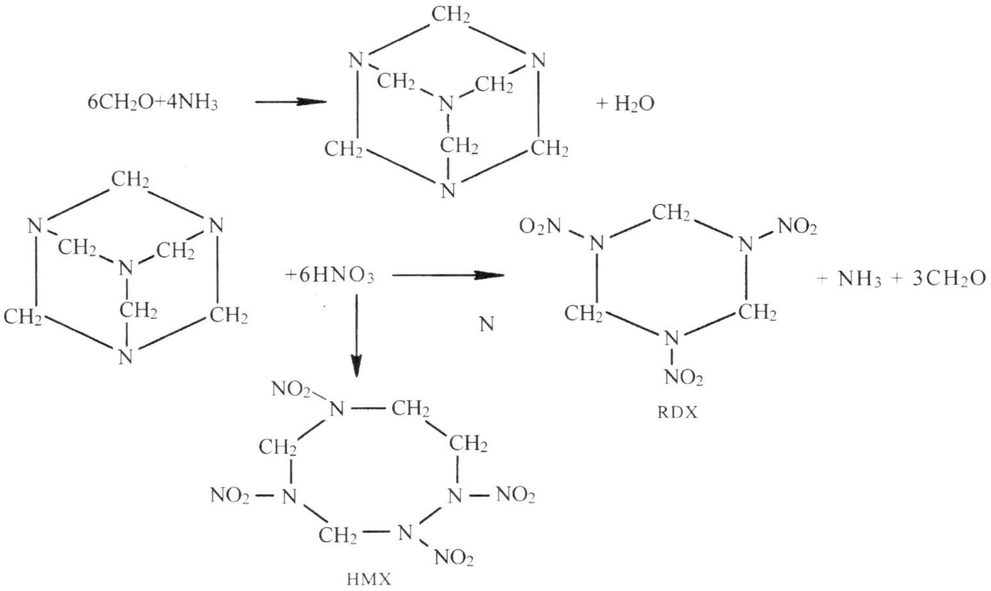

CH₂ / N N / CH₂ CH₂ / N / CH₂ CH₂ CH₂ / N

$6CH_2O + 4NH_3 \longrightarrow$ [hexamethylenetetramine structure] $+ H_2O$

[hexamethylenetetramine structure] $+ 6HNO_3 \longrightarrow$ [RDX structure: ring of N–NO₂ and CH₂ groups] $+ NH_3 + 3CH_2O$

N

RDX

↓

[HMX structure: eight-membered ring of CH₂ and N–NO₂ groups]

HMX

(14.6)

The fuels listed in Table 14.5 are practical propellants. More exotic choices such as diborane, B_2H_6; acetylene, C_2H_2; ozone, O_3; oxygen difluoride, OF_2; and nitrogen trifluoride, NF_3 are a few examples.

Solid propellants usually consist of ammonium nitrate (NH_4NO_3), ammonium perchlorate (NH_4ClO_4), and aluminum powder bound in a resin of polystyrene, polyurethane, or polybutadiene. Specific impulse values of 250–300 s are achieved depending on the choice of fuel and oxidizer.

TABLE 14.5
Some Propellants and Selected Properties

		Cryogenics		
Fuel	Oxidant	Storage temp. (°C)	ΔH^0_{comb} (kJ/mol)	I_s (sec)
H_2	O_2	−253/−183	−241.8	362
H_2	F_2	−253/−188	−271.1	352
		Storables		
Fuel	Oxidant	Storage temp (°C)	ΔH^0_{comb} (kJ/mol)	I_s (sec)
N_2H_4	H_2O_2	20/20	2686/N_2H_4	300
Ethylene oxide[a]		20	5237	170
Nitromethane[a]		20	675	220

[a]Monopropellants.

14.6. APPLICATIONS

Explosives and deflagrators are used in a variety of *applications* which include propellants, fireworks or pyrotechnics, welding, riveting, and cladding.

14.6.1. Propellants

In an age of rockets and space travel there is a constant demand for new and stronger *propellants*. These are classified into cryogenics and storable liquids, and characterized by the specific impulse (I_s) which is the thrust per unit mass flow of fuel, and by convention reported in units of seconds (see Table 14.5). The thrust depends on the combustion temperature and the exhaust and combustion chamber pressures, P_E and P_C, respectively.

RDX is often mixed with TNT (60 RDX/40 TNT) to achieve a higher VOD and thus greater strengths. When mixed with lubricating oil (12%) it became the common plastic explosive of World War II. The plastic which does not lose its flexibility, C_3, consists of 77% RDX and 23% gel composed of dinitrotoluene, nitrocellulose, and dibutyl phthalate. The putty- or doughlike material has an advantage in demolition work by forming space charges which can slice through a steel pipe or bridge support.

14.6.2. Pyrotechnics

Fireworks usually uses low explosives which are reliable, stable and of low toxicity and reasonable cost. The original oxidizer (in black powder) potassium nitrate (KNO_3) is still in use today. Potassium perchlorate ($KClO_4$), ignition temperature 560°C, is much preferred over potassium chlorate ($KClO_3$) which ignites at 220°C when mixed with sulfur. The colors are obtained from the various salts: yellow is readily obtained from sodium salts (such as Na_3AlF_6 or $Na_2C_2O_4$) due to the 589 nm line. Red is due to the strontium salts such as $Sr(NO_3)_2$ and $SrCO_3$ which emits the 606 nm band from SrO whereas SrCl gives the 636–688 nm red bands. Green is produced from barium as the nitrate or chloride. The 505–535 nm band is due to emission by BaCl. Blue is due to cuprous chloride (CuCl) which has a flame emission in the 420–460 nm range. Since $KClO_3$ is unstable in the presence of copper salts it is customary to use NH_4ClO_4 with $CuCO_3$.

The fuels consist of resins, charcoal, and sulfur. A typical aerial display shell used in fireworks is shown in Fig. 14.1. The black powder is the propellant which fires the shell to suitable heights. A fuse ignites and fires the various compartments to provide the colors and flashes in sequence.

14.6.3. Metalworking

The use of explosives in *metalworking* was initially employed in 1880 to make spittoons. Today the missile and rocket industry use explosives to shape bulkheads, nosecones, and even large rocket sections. A small 50 g charge can do the work of a

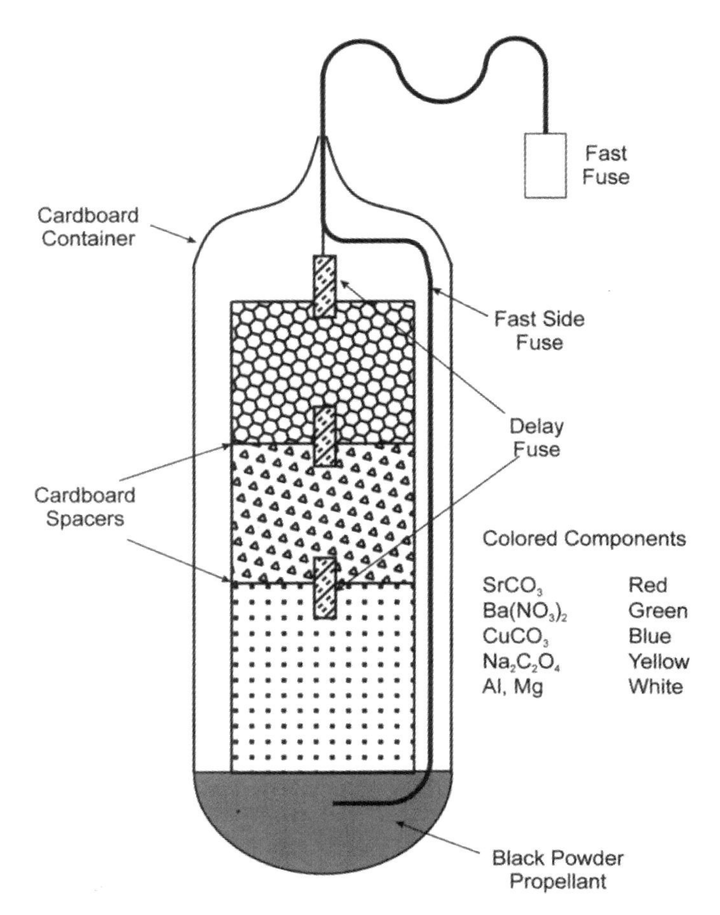

FIGURE 14.1. Typical structure of an aerial shell for fireworks display. The compartments of the rocket usually produce different colored fireworks and patterns.

1000 tonne press shaping a thick metal plate 2–3 m in diameter. An example of a metal forming system is shown in Fig. 14.2. The dye is usually cheap material of concrete or plaster which can be evacuated. The sheet metal is held in place by a bed of water in which the explosive charge is detonated. The pressure and shock wave forces the metal into the evacuated mold in μsec.

A less hazardous method is the use of an exploding wire to generate the shock wave. A thin resistive wire 0.03 mm in diameter is connected to a condenser (1–10 μF) is charged to high voltage. When the condenser discharges through the wire it explodes in microseconds creating the shock wave through the water bed. The method has been used in cladding and welding. For example, it is possible to apply a thin layer of platinum on a steel crucible which will have most of the properties of a 100% Pt crucible but which contains less than 5% Pt. Normally the metal oxides on the surface prevent the two metals from bonding. However, the shock of the explosion forces the cladding metal onto the base metal under plastic flowing conditions, producing a

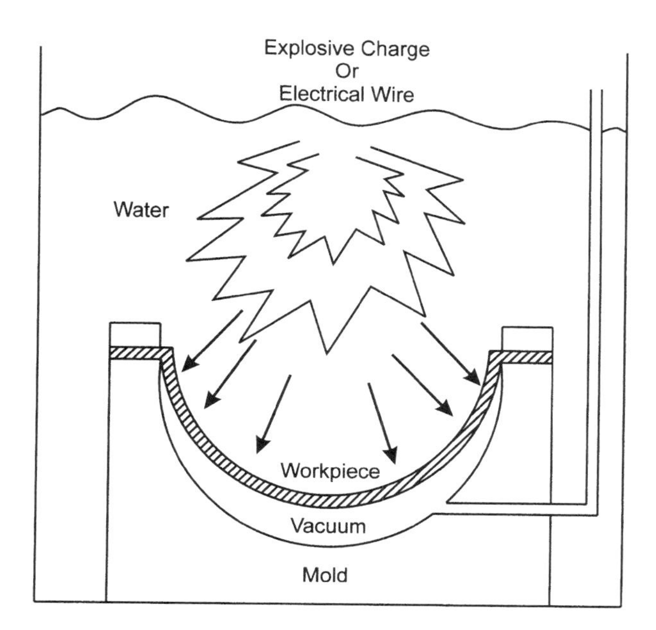

FIGURE 14.2. Typical arrangement for explosive metal forming.

metallic fluid jet at the point of contact (see Fig. 14.3). This removes the surface oxides leading to intimate welding of the two metals. The critical speed of the explosive weld is 200–300 m/sec and granular explosives are formulated to meet these requirements.

14.6.4. Riveting

Explosive *riveting* has been used for over 50 years and was originally designed for rivets where backup space was insufficient for normal processing. Applications include automobile brake shoes, aircraft industry, tanks, and other military vehicles.

The typical rivet is hollow and filled with explosive which can be detonated either thermally or by an electric charge. A typical sequence of explosive riveting is shown in Fig. 14.4.

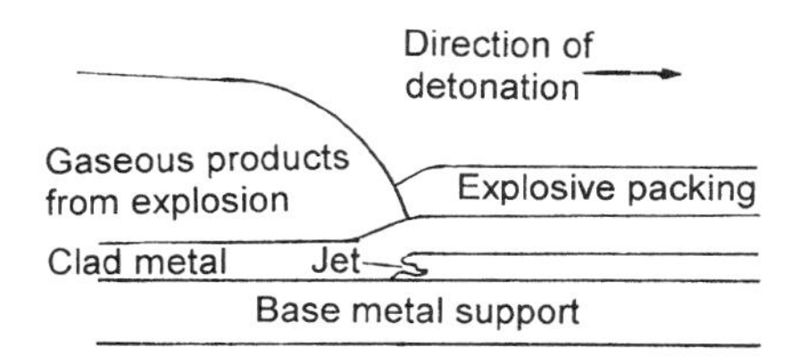

FIGURE 14.3. Schematic diagram of the explosive cladding of a metal.

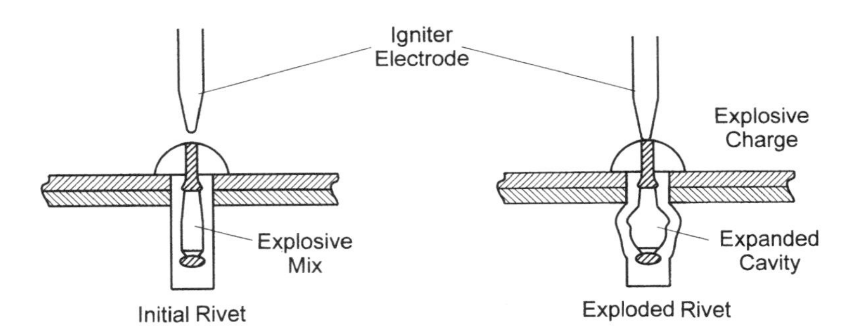

FIGURE 14.4. Typical sequence of a riveting system.

14.7. ACCIDENTAL EXPLOSIONS

From the early days of black powder usage to the industrial revolution and the extensive mining of coal, *accidental explosions* have been a constant reminder of technological development. The dust of grain causes 30–40 grain elevators to explode each year in the USA. Coal mine explosions caused about 100 deaths/year during the period 1931–1955 and in spite of precautions and improved safety regulations in coal mines and grain elevators, accidents still occur which often take lives and cause millions of dollars in losses.

There are many varied types of explosions which are accidental. A common classification is given in Table 14.6.

TABLE 14.6
Classification of Accidental Explosions

Type of explosion	Typical system and results
1. Detonation of condensed phase systems	
(a) Light or no confinement	(a) Blastwave: Manufacture, transport and handling of explosive chemical reactors, distillation columns
(b) Heavy confinement	(b) Less damaging because of the design
2. Combustion explosions in enclosures	
(a) Fuel vapor	(a) Ignition source: building, shops, boilers, low energy blast when length–diameter $(L/D) = 1$
(b) Dust ($d < 75\ \mu m$)	(b) Usually ignited by another explosion, coalmines, grain elevators, pharmaceutical industries
3. Pressure vessels	
(a) Simple failure	(a) Vessel fails due to heat or corrosion, explosion in boiler due to steam or combustion
(b) Combustion generated failure	(b) Contaminated compressed air lines
(c) Combustion and failure	(c) Fireball due to stored fuel
(d) Failure after runaway chemical reaction	(d) Exothermic reaction (Bhopal)
(e) Failure after runaway nuclear reaction	(e) Chernobyl
4. Boiling liquid expanding vapor explosion (BLEVE)	Ductile vessel (tank car) with high vapor pressure combustible liquid. Fireball
5. Unconfined vapor cloud explosion	Ignited spills of combustible fuels
6. Physical vapor explosion	Island volcanoes, liquid propane on water

TABLE 14.7
Estimation of Fuel Mass from Spherical Fireball Size

Fireball diameter D(m)	Fuel mass M (kg)	Duration t (sec)	Power (GW)
25	80	2	1.6
50	640	4	6.9
75	2160	5.8	15
100	5125	7.8	28
125	10,000	9.7	43
150	17,300	11.6	62
200	41,000	15.5	110

A fireball is the result of a fire causing a container, holding a large quantity of flammable liquid, to burst. The sudden release of the liquid which is vaporized by the heat, results in a fireball of diameter, D, lasting for a duration of time, t. Independent of the fuel or heat of combustion, the relation between the size of the spherical fireball, (diameter D in meters), the mass of combustible substance (mass M in kg), and the duration, (time t in sec), is given by

$$D = 5.8M^{1/3} \tag{14.7}$$

and

$$t = 0.45M^{1/3} \tag{14.8}$$

which are based on previous accidental explosions. Some calculated values are shown in Table 14.7 and plotted in Fig. 14.5 on a semi-log scale. The progress of the fireball and the structure, is shown in Fig. 14.6. As the pressure of the detonation products

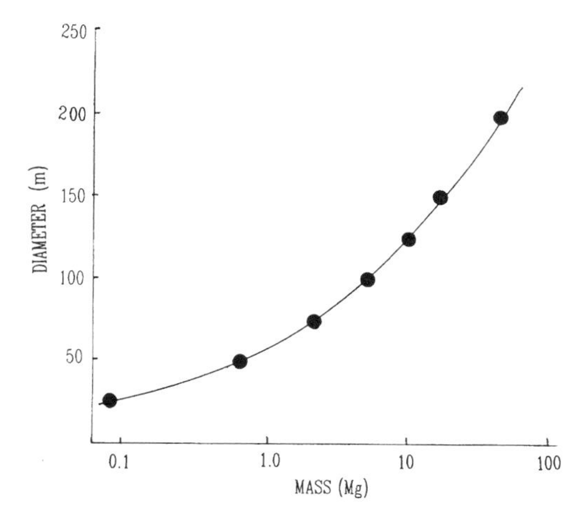

FIGURE 14.5. Semi-log plot of Eq. (14.7) from the data in Table 14.7 showing the diameter of a fireball as a function of the mass of fuel ignited.

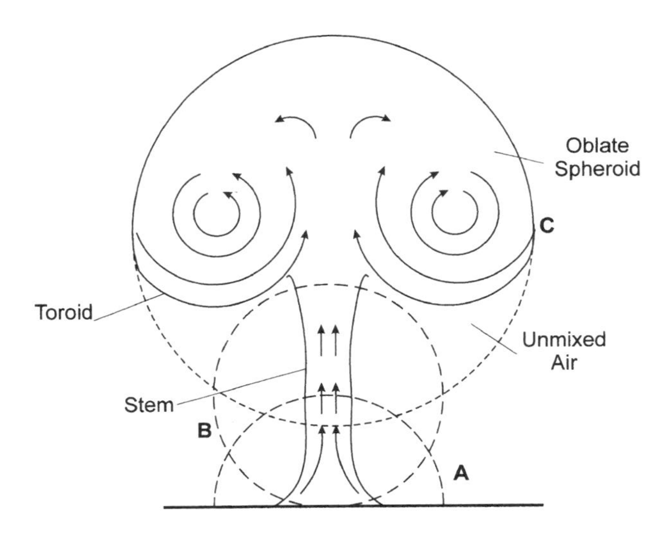

FIGURE 14.6. Typical development of a fireball from *A* to *B* to *C* showing the stem through which the fuel is funnelled.

decreases to ambient pressure, the density of the gas decreases to less than that of the surrounding air, resulting in a buoyant force causing the fire to rise. The spherical shape is formed as the fuel continues to burn and becomes more buoyant. The spherical fireball then lifts from the ground, bringing in more air by convection and vortex motion. A stem is formed from the fuel spilled on the ground, creating a mushroom-type cloud. The spherical fireball changes into an oblate spheroid, and then into a toroid, eventually consuming all the fuel and ultimately dissipating its energy.

Combustible vapors must be treated with care and respect. Standard precautions in design and construction must be followed, accidents anticipated, and provisions provided.

EXERCISES

1. Distinguish between an explosive and a detonator.
2. What chemical functional groups impart explosive properties to a compound?
3. Distinguish between a propellant and an explosive.
4. What is meant by the sensitivity of an explosive?
5. How is sensitivity to impact determined?
6. Define brisance and VOD.
7. How may the strength of an explosive be determined?
8. Write a short note on the application of explosives to metalworking.
9. Gunpowder consisted of 75% KNO_3, 15% charcoal, and 10% sulfur. Based on these properties write a chemical reaction in which K_2SO_4 is formed instead of K_2S_2 in reaction.

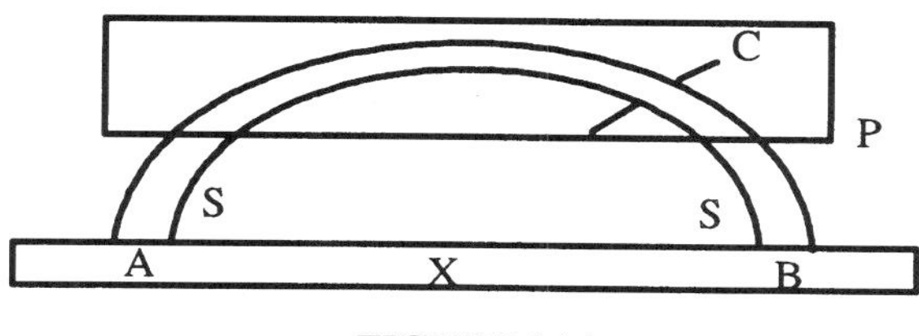

FIGURE 14.A.

10. The VOD of explosive X is determined relative to the standard explosive S (see Fig. 14.A). The explosive is initiated at O. At A the standard explosive ignites as X continues to burn. When the burn of X has proceeded to B then S ignites to meet the other flame front at C which is marked in a lead plate P. From the distances AB, AC, and BC.

    Show that $$(\text{VOD})_X = \frac{AB(\text{VOD})_S}{AC - BC}$$

11. Calculate the specific volume of gas produced at STP per g of AgN_3.
12. The explosive HMX (cyclotetramethylene tetranitramine) $C_4H_8N_8O_8$, has molecular mass of 296.17 g/mol. Calculate its oxygen balance.
13. Compare relative calculated brisance from Eq. (14.2) with the values in Table 14.4.
14. What blend of RDX and NH_4NO_3 would give a positive value for an overall oxygen balance?
15. Construct a log–log plot of Eq. (14.8) using the data in Table 14.7.
16. A recent fireball due to the explosion of a solvent recycling plant was estimated to have a diameter of 75 m, and to have lasted about 6 sec. Estimate the mass of solvent ignited.
17. A train derailment in which liquid propane exploded, formed a fireball of 250 m in diameter. Estimate the mass of propane burned and the duration of the fireball.
18. Calculate the standard enthalpy of combustion of liquid by mass and by volume. (a) Hydrogen, (b) Fluorine (see Table 14.5).

FURTHER READING

A. Bailey, *Explosives, Propellents, and Pyrotechnics*, 2nd Ed., Brasseys, Dulles, Virginia (1998).
P. W. Cooper, *Explosives Engineering*, Wiley, New York (1997).
P. W. Cooper, *Basics of Explosives Engineering*, VCH, New York (1996).
P. W. Cooper and S. R. Kurowski, *Introduction to the Technology of Explosives*, VCH, New York (1996).
M. Suceska, *Test Methods for Explosives*, Springer-Verlag, New York (1995).
S. M. Grady, *Explosives: Devices of Controlled Destruction*, Lucent Books, San Diego, California (1995).
R. Cheret, *Detonation of Condensed Explosives*, Springer-Verlag, New York (1992).
R. Meyer, *Explosives*, 4th Ed., VCH, New York (1992).
G. W. MacDonald, *Historical Papers on Modern Explosives*, Revisionist Press, Brooklyn, New York (1991).
A. Barnard and J. N. Bradley, *Flame and Combustion*, 2nd Ed., Chapman and Hall, London (1984).

K. O. Brauer, *Handbook of Pyrotechnics*, Chemistry Publication Co., New York (1974).
P. Tooley, *Fuels, Explosives and Dyestuffs*, J. Murray, London (1971).
T. Urbanski, *Chemistry and Technology of Explosives*, Eng. Transl., 3 vol., Pergamon Press, Oxford (1965).
S. S. Penner and J. Ducarme, Editors, *The Chemistry of Propellants*, Pergamon Press, London (1960).
M. A. Cook, *Science of High Explosives*, Reinhold, New York (1958).
A. M. Pennie, 1958, RDX, its history and development, *Chem. Can.* **10**:11.
International Society of Explosive Engineers, www.isee.org/
Sandia National Lab., http://www.sandia.gov/explosive/componts.htm
Can. Explosive Res. Lab., http://www.nrcan.gc.ca/mms/explosif/incerle.htm
U.S. Army Engineering Center, www.hnd.usace.army.mil
Explosives Ltd., cookbook, http://www.explosives.com/
The Ordnance shop, Types of explosives, www.ordnance.org/classifi.htm

15

Water

15.1. INTRODUCTION

In recent years, strenuous efforts have been made to conserve and recycle resources. Of major concern has been *water* due to its rising consumption and resultant shortage. Any discussion of water must include its domestic and industrial use. The treatment of waste sewage is the concern of sanitation engineers and will not be considered here.

Some selected basic properties of water are given in Table 15.1. The unique characteristics of water, although not obvious, will become apparent as we learn more about it. The three dimensional phase diagram for water, up to the pressure of 10,000 atm and for the temperature range of $-50°C$ to $+50°C$ is shown in Fig. 15.1. The density of ice (ice I) is less than that of water up to about 2200 atm. Above this pressure ice exists in various different crystalline modifications. The two dimensional phase diagram for water is shown in Fig. 15.2, where the triple point, 0.0100°C, consists of solid ice, water, and water vapor at 4.579 torr in equilibrium. Also shown is the critical point above which liquid water cannot exist in the liquid state. The negative slope of the $P-T$ line is due to the difference in molar volumes of liquid and solid, i.e. $(V_l - V_s) < 0$ and because

$$dP/dT = \Delta H_{\text{fus}}(V_l - V_s)/T \qquad (15.1)$$

This situation exists for only four known substances, water, bismuth, iron, and gallium, where the solid floats on the liquid at the melting point. Water is unique, however, since its maximum density is not at 0°C but at 4°C. This is illustrated in Fig. 15.3 where the partial structure of dimers and trimers due to hydrogen bond formation allows for a more structured arrangement of the water clusters. This structure is disrupted as salt is added as shown in Fig. 15.4, where the maximum density drops from 4° to $-1.33°C$ at 24.7 ppt (parts per 1000). At higher salt concentrations, the density of water increases with decrease in temperature. Ice formed from salt water is normally free of salt and has the normal density of ice, i.e. 0.9170 g/mL.

Water is essential to humans. The average consumption per person is 2 L/day. The average household uses about 200 L/day. Water is also necessary for many industries, for example 250 L is required for the manufacture of 1 kg of steel, and 1 kg of aluminum requires about 1300 L of water. Ten liters of water is needed for the production of each liter of gasoline, 250 tonnes for each tonne of paper, 450 L to grow enough wheat to bake a loaf of bread, grow 1 kg of potatoes, or produce one weekend newspaper.

TABLE 15.1
Selected Properties of Water

Molar mass (g/mol)	18.0153	
Melting point (°C)	0.0000	
Boiling point (°C)	100.000	
Triple point (°C)	0.0100	
Density liquid @ 25°C (g/mL)	0.997044	
Density liquid @ 0°C (g/mL)	0.999841	
Density solid @ 0°C (g/mL)	0.9170	
ΔH_f° (liquid) (kJ/mol)	−285.8	Standard enthalpy of formation
ΔH_f° (gas) (kJ/mol)	−241.8	Standard enthalpy of formation
ΔG_f° (liquid)	−237.2	Standard free energy of formation
ΔG_f° (gas) (kJ/mol)	−228.6	Standard free energy of formation
S° (liquid) (J/mol K)	69.9	Standard entropy
S° (gas) (J/mol K)	188.7	Standard entropy
ΔH_{fus} (kJ/mol)	5.98	Enthalpy of fusion
ΔH_{vap} (kJ/mol)	40.5	Enthalpy of vaporization
Heat capacity (solid)(J/mol K)	37.6	
Heat capacity (liquid)(J/mol K)	75.2	
Thermal conductivity (solid) (J/s m K)	2.1	
Thermal conductivity (liquid) (J/s m K)	0.58	
Dipole–moment (Debye)	1.84	
H—O—H angle	104°	
Bond length, H—O (pm)	96	
γ, Surface tension (mJ/m^2) @ 25°C	72	
η, Viscosity (poise, 10^{-1} kg m^{-1} s^{-1}) 25°C	0.01	
T_c, Critical temperature (°C)	347.15	
P_c, Critical pressure (atm)	218.3	
V_c, Critical volume (cm^3/mol)	59.1	
E (H—O) (kJ/mol)	464	Bond energy
D (HO—H) (kJ/mol)	498.7	Bond dissociation energy
D (O—H) (kJ/mol)	428.0	Bond dissociation energy
K_{ion} @ 25°C	1.002×10^{-14}	Ionization constant
pK_w @ 25°C	13.999	

Vapor pressure

T (°C)	0	10	20	25	50	75	90	100
P (torr)	4.6	9.2	17.5	23.8	92.5	289.1	525.8	760

The amount of water on the earth's surface is about 1.4×10^9 km^3 of which only 3% is nonocean water and of which two thirds is in the form of ice on the polar ice caps. The remaining 1% is surface and underground water in a relative ratio of about 1:3.

Seawater consists of 3.5% by weight of dissolved solids, mostly NaCl with Mg^{2+}, Ca^{2+}, K^+, and SO_4^{2-} as the main ions with many more trace components in the μg/mL and pg/mL range. For example, uranium is present in seawater at around 3 pg/mL, and gold is about 1000 times less concentrated (i.e., 3 pg/L). Distribution of water on the earth's surface is shown in Table 15.2. Water quality (as determined by standard chemical analysis) of the drinking water from some Canadian cities, and some commercially available spring waters are listed in Table 15.3. Also included are the guidelines recommended by the World Health Organization (WHO). Table 15.4 lists the WHO guidelines for the limits of organic substances which are considered a health

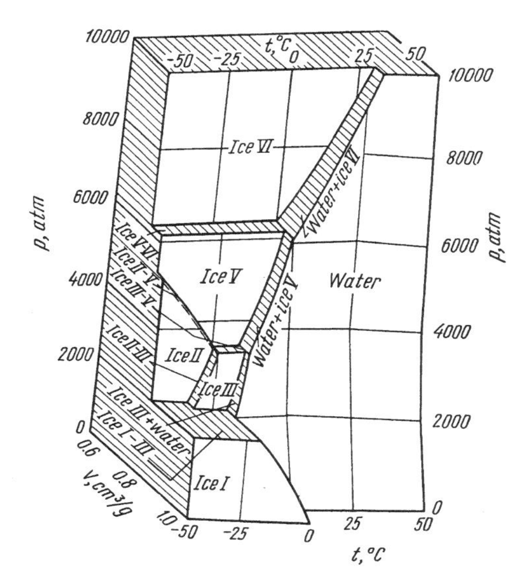

FIGURE 15.1. The 3-D phase diagram of water showing the various modifications of ice. (Ice IX is not shown in the diagram.)

hazard in potable water. These substances are seldom analyzed, but readily show their presence by ultraviolet absorption spectrophotometry.

Attempts to alleviate the water shortage which exists in many parts of the world have included extensive efforts to desalinate seawater. This continuing effort will be discussed later.

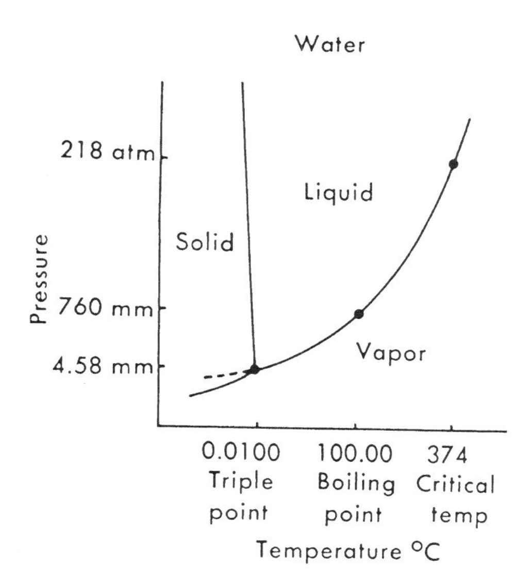

FIGURE 15.2. The 2-D phase diagram of water showing the triple point, critical temperature, and critical pressure.

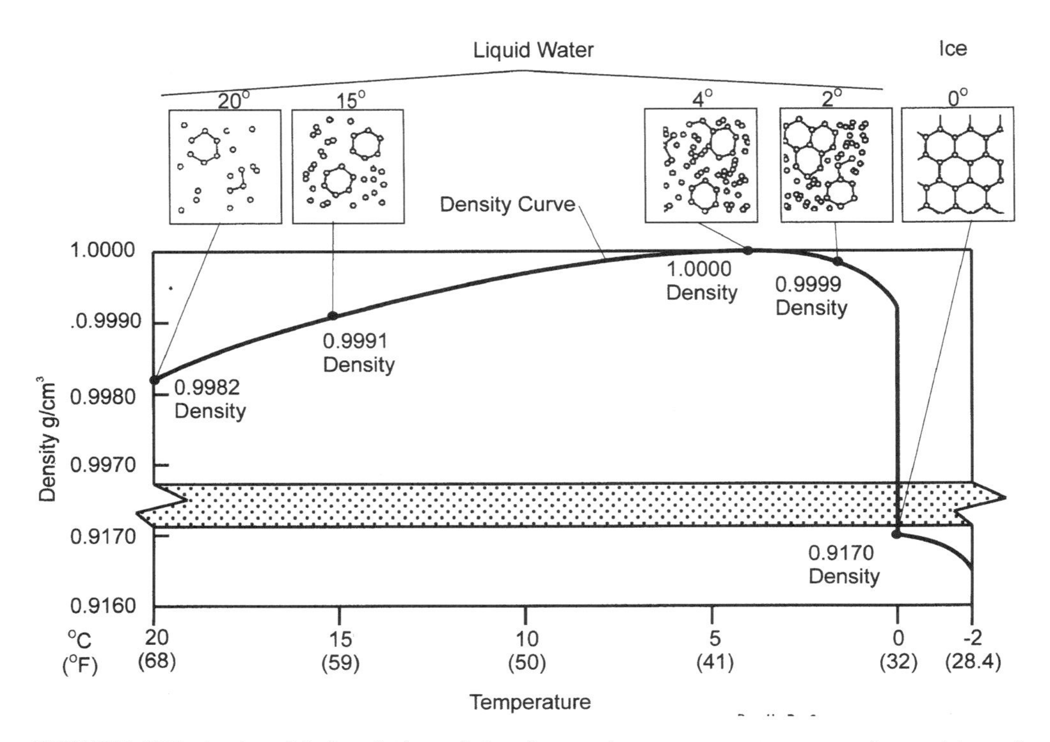

FIGURE 15.3. A pictorial description of the changes in water structure near the position of maximum density and as water changes to ice it shows the formation of ice clusters in fresh water. The density of fresh water ranges from 0.9982 g/cm^3 at 20°C to a maximum of 1.000 g/cm^3 at 4°C. However, the density of ice (solid water) is only 0.9170 g/cm^3, so it floats on water.

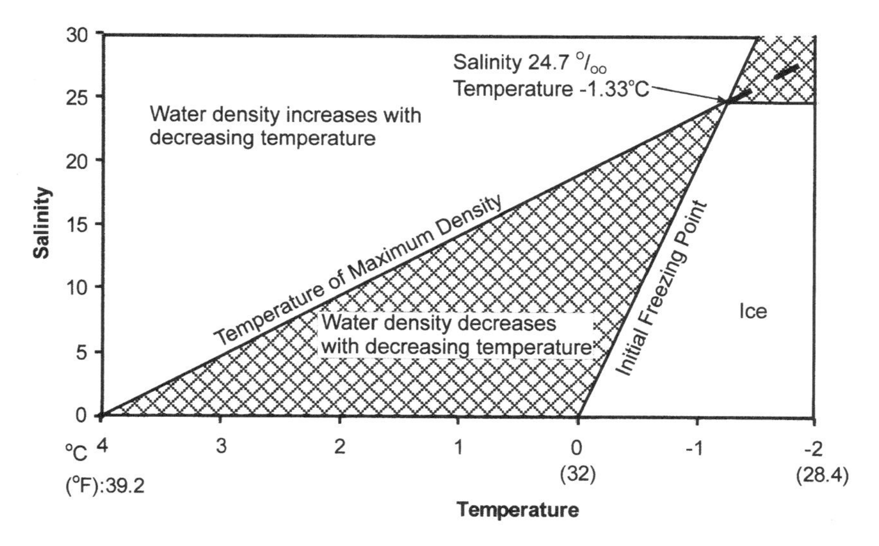

FIGURE 15.4. Freezing point and temperature of maximum density as a function of salinity, (grams of salt per 1000 g of solution).

TABLE 15.2
Approximate Water Distribution on the Earth's Surface

Source	(%)
Oceans	96.1
Ice and snow	2.2
Underground	1.2
Surface and soil	0.4
Atmosphere	0.001
Living species	0.00003
Total = 1.36×10^9 km^3	100

One elaborate scheme, proposed by Prince Mohamed Al Faesal of Saudi Arabia, involves the towing of icebergs, which are essentially salt free, from the Antarctic to desert countries. It is calculated that at a speed of 3 knots the trip could take over 100 days, and that a 1-km long iceberg would melt before reaching the Gulf of Arabia. Hence, it was proposed to insulate the iceberg to reduce its melting rate. The ice could also be used for air conditioning while being converted into potable water. The estimated cost of the water was determined to be about $0.06 per 1000 US gal.

Another recent scheme involved the transportation of water by sea from areas of plentiful supply in floating inflatable bags—called the *Medusa bag*—to areas where water is in short supply. This is considered feasible because fresh water is less dense than seawater, and will float on the ocean surface.

15.2. NATURAL WATER

The main source of domestic water is *natural water* which include lakes, rivers, and wells. It is characterized by a variety of measurable quantities which include color, turbidity, odor, bacteria, suspended solids, total solids, pH, conductivity, and most importantly, hardness.

15.2.1. Turbidity

Suspended solids and colloids cause water to be turbid. Various turbidity scales have been established for the determination of the relative *turbidity* of a water sample. Active carbon filtration can often clarify water by adsorbing colored components and removing colloids.

15.2.2. Color

The presence of dissolved organic substances in water gives rise to coloration. The *color* of water can be determined by various methods that either involve a comparison with standards which are either colored glass or standard solutions of a blend of K_2PtCl_6

TABLE 15.3

A Comparison of Various Municipal Water Supplies (T, M, W) with Selected Spring Bottled Waters (A, P, E, C) with Snow and WHO Guidelines

	T	M	W	A	P	E	C	Snow	WHO
Contents (mg/L)	1989	9/91						3/92	
Specific conductivity (μmho/cm)	328	180	181	60.5	744	576	822	28	
pH	7.6	7.4	7.45	7.14	5.58	7.63	5.58	7.18	6.5–8.5
Alkality (total) ($CaCO_3$)		70–80	72.9	24.2	300	294	328	14	
Alkalinity (bicarbonate)		80–100	88.9	29.5	367	358	400	17.1	
Alkalinity (carbonate)	84	—	<0.06	<0.06	<0.06	<0.06	<0.06	Nil	
Alkalinity (hydroxide)		—	<0.34	<0.34	<0.34	<0.34	<0.34		
Hardness ($CaCO_3$)	132	80	81.6		326	298	469	8	500
Fluoride (F^-)	1.14		1.05		0.14	<0.01	0.44	0.1	1.5
Chloride (Cl^-)	26	5–10	7.66	<0.1	23	4.9	2.6	<1	250
Nitrate (nitrite-N)	0.46	0		0.37	3.64	0.74	0.37	0.25	10
Sulfate (sol)	30	<5	2.15		24	17	164	<0.5	400
Calcium (ext)	40	20	21.5	0.89	125	76.3	100	3.02	
Magnesium (ext)	8.4	6–10	6.75	7.97	3.37	26.1	52.9	0.32	CN
									0.1
Aluminum	0.09							0.095	0.2
Iron (ext)	0.004		0.07		<0.02	0.02	0.04	0.095	0.3
Manganese (ext)	0.001		<0.02	0.02	<0.02	<0.02	<0.02	0.005	0.1
Sodium (ext)	12		2.08	0.02	8.59	6.1	21.7	1.17	200
Copper (ext)	0.0045		1.09	1.06	0.19	0.05	0.32	0.161	1
Zinct (ext)	<0.001		0.01	0.06	0.01	0.01	0.05	0.019	5
Lead (ext)	<0.005		<0.003	<0.00	<0.00	<0.00	<0.00	<0.01	0.05
				3	2	2	2	5	
Boron (sol)			<0.05	<0.05	0.07	<0.05	<0.05	0.018	
Phosphorus (total)	<0.15			<0.01	<0.01	<0.01	0.035	<0.05	Cr.
									0.05
Arsenic total			<0.001	<0.00	<0.00	<0.00	<0.00	<0.03	0.5
				1	1	1	1		
Cadmium (ext)	<0.001		<0.001	<0.00	<0.00	<0.00	<0.00	<0.00	0.05
				1	1	1	1	3	
Nickel (ext)				<0.00	<0.00	<0.00	<0.00	0.007	0.005
				5	5	5	5		
Potassium (ext)	1.5			<5	<5	<5	<5	0.54	Se
									0.01
Dissolved minerals (est)	197		136	45	558	432	617		1000
Mercury	$<5 \times 10^{-5}$								0.001

(ext, extractable metals; est, estimated concentrations.

TABLE 15.4
WHO Guidelines for Organic Constituents in Drinking Water (1984)

Constituent	Concentration (μg/L)
Aldrin and dieldrin	0.03
Benzene	10
Benzo(a)pyrene	0.01
Carbon tetrachloride	3
Chlordane	0.3
Tetrachlorethane	10
Trichloroethene	30
Chloroform	30
1,2-Dichloroethane	10
1,1-Dichloroethane	0.3
2,4-D	100
Heptachlor and heptachlorepoxide	0.1
Hexachlorobenzene	0.01
Lindane	3
Methoxychlor	30
Pentachlorophenol	10
2,4,6-Trichlorophenol	10

and $CoCl_2$ or by the more comprehensive approach, the spectrophotometric method where the transmission of visible light is determined as a function of the wavelength of light.

The color value of water depends on pH, and usually increases as the pH increases. Hence, the specification of color must also include the pH value of the water.

15.2.3. Odor and Taste

The presence of both organic and inorganic compounds in water contribute to the "chemical" senses—*odor* and *taste*. Pure distilled water is free of contaminants and has neither odor nor taste. The odor and taste of water can be removed by passing water through a column of activated carbon. Good grade drinking water is free of odor and toxic substances and has a pleasant taste.

Aeration also improves odor and taste and is often the standard treatment during the conversion of natural waters to potable grade. This full treatment involves filtering the intake water to remove debris, fish, and other large items. The water is then treated with trivalent cations such as alum $[KAl(SO_4)_2 \cdot 12H_2O]$ or ferric chloride ($FeCl_3$) that hydrolyze to amorphous, gelatinous precipitates, $Al(OH)_3$ and $Fe(OH)_3$, respectively, and as the floc settles by gravity it clarifies the water by trapping and adsorbing the suspended impurities. Aeration is used to improve the color, odor, and taste of water. The soluble organic compounds and salts are adsorbed on activated carbon followed by disinfection by the addition of chlorine.

15.3. WATER STERILIZATION

15.3.1. Chlorine

Water for drinking purposes—potable water—is treated to make it acceptable and free of harmful bacteria. This is most often accomplished by adding *chlorine* to the water which forms the strong oxidizing hypochlorous acid, HOCl.

$$Cl_2 + H_2O \rightleftharpoons HOCl + H^+ + Cl^- \tag{15.2}$$

$$HOCl \rightleftharpoons H^+ + OCl^- \tag{15.3}$$

Depending on pH and products formed, the effective residual concentration of chlorine [free available chlorine (FAC)] HOCl or OCl^- is 0.1–0.2 mg/L. Higher concentrations tend to give water a definite taste.

A more persistent source of the hypochlorite ion is the chloramines (made from the reaction of chlorine with ammonia at pH 4.5–8.5)

$$Cl_2 + NH_3 \rightleftharpoons NH_2Cl + HCl \tag{15.4}$$

$$NH_3 + HOCl \rightleftharpoons NH_2Cl_2 + H_2O \tag{15.5}$$

The dichloramine, pH 4.5, and trichloroamine, pH 4.4, are formed by successive additions of Cl_2 or HOCl to the monochloroamine. Chloramines impart a green color to water. As the hypochlorite ion is consumed by oxidizing the contaminants the chloramines supply more HOCl and so maintain a "safe" level of primary disinfectant. Excess chlorine is removed by reaction with sodium bisulfate

$$NaHSO_3 + Cl_2 \rightarrow NaCl + H_2SO_4 + HCl \tag{15.6}$$

The chlorine demand is the difference between the chlorine added and the residual concentration after a designated reaction time of approximately 10 minutes. Total residual chlorine is determined by the oxidation of *N*,*N*-diethyl-*p*-phenylenediamine (DPD) to produce a red colored product. Addition of iodine then catalyzes further reaction with chloramines and it is possible to obtain values for all chloramines and free chlorine.

The extensive use of chlorine to purify water has recently been shown to result in the formation of chlorinated hydrocarbons. Low molecular weight compounds, such as the haloforms (HCX_3), also called trihalomethanes (THM), are volatile, and have been shown to be carcinogenic. They have been detected in drinking water and in the air of enclosed swimming pools. Thus, several alternate disinfectants (such as ozone, chlorine dioxide, UV, and ferrates) have been considered as alternates to chlorine. Of these, the use of ozone has been most developed.

15.3.2. Ozone

Ozone is a triatomic molecule of oxygen, O_3, with a triangular structure where the O—O—O angle is 116° 49′. It has been used as a water disinfectant since 1903 and water treatment is the single major use of ozone. Next to fluorine (F_2) it is the most powerful oxidizing agent available. Its widespread use is due to its ability to add readily across a carbon–carbon double bond to form an ozonide

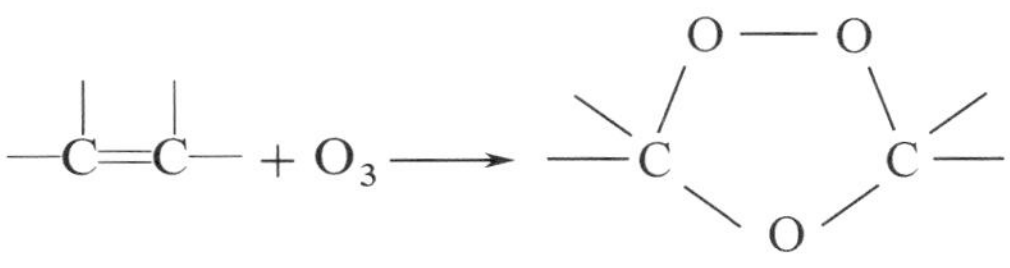

$$-\overset{|}{C}=\overset{|}{C}- + O_3 \longrightarrow \text{azonide (}{>}C(-O-O-)(-O-)C{<}\text{)} \quad (15.7)$$

azonide

which subsequently decomposes to form aldehydes, RCHO, and ketones

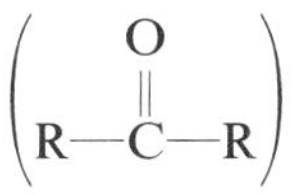

$$\left(R-\overset{\overset{O}{\|}}{C}-R\right)$$

Ozone also decomposes to form an oxygen atom that is also a powerful oxidizing agent.

Ozone is prepared either by the photodissociation of oxygen (O_2)

$$O_2 + h\nu \xrightarrow{\lambda < 200\,\text{nm}} 2O \quad (15.8)$$

$$O + O_2 \longrightarrow O_3 \quad (15.9)$$

or by a high voltage ac electrical (silent) discharge through oxygen (or air). A typical ozone generator is shown in Fig. 15.5.

A mixture of ozone and air can be bubbled through water to oxidize impurities. Ozone is toxic, and can be detected by its odor at about 0.01 ppm whereas its TLV is 0.1 ppm.

Unlike chlorine, ozone does not produce known carcinogens as a byproduct of its water treatment and, therefore, is gaining increased use for domestic as well as industrial water supplies. In North America, chlorination must follow ozonation in public water supply because ozone decomposes rapidly, and a chlorine residual may be carried throughout a distribution system. Chloramines are often used.

15.4. INFECTIOUS AGENTS

The final criterion of satisfactory sterilization of domestic water is the reduction in bacterial concentration to very low values. Bacteriological examination of drinking water uses the coliform bacteria (*Escherichia coli*—often referred to as *E. coli*) as an indication of the purity of the water since these bacteria are the normal inhabitant of the intestinal tract, and constitute about 30% of the dry weight of adult human feces. Water suitable for human consumption should contain less than one viable coliform per 100 mL.

Three different types of standard tests are used.

1. *Multiple-type fermentation method.* The various aliquots of samples are incubated for 48 hr at 35°C in a culture medium. The absence of gas formation indicates a negative test for coliform.

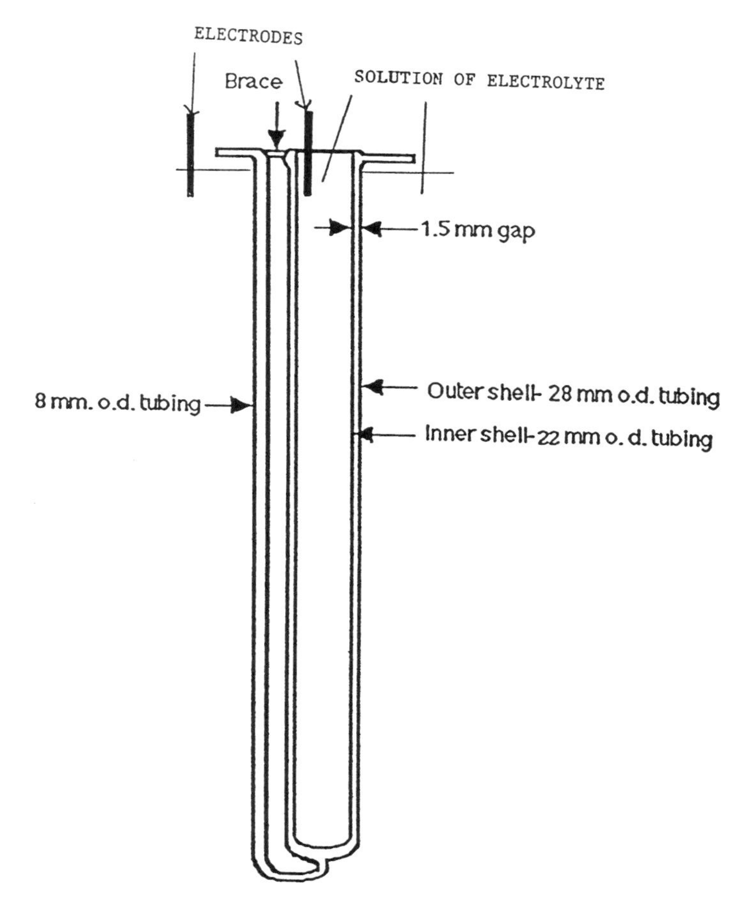

FIGURE 15.5. Silent discharge reactor tube for ozone production. The tube is filled with electrolyte and placed in a container full of electrolyte solution which hold the high voltage ac electrodes. The electrolyte is usually copper sulfate or sodium sulfate in water. The electrodes are usually copper.

2. *Membrane filter technique.* Water is filtered through a 0.45 μm sterile filter which is then placed in a culture medium and incubated at 35°C for 24 hr. The growth of visible colonies on the filter indicate the number of bacteria present.
3. *Standard plate count method.* Various diluted volumes of the sample are added to a solid agar culture medium and incubated at 35°C for 72 hr. The colonies are counted and recorded as the number per milliliter.

Common bacterial concentration is designated as the "most probable number" (MPN) per 100 mL, and represents a statistical interpretation of the results of replicate analysis.

Prior to the 1950's, most houses and buildings used galvanized iron pipes to distribute water. However, these pipes corroded and usually started leaking after about 25 years.

Copper has since replaced iron and because of the solder joints which contain lead, copper is being replaced by plastic for cold-water systems and CPVC and polybutylene are commonly used for both hot and cold water lines. There are, however, a large number of older houses which still have lead water pipes or lead connectors from the city water line. This metal is appreciably soluble in water containing carbon dioxide or organic acids. Chloride of lead ($PbCl_2$) is also soluble to a considerable extent. Cases of chronic lead poisoning have been observed, due in some instances to the habitual use of water that remained in the lead pipes overnight. The fall of ancient Rome is attributed by some scholars to the use of lead casks and vessels to store wine.

It was shown in the 1930's that fluoride in natural waters in concentrations greater than 4 ppm caused a brownish discoloration or mottling of tooth enamel. Prior to World War II it was demonstrated that such teeth were free from cavities. In 1945, fluoride was for the first time, added to the public water supply in Grand Rapids, Michigan. Since then it has been confirmed throughout the world that fluoride added to drinking water reduces cavities in tooth enamel. Fluoride is now usually added to domestic water supplies as Na_2SiF_6 to bring the F^- level to about 1 μg/mL. Though fluoride in high concentrations is harmful, the benefits of reduced dental cavities offset the potential hazards.

15.5. WATER QUALITY—HARDNESS

The quality of water varies considerably from one city to another, and from one natural source to another. The results of analysis of several water supplies are given in Table 15.3. The water for three Canadian cities and some common spring bottled waters are compared with freshly fallen snow. Specific conductivity is a rough measure of the total ion concentration in the sample. Metallic ions are determined simultaneously by their emission spectra in a plasma or flame. Anions usually require special colorimetric methods. Spring waters listed in Table 15.3 show differences which do not truly demonstrate the organic content of the water. Only by a measure of the UV transmission through at least a 10 cm path length of water over the wavelength range of 200–360 nm, is it possible to show which of the waters is the cleanest. Surprisingly, the fresh fallen snow is not the purest water.

One of the most important characteristics of domestic and industrial water, especially in boiler water systems, is a measure of its *hardness*. The hardness of water is due to the bicarbonate ion (HCO_3^-) which, in the presence of calcium and magnesium ions (Ca^{2+}, Mg^{2+}), will form insoluble carbonates when heated:

$$Ca^{2+} + 2HCO_3^- \rightarrow CaCO_3\downarrow + CO_2 + H_2O \tag{15.10}$$

$$Mg^{2+} + 2HCO_3^- \rightarrow MgCO_3\downarrow + CO_2 + H_2O \tag{15.11}$$

Hardness due to carbonate is formed when carbon dioxide (CO_2) in air is dissolved in water which is in contact with calcium carbonate ($CaCO_3$ or limestone)

$$CaCO_3\downarrow + CO_2 + H_2O \rightarrow Ca^{2+} + 2HCO_{3-} \tag{15.12}$$

This reaction is reversed at high temperatures, precipitating the $CaCO_3$, which forms as scale on the surfaces of heaters or other hot surfaces.

Hardness of a water supply is measured in terms of the equivalent amount of $CaCO_3$ that would precipitate. Thus, 40 mg/L of Ca^{2+} (1 mMolar) would form 100 mg/L of $CaCO_3$ (1 mMolar). Similarly, 24.3 mg/L of Mg^{2+} (1 mMolar), would also produce 1 mM equivalent of $MgCO_3$, and its hardness (in mg/L) would also be 100.

Soap consists of the sodium salt of a fatty acid that ionizes in water.

$$RCOONa \rightarrow RCOO^- + Na^+ \tag{15.13}$$

The cleaning agent is the fatty acid anion $RCOO^-$ which is precipitated by divalent or trivalent metal ions such as Ca^{2+}, Mg^{2+}, and Al^{3+}.

$$2RCOO^- + Ca^{2+} \rightarrow (RCCO)_2Ca\downarrow \tag{15.14}$$

The presence of calcium or magnesium bicarbonate in water is called temporary hardness, and requires the use of excess soap to precipitate the free Ca^{2+} and Mg^{2+} before the soap can work. Detergents are synthetic ionic compounds, e.g., sodium salts of sulfonic acids,

$$(C_nH_{2n+1})\text{—}C_6H_4\text{—}SO_3^- + Na^+$$

where $n = 10$ to 20.

The detergent anion does not form an insoluble salt with divalent or trivalent cation and hence it can function in hard water almost as efficiently as in soft water, if in very hard water, the Ca^{2+} and Mg^{2+} is complexed by "builders" such as sodium pyrophosphate ($Na_4P_2O_7$) or sodium tripolyphosphate ($Na_5P_3O_{10}$). The extensive use of phosphates with detergents has resulted in contamination of rivers and lakes since the detergents are not as readily biodegradable as fatty acids, though recent modification has improved this. Phosphates which are released, supply a required nutrient to aquatic plants, namely algae, which can overrun a lake with algal bloom (eutrophication), depriving the water of oxygen needed by living organisms. Lakes can die if corrective measures are not instituted. Thus nonphosphate detergents have been developed which now include "builders", such as sodium citrate, to complex the calcium and magnesium ions in hard water.

Hardness is determined by an analysis of the Ca^{2+} ion Mg^{2+} ions in solution, together with the remaining multivalent ions. Titration can be easily performed using a chelating reagent ethylene diaminetetraacetic acid (EDTA).

$$(HOOC\text{—}CH_2)_2N\text{—}CH_2\text{—}CH_2\text{—}N(CH_2\text{—}COOH)_2$$

EDTA combines with many multivalent cations such as Ca^{2+} to form an octahedrally coordinated chelate complex

CaH_2EDTA

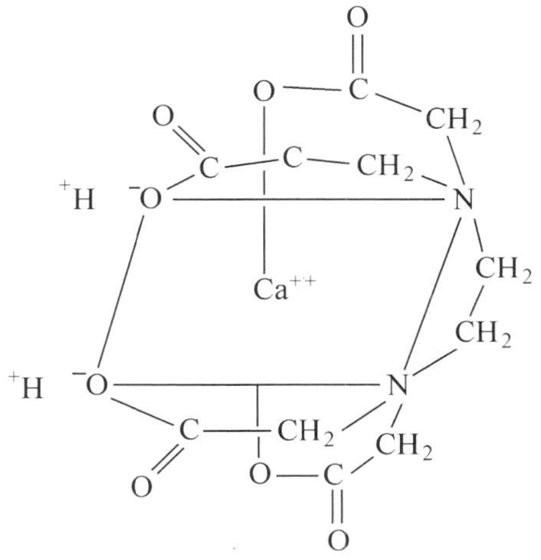

15.6. WATER SOFTENING

Hard water can be softened by several methods which vary in efficiency and cost. Distillation will result in water free of dissolved salts and nonvolatile organic substances. This can be carried out using solar energy, and many devices have been described whereby seawater can be converted into potable drinking water, usually on a small scale.

Water frozen slowly will be free of dissolved organic and inorganic compounds. The freeze–thaw cycle is used to purify semiconductors by the process called *zone refining*. Fog and dew have also been used as sources of fresh water.

The removal of the Ca/Mg hardness can be effected by the lime process. In this process, lime (as CaO or $Ca(OH)_2$) is reacted with $Ca(HCO_3)_2$ or $Mg(HCO_3)_2$ to form $CaCO_3$ and $Mg(OH)_2$, respectively.

$$Ca(HCO_3)_2 + Ca(OH)_2 \rightarrow 2CaCO_3\downarrow + 2H_2O \tag{15.15}$$

$$Mg(HCO_3)_2 + Ca(OH)_2 \rightarrow Mg(OH)_2\downarrow + Ca(HCO_3)_2 \tag{15.16}$$

Lime to be added is calculated from the measured Ca/Mg hardness and adjusted to a pH of 9–9.5 for calcium removal, but a pH of about 12 is needed to remove most of the Mg^{2+}.

The $CaCO_3$ can be used to regenerate the lime by the reaction

$$CaCO_3 - \text{heat} \rightarrow CaO + CO_2\uparrow \tag{15.17}$$

Lime treatment does not remove all of the hardness. However, should complete removal be required, then the final treatment involves ion-exchange resins.

15.6.1. Ion-Exchange

Naturally occurring minerals, such as *zeolites* or *permulites* are sodium aluminum silicates ($NaAlSiO_4$) which can exchange sodium for calcium or magnesium:

$$2NaAlSiO_4 + Ca^{2+} \rightarrow 2Na^+ + Ca(AlSiO_4) \qquad (15.18)$$

Zeolite can be regenerated by the addition of salt (NaCl) which reverses reaction (15.18).

More efficient ion exchangers such as synthetic polymers of substituted polystyrene, are referred to as *ion-exchange* resins. These are either cation or anion exchangers, and when mixed together in a single bed, will deionize water to a distilled water grade if the cation exchanger is initially in the protonated or acidic form (H^+) and the anion exchanger is in the basic or hydroxide (OH^-) form. If the cation exchanger is in the sodium form (R.Na), then the resin will exchange Ca^{2+} and Mg^{2+} for Na^+

$$2R.Na + Ca^{2+} \rightarrow (R)_2Ca + 2Na^+ \qquad (15.19)$$

Regeneration of the single bed ion exchangers is the reverse reaction, i.e., adding a solution of NaCl for reactions (15.17) and (15.18). In the case of the complete deionization process of water, acid (HCl) must be used to regenerate the cation exchange resin and base (NaOH) is used to regenerate the anion exchange resin.

15.6.2. Reverse Osmosis

Another popular method which gives a high quality water is *reverse osmosis* (RO) often called, *ultrafiltration* or *hyperfiltration.* Though often considered too expensive for industrial use, RO has found extensive applications in domestic water supplies (see Chapter 1). The production of highly efficient osmotic membranes has made RO competitive with distillation for the production of saltfree water. RO does not, however, remove volatile organic compounds (VOC) from the water supply. Treatment with granulated activated carbon (GAC) can be very effective for this purpose (see Appendix C).

15.6.3. Electrocoagulation

Recently, it has been reported that a simple patented process can convert raw water into potable water. Raw water is passed through a specially designed electrolysis cell shown in Fig. 15.6. Anodes and iron rods are centered in perforated cylindrical stainless steel cathodes with a spacing gap of about 1.5 mm between anode and cathode. The DC voltage across the electrodes is 3 V, drawing a current of about 0.2 A, producing an electrical field of up to 2000 V/m. Raw water flows upward carrying the evolved gases, namely oxygen and hydrogen (though ozone has also been reported) to an exhaust vent. Single units have 10–20 cells in parallel.

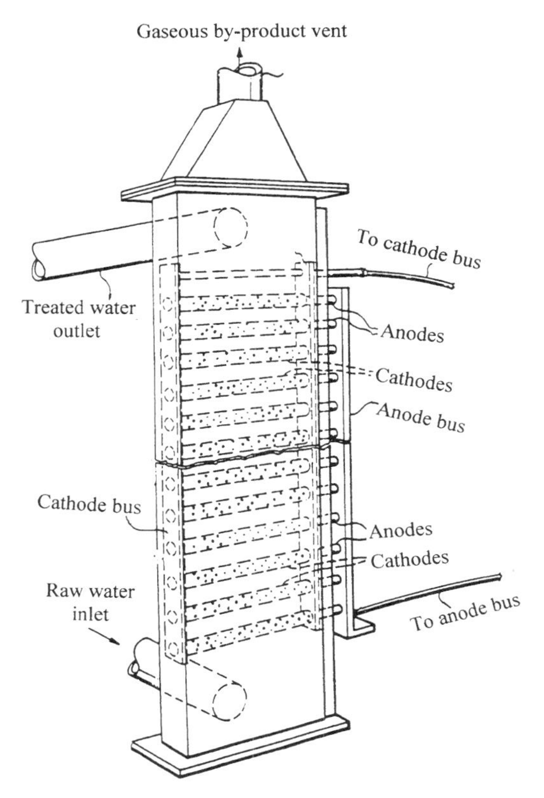

FIGURE 15.6. Electrolysis cell used in the electrocoagulation process of converting raw water into potable water.

This process has been used to prepare potable water from natural, industrial contaminated, and sewage effluent water. Bacteria such as *E. Coli* and Coliform are removed, organic material is oxidized, and heavy metals are plated out or precipitated. The chemical oxygen demand (COD)* and biochemical oxygen demand (BOD) are reduced, as well as the hardness of the water. How does the system work? It has been proposed that the highly reactive hydroxyl (OH) radical is responsible for the strong oxidizing power. The OH radical could be formed by Fenton's reaction:

$$Fe^{2+} + H_2O_2 \rightarrow Fe^{3+} + OH^- + {}^{\cdot}OH \quad (15.20)$$

Since O_3 may be formed, it is likely that hydrogen peroxide may also be produced. Fe^{2+} is readily formed from the iron anode.

Another possible process which may be occurring, depends on the presence of chloride (Cl^-) ions in the treated water. Chloride could be oxidized to chlorine (Cl_2),

*Biochemical oxygen demand (BOD) is defined as the amount of oxygen needed by bacteria in aerobic stabilization of organic matter in water. Chemical oxygen demand (COD) is the oxygen needed to completely oxidize the organic matter in the water.

which can oxidize the organic matter and kill the pathogenic bacteria. In either case, the electrolysis relies on reactants which are formed during the electrolysis. The simplicity of this process and its reported effectiveness makes it an attractive method of recycling water.

15.6.4. Electrodialysis

An electrode potential can direct the motion of ions that depends on the polarity of the voltage and the sign of the charge of the ions. This is illustrated in Fig. 15.7 where the flow of cations are directed through cation exchange membranes (C), moving towards the negative terminal, and anions will move in the opposite direction through the anion exchange membrane (A), depleting the solution of ions. The alternate channels, however, become enriched in salt. The flowing current is an indication of the desalination process.*

Some water softening (conditioning) units which are sold for domestic use, are advertized as "no saltwater conditioners." These units are reported to remove $CaCO_3$ from hard water by using a catalyst, which by epitaxial nucleation, and the reduction of pressure by virtue of a change in water velocity, converts the $Ca(HCO_3)_2$ into $CaCO_3$ and CO_2. These units are very attractive and are advertised to work with detergents but are not intended for use with soap. This can be interpreted to mean that

*When fresh water is alternated with saline water and separated by the anion and cation exchange membranes the movement of ions from the saline to the freshwater channels will result in a current flow and represent a means of generating electrical energy. This is called *reverse electrodialysis*.

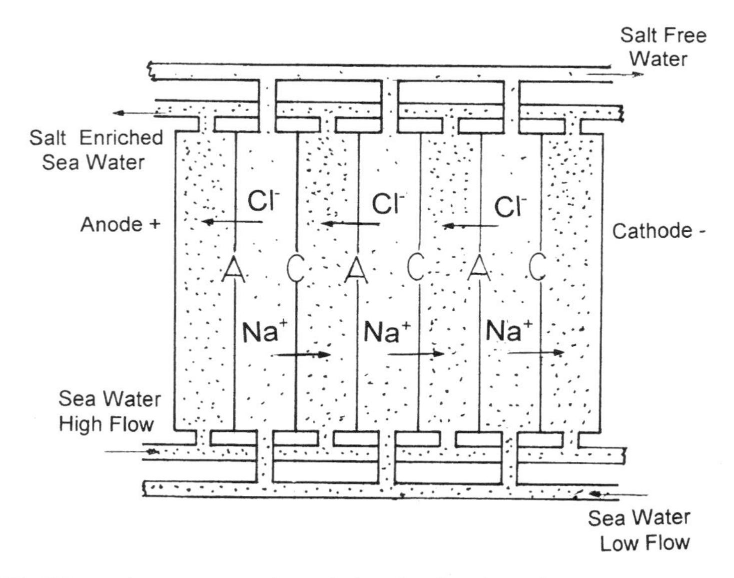

FIGURE 15.7. Schematic representation of the desalination of water by electrodialysis. "A" refers to the anion exchange membrane and "C" refers to the cation exchange membrane. The applied voltage attracts the ions from the low flow water stream into the high flow wastewater.

the calcium ions are not removed from the water system, and that a precipitate will form from the calcium salt of the fatty acid from the soap [Eq. (15.14)].

15.7. BOILER SCALE

There are several types of scale that deposit on the hot surfaces in a boiler. The most common is calcium carbonate, which is formed from the hardness of water. Scale is also formed from the deposition of insoluble salts such as calcium sulfate, calcium phosphate, or insoluble silicates. The relation used to calculate the influence of scale on the temperature drop across a boiler tube is

$$\Delta T = QL/K \tag{15.21}$$

where Q is the heat transferred in W cm^{-2}, L is the thickness of the scale in cm and K is the thermal conductivity in W cm^{-1} K^{-1}, and ΔT is the temperature drop across the pipe in K or °C.

The thermal conductivity of scale such as $CaCO_3$, is approximately 0.03 W cm^{-1} K^{-1}, and that of $CaSO_4$ is around 0.003 W cm^{-1} K^{-1}. Production of steam at 600 psi (40 atm) in a 4 cm OD tube of SA 210 carbon steel 3.4 mm thick ($K = 0.41$ W/cm K), that requires a heat flux of 40 W/m^2 has a temperature gradient across the steel tube (32 K), and a water boundary film of about 2 mm (40 K). Water temperature is approximately 250°C, and outside tube temperature is around 325°C, both well below the safe limit of 525°C.

With a $CaSO_4$ scale of 0.15 mm thick, the temperature gradient across the scale is approximately 200 K in order to maintain the rate of formation of steam. This raises the external temperature of the steel from 325°C in the absence of any scale to about 525°C. This exceeds the safe temperature for steel in an air atmosphere, and at the steam pressure generated, will result in tube failure.

Scale formation is thus a major problem of failure in steam boilers where hard water is used. Thermal conductivity of silicate scale is so low (0.0008 W/cm K) that a buildup of as little as 0.05 mm can cause boiler failure.

Scale formation can be prevented by removing salts which cause the scale or by adding substances which prevent the formation of scale in the boiler. A common additive is Na_3PO_4 though other phosphates such as NaH_2PO_4, Na_2HPO_4, and $Na_2P_2O_7$ are also used depending on the acidity of the water. Phosphate is added to the water at 25–50 mg/L to precipitate the calcium as the insoluble $Ca_3(PO_4)_2$. In small plants, additives such as tannin or starch, are added to soften the scale, making it easier to remove after it has formed. EDTA can also prevent calcium and magnesium scale by complexing the cations. Similarly, EDTA can remove scale once formed.

A simple, but as yet unexplained process of reducing scale formation is the use of a magnetic field of up to 7000 G. The method was initially proposed in 1865 and several patents have been issued for what is also referred to as a magnetohydrodynamic effect. The most successful applications are those in which the magnetic field is at 90° (orthogonal) to the water flow. The crystal size and morphology of the $CaCO_3$ formed are influenced by the magnetic field. Various claims have been made by manufacturers of magnetic treatment devices, and these include not only the reduction in scale formation but also its removal.

Water treated with ozone tends to reduce scale formation because of the chelating compounds formed by the oxidation of the organic substances in the water. This aspect, though significant, is often of secondary consideration in choosing ozone for water treatment.

The cost of scale formation to USA industry is estimated to be $10b annually, and even a small reduction in boiler scale will result in large savings.

15.8. WASTEWATER

Industrial and municipal *wastewater* treatment is designed to permit the safe disposal of the discharged water. The important components of the contaminants are:

1 suspended solids,
2 biodegradable organics,
3 pathogens,
4 nutrients,
5 industrial metals and organics.

The cleanup is classified into primary and secondary treatment.

Primary treatment includes sedimentation and filtration which reduces the BOD by about 30%. The secondary treatment contains about 40% of the suspended solids and all of the dissolved metals and organic substances. Microorganisms can be used to remove the organics by aerobic digestion where pure oxygen is often used to accelerate the biological rate. The remaining dissolved organics and metals are removed by physical processes such as adsorption, microporous membrane filtration, and by oxidation and precipitation. In some cases, further tertiary treatment is used to remove nitrogen and phosphorus compounds and other plant nutrients.

An interesting and novel approach to the treatment of wastewater has recently been developed by Delta Engineering (Ottawa) called the *Snowfluent* method. The process consists of 3 steps:

1 During freezing conditions, with outdoor temperatures below $-6°C$ the contaminated raw wastewater (from a storage lagoon) and compressed air are sprayed into the air forming fine droplets ($d <$ 200 μm) which freeze and produce snow. The volatile components, CO_2, NH_3, H_2S, and some VOCs are released into the atmosphere while the freezing of bacterial cells cause membrane ruptures, thus, sterilizing the snow. When conducted during daylight, the UV contributes to an additional disinfection effect. At distances of 30 m from the spray, the odors are significantly reduced and considered not to be a nuisance.
2 The snowpack is aged during which time the BOD decreases further, and the pH increases while nutrients such as phosphorus precipitate, and do not redissolve when the snow melts. Organic particles also separate out forming a residue on melting.
3 The remaining process leaves runoff and residue as the snow melts. Meltwater can be released into neighboring freshwater systems or used for irrigation. In the case of animal waste treatment, the runoff can be stored in a holding pond for recycling as a flushing system for barns.

The few remaining surviving pathogens in the snowpack are destroyed by exposure

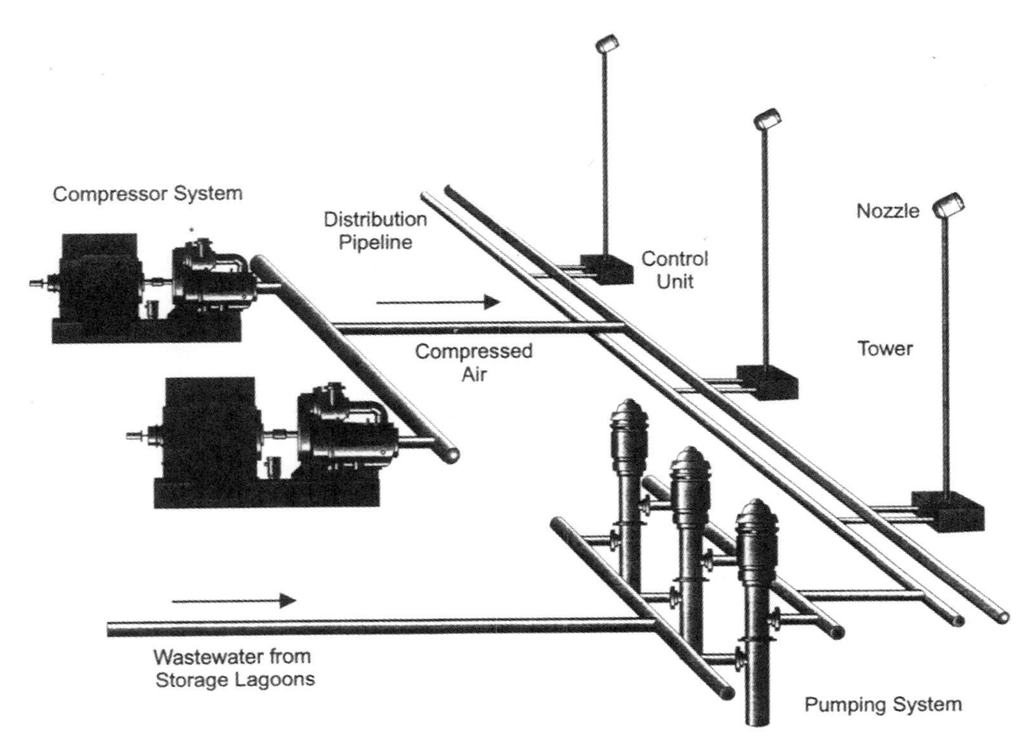

FIGURE 15.8. Diagram of the basic components of the Snowfluent process which converts wastewater into sterile snow.

to sunlight (UV) or are unable to reproduce due to ice damage. A diagram of the process is shown in Fig. 15.8. The cost of the Snowfluent process is from 10 to 25 cents/m^3 compared to \$0.50 to \$1.50/m^3 for conventional wastewater treatment.

A United Nations report has predicted that by 2025, two thirds of the world's population will be facing water shortage. The need to conserve and recycle water cannot be overemphasized if mankind is to prosper in our limited global environment.

EXERCISES

1. What are the physical characteristics that determine water quality?
2. How are the quantities in Exercise 1 evaluated?
3. How is water made safe for drinking?
4. What 6 substances (in order of decreasing importance) would you want to have analysed in your well water?
5. Write the hydrolysis reactions for alum and $FeCl_3$ used to clarify water.
6. What is hardness and how can it be determined?
7. How can the hardness of water be reduced?
8. Draw a diagram showing the EDTA complex for Al^{3+}.
9. Calculate the temperature gradient across a silicate scale of 0.1 mm thick if a heat flux of 40 W/cm^2 is required.

10. Speculate on the possible mechanism whereby a magnetic field influences the formation of scale.
11. Why is reverse osmosis so popular?
12. Calculate the amount of gold present in the oceans.
13. Explain how a nuclear powered ship could successfully transport icebergs to desert lands.
14. What is the BOD and COD of a water sample?
15. Describe how gas hydrates (see Chapter 6) can be used in the desalination of seawater.

FURTHER READING

M. Dore, Editor, *Chemistry of Oxidants and Treatment of Water*, VCH, New York (1996).
R. L. Droste, *Theory and Practice of Water and Wastewater Treatment*, Wiley, New York (1996).
S. A. Lewis, *The Sierra Club Guide to Safe Drinking Water*, Sierra Club Books, San Francisco, California (1996).
R. G. Nunn, *Water Treatment Essentials for Boiler Plant Operation*, McGraw-Hill, New York (1996).
A. E. Greenberg, Editor, *Standard Methods for the Examination of Water and Wastewater*, 19th Ed., APHA, AWWA, WPCF, Washington, DC (1995).
L. C. Wrobel et al., Editor, *Water Pollution III, Modelling, Measuring and Prediction*, Comp. Mechanics, Billerica, Massachusetts (1995).
Guidelines for Canadian Drinking Water Quality, Health and Welfare Canada (1993).
M. E. Pilson, *Introduction to the Chemistry of the Sea*, Prentice-Hall, Englewood Cliffs, New Jersey (1992).
S. Postel, *The Last Oasis: Facing Water Scarcity*, Norton, New York (1992).
J. C. Stewart, *Drinking Water Hazards*, Envirographics, Hiram, Ohio (1990).
J. D. Hem, *Study and Interpretation of the Chemical Characteristics of Natural Water*, State Mutual, New York (1990).
S. D. Faust and O. M. Aly, *Chemistry of Water Treatment*, Butterworths, Boston, Massachusetts (1983).
C. N. Sawyer and P. L. McCarty, *Chemistry for Sanitary Engineers*, 3rd Ed., McGraw-Hill, New York (1978).
F. Coulston and E. Mrak, Editors, *Water Quality*, Academic Press, New York (1977).
Water Quality Parameters, ASTM Publication #573 ASTM, Philadelphia, Pennsylvania (1975).
T. R. Camy and R. L. Meserve, *Water and its Impurities*, Dowden, Hutchinson and Ross Inc., Stroudsburg, Pennsylvania (1974).
L. L. Ciaccio, Editor, *Water and Water Pollution Handbook*, vol. 4, Dekker, New York (1971).
S. Sourirajan, *Reverse Osmosis*, Academic Press, New York (1970).
K. S. Spiegler, Editor, *Principles of Desalination*, Academic Press, New York (1966).
K. S. Speigler, *Salt-Water Purification*, Wiley, New York (1962).
American Water Works Association, www.awwa.org/
American Water Resources Association, www.awra.org/
Water Quality Association, www.wqa.org/
National Ground Water Association, www.ngwa.org/
Water Environment Federation, www.wef.org/
U.S. EPA, www.epa.gov/ow
World water forum, www.worldwaterforum.org/
Water info., www.ecoworld.com/water/ecoworld—water—home.cfm
US water resources, http://water.usgs.gov/
University Water Info. Network, http://www.uwin.siu.edu
Scale formation, http://www.fuelefficiencyllc.com/

16

Cement, Ceramics, and Composites

16.1. INTRODUCTION

Cement and its applications as concrete (a composite of cement and aggregate) is known throughout the world. The most common cement used today is Portland, named after the grey rock of Portland, England which it resembled. World production of Portland cement increased from 133 million tonnes in 1950 to about 1000 million tonnes in 1985 and close to ten times the 1950 value in 1995. The energy usage during this period dropped from 9.6 MJ/kg to about 5.7 MJ/kg in 1990. Research continues in all aspects of cement from quick setting to increase in strength—the predictability of which is still a major problem.

The history of cement starts in the earliest times when the Assyrians and Babylonians used clays to bind stones into massive walls. The Egyptians used a lime and gypsum mortar as a binding agent for the Pyramids. The Romans perfected such mortar and concrete for use in their structures, some of which still stand. They mixed slaked lime with volcanic ash from Mount Vesuvius to form a cement which hardened under water. The Mongols and Aztecs had developed a similar technology. However, the word "cement" is derived from the Roman "caemenium," meaning building stone.

This skill was lost during the Middle Ages and was not rediscovered until the scientific approach was taken by John Smeaton in 1757 when he built the Eddystone lighthouse on the southwest coast of England. He found that a good hydraulic cement was formed when the limestone used had clay impurities. We now know that aluminosilicate clays, when calcined with lime, form the desired cement. Between 1757 and 1830 the essential roles of the lime and silica were established by Vicat and Lisage in France and by Parker and Frost in England.

In 1824 Joseph Aspdin, a bricklayer and mason from Leeds obtained a patent for a superior hydraulic cement which he called *Portland Cement*. His process required the mixing and pulverizing specific quantities of limestone and clay, and heating the mixture to a required temperature forming clinkers. Unfortunately all details are missing, including the kiln Aspdin used. Two major improvements were introduced about 1890: (a) the addition of gypsum ($CaSO_4 \cdot 2H_2O$) to the clinker grinding step to act as a set retardant, (b) higher burning temperatures to permit higher lime and silicate content which results in more rapid strength development in concrete.

Portland cement presently constitutes over 60% of all cement produced and is a carefully apportioned combination of the oxides of calcium, silica, aluminum, and iron.

TABLE 16.1
Components and Nomenclature of Portland Cement

Symbol	Compound	Cement (%)	Clinker (%)
C	CaO	60–67	67
S	SiO_2	18–24	22
A	Al_2O_3	4–8	5
F	Fe_2O_3	2–5	2.6
N	Na_2O	0.1–1	0.2
K	K_2O	0.1–1.5	0.5
M	MgO	1–2	
S	SO_3	2–3	
C	CO_2	3	
H	H_2O	3	

16.2. CEMENT NOMENCLATURE

Special notation is used by cement chemists to denote the various ingredients in cement. These are listed in Table 16.1 with the approximate concentrations used. Thus tricalcium silicate is represented as $C_3S(3CaO \cdot SiO_2)$. Portland cement is a mixture of minerals: C_3S, 42–60%; C_2S, 15–35%; C_3A, 5–14%; C_3AF, 10–16%; and C and M. The minerals are formed in the kiln during the burning of limestone and the silicate clay which are usually readily available and inexpensive materials. The relative energy expended in producing various materials is given in Table 16.2. The advantages of cement in an age of rising energy costs is obvious.

16.3. MANUFACTURE OF PORTLAND CEMENT

Portland cement is made from readily available and cheap raw materials (limestone, sand, and clay). The components in an appropriate composition are mixed as a wet slurry and passed into the top end (500°C) of a rotary iron kiln which is lined with fire

TABLE 16.2
The Relative Energy Content of Various Materials

Material	Relative energy (vol.)
Portland cement	1.0
Flat glass	3.0
Polyvinylchloride	3.8
Polyethylene (low density)	4.2
Polyethylene (high density)	4.4
Polystyrene	6.0
Steel	19.2
Stainless steel	28.8
Aluminum	31.8
Zinc	34.8

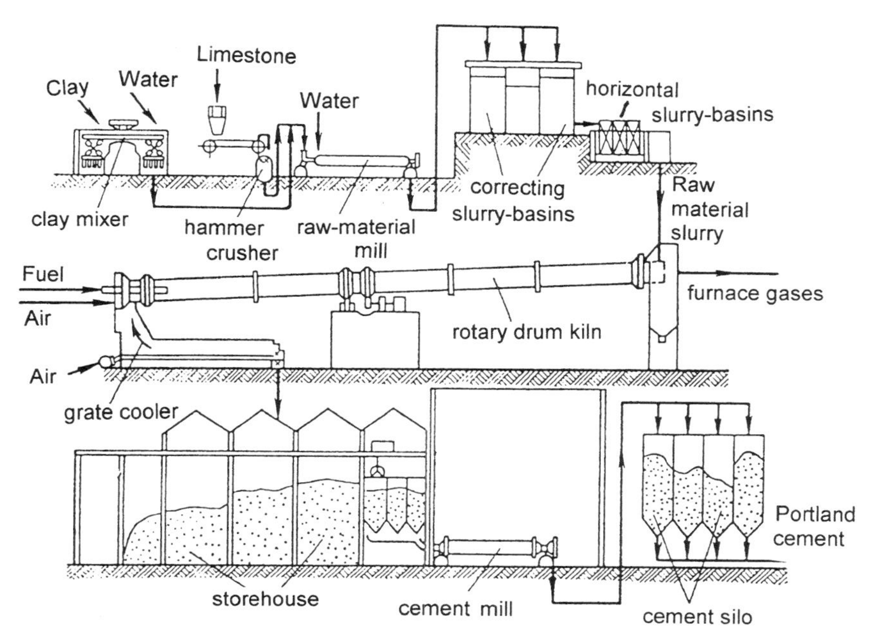

FIGURE 16.1. Flow diagram for the manufacture of Portland cement by the wet process. The limestone is crushed, mixed with wet clay, and ground to a fine slurry in a mill. This raw material is stored and corrected for composition by blending before being fired in a rotary kiln where the process of water evaporation, mineral dehydration, limestone dissociation, and chemical reaction proceed. Clinker formation forms finally at 1450°C and is cooled and ground with additives before storage.

brick, about 150 m long, and at a 15° angle to the horizontal. A diagram of the flow process commonly used in the manufacture of cement is shown in Fig. 16.1. The following reactions occur in the kiln as the temperature increases.

1 Free water is evaporated.
2 Combined water from the clay is released.
3 Magnesium carbonate is decomposed and CO_2 is released.
4 Limestone is decomposed to form lime and CO_2.
5 Lime and magnesia combine with the clays and silica to form "clinker."
6 Cooled clinker is then ground and some 20% gypsum is added to prevent the cement from setting too rapidly.

A common accelerator which speeds up the hydration process is $CaCl_2$ but its corrosiveness makes it unacceptable in steel reinforced concrete.

An alternate process which is gaining popularity is to introduce the ingredients into the kiln in the dry state and so reduce the energy required to drive off the water present in the wet slurry process.

16.4. SETTING OF CEMENT

The *setting* of Portland cement consists of the hydration of the various silicates and aluminates as well as the compound $3CaO \cdot Al_2O_3 \cdot 3CaSO_4 \cdot 32H_2O$ (a process which occurs over a long period). Attempts to reduce setting time and time to maximum strength have included resistive heating, accelerated microwave, and RF drying as well as the use of additives. Evidence of gel formation with fibrillar growth during the dehydration has also been obtained.

Hardened cement is a porous solid with a density of about 2.5 g/mL and a surface area of 100–300 m^2/g. Such pores can be fully sealed by polymer impregnation that has numerous applications.

The compressive strength of cement increases from 10^2 N/m^2 when initially set to 10^7 N/m^2 after a few days, reaching 70% of its final strength value (10^8 N/m^2) after 28 days. Further small increases in strength are observed even after 1–2 years. Thus the normal working stress, with the appropriate safety margin, is restricted to about 10 MPa. The tensile strength, σ, of a solid (measured in bending) can be given by the Griffith relationship

$$\sigma = (ER/\pi c)^{1/2} \tag{16.1}$$

where E is Young's modulus of elasticity, R is the surface fracture energy, and c is the crack length. A reduction in particle size of the cement and the removal of all the air bubbles creates a macro defect free (MDF) cement which results in a strength of up to 150 MPa, a value comparable to aluminum and its alloys. The high setting strength of MDF cement is illustrated by the coiled cement spring in Fig. 16.2 which is shown in the relaxed and extended states.

16.5. CONCRETE

Concrete is composed of cement and filler, or aggregate, which can be gravel or stones. A proper concrete is composed of aggregate of different sizes which permits better packing and fewer voids. For lightweight concrete the aggregate can be vermiculite, perlite, or other low density filler. The concrete must be vibrated as it is poured. The tensile strength of concrete can be increased by embedding iron rods, called *Rebar*, and iron wire mesh in the slurry before it sets. The coefficient of expansion of concrete and steel are similar and thus temperature changes do not disrupt the structure—though corrosion of the iron must be avoided if long term strength is to be maintained.

The strength of polymer impregnated concrete is determined by the glass transition temperature, T_g, of the polymer. Above the T_g the reinforcement is lost and the concrete is less resistant to salt penetration.

A patented invention (1995) has described how a conducting concrete can be made by adding conducting carbon fibers and particles to the cement. An electrical current through the concrete can heat the material. This can have a broad application for de-icing in areas where freezing temperatures occur and cause traffic accidents. Airport

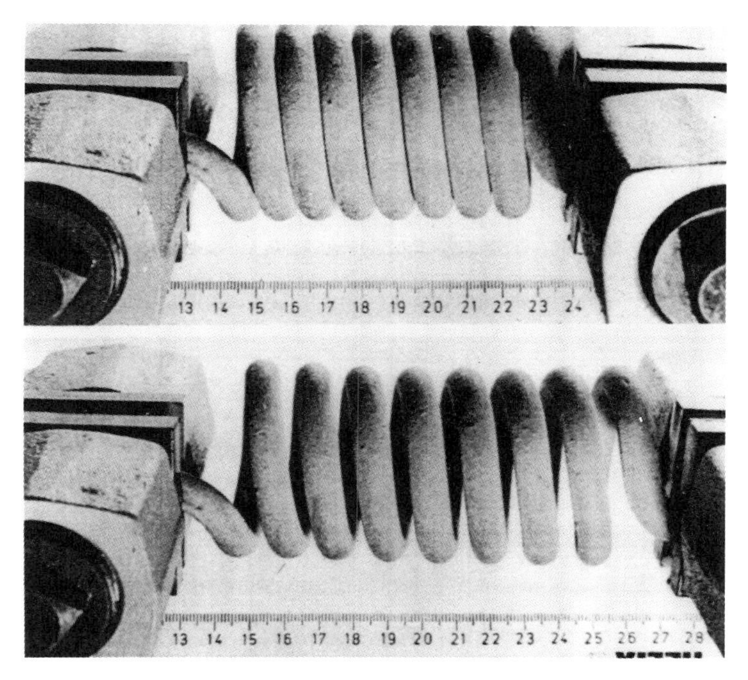

FIGURE 16.2. Coiled high strength Micro Defect Free (MDF) cement spring in relaxed and extended positions.

runways, bridges, and highway intersections are other important locations where ice free surfaces are desirable. Another application is as a secondary anode in existing cathodic protection systems.

The cement content of 28 day dried concrete is normally determined by dissolving the cement in hydrochloric acid (HCl), leaving the insoluble aggregate. A simpler and rapid determination can be carried out using differential thermal analysis (DTA) of a thoroughly ground and mixed sample (mesh <80). A typical DTA trace is shown in Fig. 16.3 where the area of the cement peak is directly proportional to the amount of cement in the concrete. It remains to determine the ultimate strength of concrete while the cement is still a slurry in the cement truck. Though several methods have been proposed none has, as yet, had any success.

16.6. CERAMICS

Ceramics are inorganic solids, usually oxides, which contain ionic and covalent bonds. The material, formed by sintering at high temperatures, range from amorphous glasslike material to highly crystalline solids, from insulators to conductors or semiconductors. They include earthenware, which is fired at 1100–1300 K and a porosity of about 8%; fine china or bone china, fired at 1400–1500 K with a porosity of less than

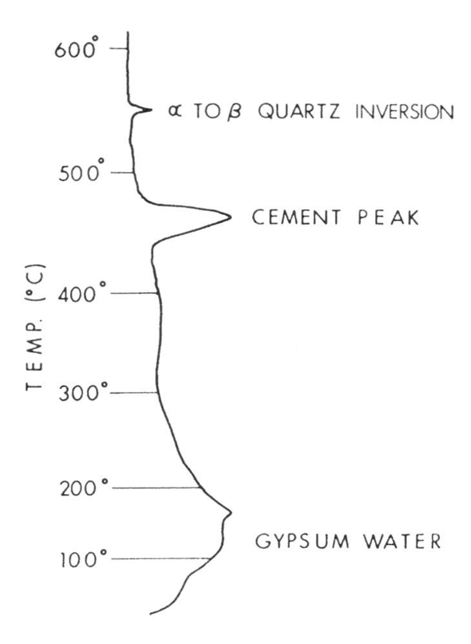

FIGURE 16.3. Differential thermal analysis (DTA) of concrete sample ground to <80 mesh. Area of cement peak is proportional to amount of cement in sample.

1%; stoneware, fired at over 1500 K with a porosity of about 1% before glazing; and porcelain which is fired at over 1600 K and has a much finer microstructure than either stoneware or bone china.

One of the most noteworthy ceramics is the high temperature superconductors that were first described is 1988, and show superconductivity at temperatures as high as 90 K.

The superconductor $YBa_2Cu_3O_7$ ($T_s < 90$ K), known as the "1-2-3" compound, is prepared by mixing powdered Y_2O_3, $BaCO_3$, and CuO in the appropriate stoichiometry and heating to 950°C. The cooled solid is then pressed and sintered again at 950°C to bond the grains and to increase the density of the pellet. Final heating is in the presence of oxygen at 500–600°C followed by slow cooling to room temperature.

The glass or vitreous state of matter is a solid with the molecular random structure similar to that of a liquid. Such a state is not limited to inorganic substances since many organic compounds and mixtures can form transparent glasses. Glass can be translucent or opalescent due to the presence of small crystalline material in the glass. However, the term "glass" normally refers to a solution of inorganic oxides with silica (SiO_2) as the basic material.

Ceramics can be cast prior to heat treatment or, like glass, can be formed by the sol–gel process, a recent development whereby a glass can be formed without first forming the melt. The oxides are converted into a colloidal gel which can also be formed from the organic alkoxide, $M(OR)_n$ where n is the valency of the element M. The controlled hydrolysis reaction in an organic solvent such as methanol forms the

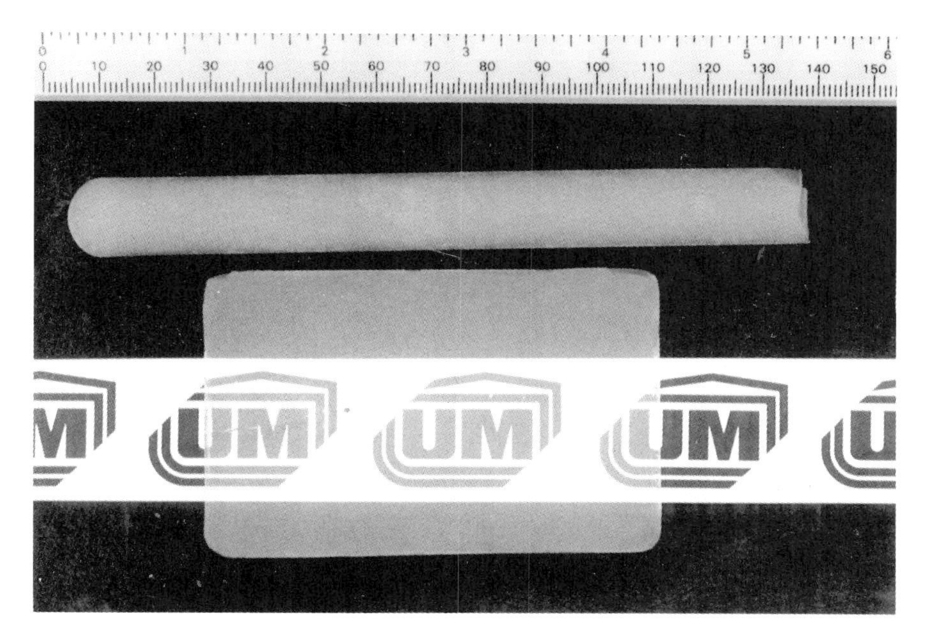

FIGURE 16.4. Photograph of two aerogels prepared from the hydrolysis of tetraethylortho silicate in methanol and dried under supercritical conditions.

hydroxide gel:

$$M(OR)_n + nH_2O \rightarrow M(OH)_n + nROH \tag{16.2}$$

When hydroxide gel is dried slowly in air a xerogel is formed which is porous, often transparent, and usually smaller and more dense than the original gel. If the solvent is removed under supercritical conditions (temperature and pressure above the critical point), the shrinkage is minimal and the solid is called an *aerogel*. This is shown in Fig. 16.4. Aerogel has a very low thermal conductivity (10 mW/m, K) which is slightly lower than that of a xerogel and much lower than polyurethane foam (28 mW/m, K). A recent silica xerogel produced in Norway had a bulk density of 0.26 g/cm^3 and a thermal conductivity of 18 mW/m, K. These materials are being studied for possible use as insulating windows because of their high degree of transparency.

The possibility of blending, doping, or mixing various materials in the gel state make the resulting homogeneous solids very easy to prepare to exact specifications. When a xerogel or aerogel is heated the pores collapse, forming the oxide solid at a temperature much lower than that required if formed from the melt.

An interesting glass ceramic, Macor, manufactured by Corning, is a two-phase machinable crystalline solid composed of mica and glass. Heat treatment converts the amorphous glass into a crystalline structure, (Fig 16.5). A comparison of the properties relative to other ceramic materials is given in Table 16.3. Its machining characteristics are compared to some metals and nonmetals in Table 16.4. Such excellent properties are unique to ceramics and ceramiclike materials.

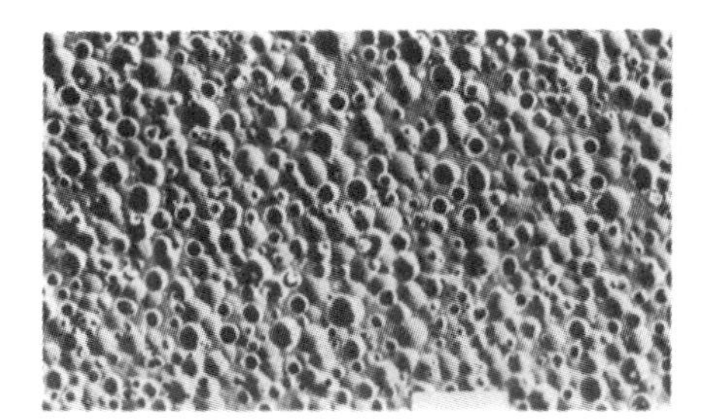

A **Droplet-imbedded** parent-glass material before heat treatment

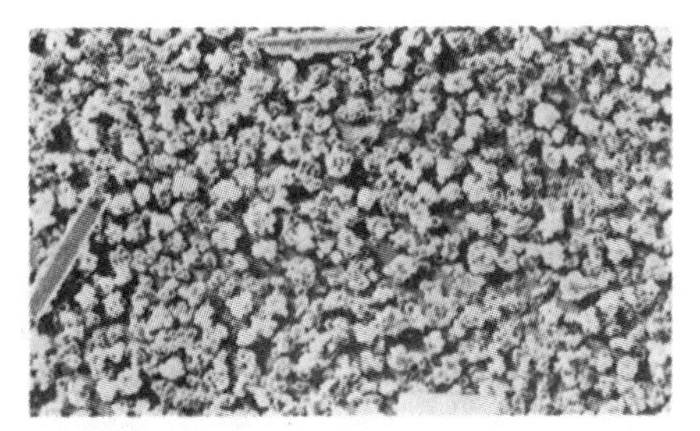

B **Second** intermediate crystal phase seen after heating to 825C

C **Beginning** of mica-crystal formation seen after heating to 850C

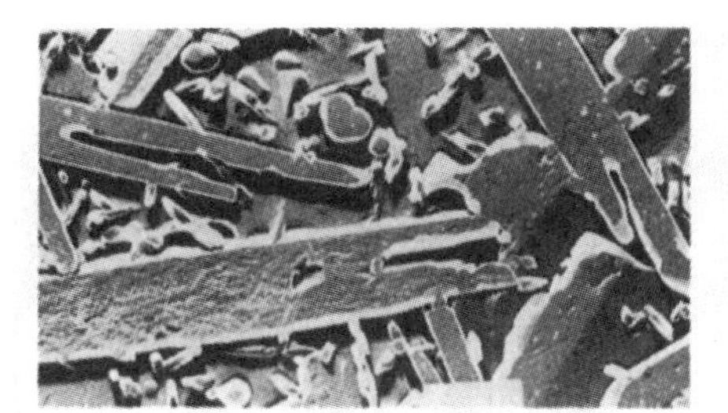

D **Fully crystallized** Macor glass-ceramic seen after heating to 950C

FIGURE 16.5. Surface photomicrographs of Macor at various stages of heat treatment and formations. (A) Droplet-imbedded parent–glass material before heat treatment, (B) second intermediate crystal phase after heating to 825°C, (C) beginning of mica-crystal formation seen after heating to 850°C, (D) fully crystallized (zero porosity) Macor glass–ceramic seen after heating to 950°C with mica crystals occupying about 55% and glass matrix the remaining 45% of the material.

16.7. COMPOSITES

Composite materials were described in the Bible (Exodus), where straw was required and used in the preparation of reinforced bricks. The Inca and Maya people mixed plant fiber with their pottery to reduce cracking while being dried. Concrete is a composite structure and most children are familiar with papier maché (a composite

TABLE 16.3
A Comparison of Macor with Other Common Ceramic Materials

Property	Units	Macor™ machinable glass ceramic	Boron nitride 96% BN	Alumina nominally 94% Al_2O_3	Valox thermoplastic polyester
Density	g/cm^3	2.52	2.08	3.62	1.31
Porosity	%	0	1.1	0	0.34
Knoop hardness	NA	250	<32	2000	NA
Maximum use	°F	1832	5027	3092	204°
temp. (no load)	°C	1000	2775	1700	140°
Coefficient of thermal	in/in (°F)	52×10^{-7}	23×10^{-7}	39×10^{-7}	530×10^{-7}
expansion	in/in (°C)	94×10^{-7}	41×10^{-7}	71×10^{-7}	934×10^{-7}
Compressive strength	psi	50,000	45,000	305,000	13,000
Flexural strength	psi	15,000	11,700	51,000	128,000
Dielectric strength (ac)	Volts-mil	1000	950	719	590
Volume resistivity	Ω-cm	$>10^{14}$	$>10^{14}$	$>10^{14}$	$>10^{14}$

of paper and a glue made of flour and water). The Egyptians mixed their old papyrus manuscripts with pitch to wrap their mummies.

A durable and popular composite, *Transite*, is asbestos with up to 15% cement. In the form of boards and sheets it is a substitute for wood, and in severe climates is used for roofing, fence materials, and other structures because it weathers extremely well. Other *composites* include linoleum (linseed oil and jute), Bakelite (phenol–formaldehyde resin and cellulose fiber), plywood, and vehicle tires.

The more recent composites are the fiber-reinforced plastics and resins. The fibers include glass, Kevlar, and carbon fibers. Glass fibers used to reinforce plastics were introduced during World War II. The fibers are from 5–10 μm in diameter and have a

TABLE 16.4
Machinability Index for Various Materials[a]

Material	Machinability Index (MI)
Graphite	1
Teflon TFE	2.5
Macor (glass ceramic)	25
Free-machining brass	36
Aluminum 2024-T4	50
Copper alloy no. 10	97
Cold-rolled 1018 steel	111
AISI 4340 steel (Rc46)	206
304 Stainless steel	229

[a]The index unit (MI) is arbitrary and increases with difficulty of machining. Its approximate value is 1 HP/in^3/min = 200 MI.

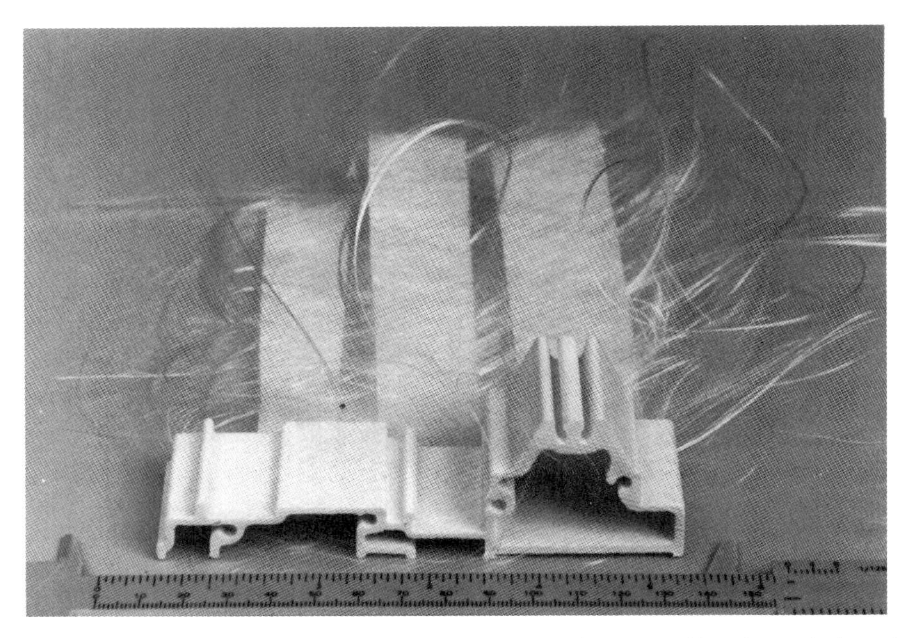

FIGURE 16.6. Photograph of a short section of a polystyrene–fiberglass window frame prepared by the continuous exuding process of styrene–fiberglass and fiberglass mats.

tensile strength of about 3000 MN/m^2. The glass fibers are treated with coupling agents to prevent the fibers from bundling, and to bond the fibers to the plastic or resins. Because the fibers are about 100 times stronger than the plastic, the strength of the composite is proportional to the fiber content. An example of an extruded section of a polystyrene window frame which is reinforced with glass fiber is shown in Fig. 16.6. Molded fiberglass reinforced polyester is the material of choice for boats though reinforced cement boats have also been built.

Carbon fibers were used about 100 years ago by Edison for his electric lamps. In 1964 the carbonized polyacrylonitrile (Orlon) fibers were first produced with a tensile strength of about 2000 MN/m^2 and a high modulus of over 400,000 MN/m^2. Thus, carbon fiber reinforced resins are very stiff and have found wide application in artificial limbs, golf clubs, tennis rackets, skis, and many aircraft parts. However, the composites are not especially strong in tension.

Vehicle tires are another example of a composite which are composed of rubber cord plies and beads which hold the tire to the wheel. The cord plies are usually nylon, rayon, polyester, fiberglass, and steel wire. The construction of the tire was initially bias belted, with the cords running from bead to bead, crossing the tread at an angle with the number of plies determining the strength of the tire. Addition of two stabilizing belt plies below the tread increased traction, and resistance to punctures. The radial-ply tires, manufactured first by Michelin in 1948, had cords which ran from bead to bead with no bias angle (i.e., 90° to the longtitudal tread). This is shown in Fig. 16.7. Tread design has reduced planing on wet pavement and increased the life of a tire to approximately 80,000 miles.

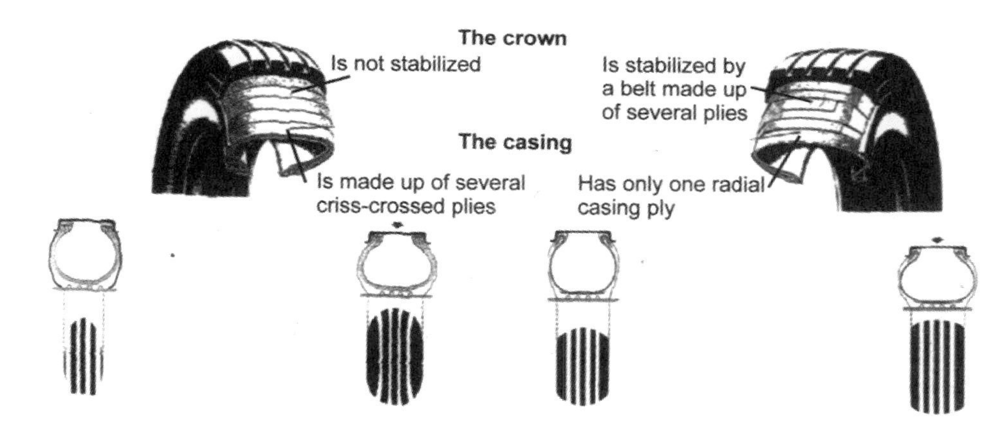

FIGURE 16.7. A comparison of the diagonal ply construction and the Michelin X radial tire.

New composites are constantly being developed to meet specific requirements with specified strengths, many to satisfy the aerospace industry.

EXERCISES

1. What are the primary components of cement?
2. What reactions occur in the kiln?
3. What reactions occur during the hydration process?
4. The lime (C), silica (S), and alumina (A) phase diagram is shown in Fig. 16.1A as mol% and where the area of Portland cement and high alumina cement are indicated. Identify on the phase diagram the location for the following minerals

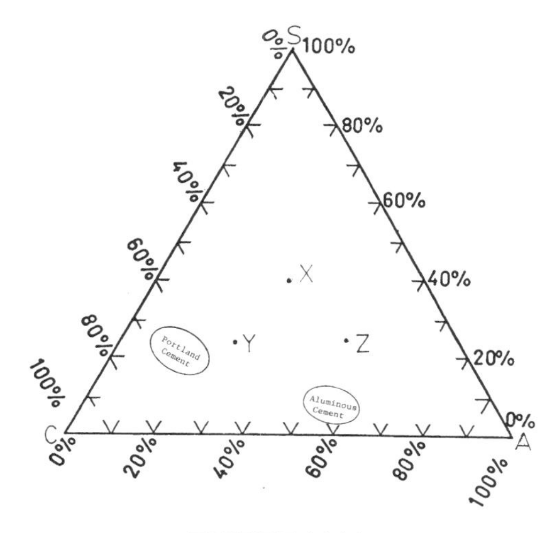

FIGURE 16.1A.

found in: (a) Portland cement clinker, C_3S, C_2S, and C_3A; (b) found in high alumina clinkers, CA, and CA_2.
5. In Fig. 16.1A locate the positions of the substances listed in Exercise 4a and 4b assuming that the scale of the diagram is in units of wt.%.
6. Identify the composition of points *x*, *y*, and *z* on the phase diagram (Fig. 16.1A) for both a mol% and wt.% scale.
7. The silicate garden in which metallic salts grow colored "trees" when crystals are dropped into water glass (aqueous sodium silicate), is analogous to the fibrillars developed during the hydration process in cement. Explain.
8. In what way is polymer impregnated concrete an important improvement?
9. It has recently been shown that the treatment of concrete with linseed oil (in dilute solutions of an organic solvent) extends the life of the concrete. Explain.
10. What advantages would there be in using fatty acid methyl esters of linseed oil instead of oil (triglyceride) to treat concrete?
11. Distinguish between cement, ceramics, and composites.
12. What is the difference between xerogels and aerogels and how are these materials used?
13. How is a ceramic or glass formed from the xerogel or aerogel?
14. What are the ideal properties of a fiber for reinforcement used in a composite?
15. Explain how the modern radial tire is an example of a composite structure.
16. Why is the addition of $CaCl_2$ to steel reinforced concrete not to be recommended?

FURTHER READING

D. Munz and T. Fett, *Ceramics: Mechanical Properties, Failure Behavior, Materials Selection*, Springer Verlag, New York (1999).

F. L. Matthews, *Composite Materials*, CRC Press, Boca Raton, Florida (1999).

K. K. Chawla, *Composite Materials: Science and Engineering*, 2nd Ed., Springer Verlag, New York (1998).

H. F. Taylor, *Cement Chemistry*, 2nd Ed., T. Telford, London (1997).

H. Bach, Editor, *Low Thermal Expansion Glass Ceramics*, Springer Verlag, New York (1996).

L. Dean-Mo, Editor, *Porous Ceramic Materials: Fabrication, Characterization, Application*, LPS District Center, Lebanon, New Hampshire (1996).

V. M. Mallhotra, *Pozzolanic and Cementitious Materials*, Gordon and Breach, Newark, New Jersey (1996).

J. B. Watchman, *Mechanical Properties of Ceramics*, Wiley-Interscience, New York (1996).

H. Yanagida, *The Chemistry of Ceramics*, Wiley, New York (1996).

C. A. Harper, Editor, *Handbook of Plastics, Elastomers and Composites*, 3rd Ed., McGraw-Hill, New York (1996).

D. Hull and T. W. Clyne, *An Introduction to Composite Materials*, Cambridge University Press, New York (1996).

A. M. Brandt, *Cement-Based Composites*, Chapman and Hall, New York (1995).

E. J. Pope et al., Editors, *Sol-Gel Science and Technology*, Proc. Internat. Sympos. on Sol-Gel, Am. Ceramic, Westerville, Ohio (1995).

R. J. Struble, *Cement Research Progress*, Am. Ceramic, Westerville, Ohio (1994).

A. Kelly et al., Editors, *Concise Encyclopedia of Composite Materials*, revised ed., Elsevier, New York (1994).

K. Ashbee, *Fundamental Principles of Fiber Reinforced Composites*, 2nd Ed., Technomic, Lancaster, Pennsylvania (1993).

P. Balaguru and S. P. Shah, *Fiber-Reinforced Cement Composites*, McGraw-Hill, New York (1992).

R. J. Brook, Editor, *Concise Encyclopedia of Advanced Ceramic Materials*, Elsevier, New York (1991).

V. K. Chao, *Fundamental of Composite Materials*, Knowen Acad., La Cruces, New Mexico (1990).

H. F. W. Taylor, *Chemistry Cement*, Academic Press, London (1990).
K. Perry, *Rotary Cement Kiln*, 2nd revised ed., Chemistry Publishing, New York (1986).
G. C. Bye, *Portland Cement*, Pergamon Press, New York (1983).
G. Lubin, Editor, *Handbook of Composites*, Van Nostrand Reinhold, New York (1982).
R. W. Davidge, *Mechanical Behavior of Ceramics*, Cambridge University Press, Cambridge (1979).
Portland Cement Association, http://www.portcement.org/
Cement and Concrete Instutute, cnci, http://www.cnci.org/
Cement & Concrete Association, NZ, http://www.cca.org.nz
Comprehensive Procurement Guideline Program,
http://www.epa.gov/cpg/products/cement.htm
Corning Ltd, http://www.corning.com/lightingmaterials/products/macor
Ceramic sources, http://www.ceramics.com
American Ceramic Society, www.acers.org/
Silica Aerogels, a history, http://www.eetd.lbl.gov/ECS/aerogels/satoc.htm
Aerogels, Lawrence Livermore Nat'l Lab., http://www.llnl.gov
New Virtual Library, http://www.ikts.fhg.de/VL.ceramics.html
All about composites, http://www.owenscorning.com/owens/composites/about/
Honeycomb composites, http://www.euro-composites.com/
Composites Fabricators Assoc., www.cfa-hq.org/
Aerospace Composite Products, www.acp-composites.com/
Composite applications, http://www.netcomposites.com/
Worldwide composites search engine, www.wwcomposites.com/
Composites Registry, http://www.compositesreg.com/

Appendix A
Fundamental Constants and Units

Speed light (in vacuum)	$C = 2.99792458 \times 10^{8}\ \mathrm{m\,s^{-1}}$
Planck's constant	$h = 6.6260755 \times 10^{-34}\ \mathrm{Js}$
Gas constant	$R = 0.0820584\ \mathrm{L\ atm\ mol^{-1}\ K^{-1}}$
	$R = 8.314510\ \mathrm{J\ mol^{-1}\ K^{-1}}$
Gravitational constant	$g = 9.8066\ \mathrm{m/s^2}$
Faraday constant	$F = 9.6485309 \times 10^{4}\ \mathrm{C\ mol^{-1}}$
Avogadro number	$N_A = 6.0221367 \times 10^{23}\ \mathrm{mol^{-1}}$
Electron charge	$e^- = -1.60217733 \times 10^{-19}\ \mathrm{C}$
Electron mass (rest)	$m_e = 9.1093897 \times 10^{-31}\ \mathrm{kg} = 0.00054858\ \mathrm{amu}$
Proton mass (rest)	$m_p = 1.67262 \times 10^{-27}\ \mathrm{kg} = 1.007276\ \mathrm{amu}$
Neutron mass (rest)	$m_n = 1.67493 \times 10^{-27}\ \mathrm{kg} = 1.008665\ \mathrm{amu}$
Solar constant (sea level)	$1370\ \mathrm{W/m^2}$
Solar luminosity	$3.85 \times 10^{26}\ \mathrm{W}$

Nomenclature for powers of 10

Prefix	Symbol		Prefix	Symbol	
atto	(a)	10^{-18}	kilo	(k)	10^{3}
femto	(f)	10^{-15}	mega	(M)	10^{6}
pico	(p)	10^{-12}	giga	(G)	10^{9}
nano	(n)	10^{-9}	tera	(T)	10^{12}
micro	(μ)	10^{-6}	peta	(P)	10^{15}
milli	(m)	10^{-3}	exa	(E)	10^{18}

Length

$\text{Å} = 10^{-10}\ \mathrm{m}$

inch = 2.54 cm

mile = 1.609 km

knot = 1 nautical mile/hr
= 1.1516 mile/hr
= 1.853 km/hr

Area

acre = 43,560 $\mathrm{ft^2}$
= 4047 $\mathrm{m^2}$

hectare = 10^4 m^2
square mile = 640 acres
= 259 hectares

Volume
1 ft^3 = 28.317 L
1 m^3 = 1000 L
1 gal (US) = 3.785 L
= 4 quarts
= 128 fluid oz.
1 gal (Imp) = 4.546 L
= 4 quarts
= 160 fluid oz.
1 barrel (US) = 31.5 US gal
1 barrel (Imp) = 36 gal (Imp)
1 barrel (oil US) = 42 US gal
= 158.9873 L

Mass
1 lb. = 453.6 g
1 ton = 2000 lb.
1 tonne = 1000 kg
= 2204.6 lb.
1 kg = 2.2046 lb.
1 amu = 1.66054×10^{-24} g
= 931.4874 MeV

Pressure
1 atm = 1.01325 bars
= 760 torr
= 101,325 Pa

Energy
Joule (J) = 1 volt (V) × 1 amp
1 eV = 1.6021×10^{-19} J
= 4.450×10^{-26} kWh
1 MeV = 9.65×10^{10} J/mol
Watt (W) = 1 J/sec
Curie (Ci) = 3.7×10^{10} disintegration/sec
Becquerel (Bq) = 1 disintegration/sec
1 calorie (cal) = 4.183 J
1 horsepower (HP) = 745.7 W
1 kWh = 3412 Btu
1 Btu = 1055.06 J
$\simeq$ 1 kJ = 2.931×10^{-4} kWh
1 Quad $\approx 10^{15}$ Btu
$\approx 10^{15}$ kJ
= 10^{12} ft^3 (CH_4)
= 2.93×10^{11} kWh

$\simeq 40 \times 10^6$ tons coal
$= 170 \times 10^6$ bbl crude oil
$= 8.0 \times 10^9$ US gal gasoline
1 bbl oil = 42 gal (US)
$= 5.8 \times 10^6$ Btu
$= 1.65 \times 10^3$ kWh
= 159 L
= 136 kg
= 5.8 M Btu
1 ft^3 Natural gas (CH_4) = 1035 Btu
= 0.310 kWh
1 lb. coal = 3.84 kWh
1 gal (US) gasoline = 36.7 kWh
1 ton oil equiv. (toe) $= 4.19 \times 10^{10}$ J
$= 10^7$ kcal
1 ton coal equiv. (toe) $= 2.93 \times 10^{10}$ J
$= 7 \times 10^6$ kcal
1 therm $= 10^5$ Btu
1 kg TNT $= 10^9$ J

On July 23, 1983 a new Boeing 767 (Air Canada Flight No. 143) refueled in Montreal. The fuel gauge was not working so it was decided to refuel by the manual use of a dipstick which correctly showed the fuel in the aircraft to be 7682 L. The required fuel for the trip was 22,300 kg. The mechanics, using 1.77 as the density conversion factor, calculated the required fuel necessary.

$$7682 \times 1.77 = 13{,}597 \text{ kg on board}$$

$$22{,}300 - 13{,}597 = 8703 \text{ kg to be added}$$

$$8703 \div 1.77 = 4916 \text{ L of fuel to be added}$$

They believed 1.77 represented the conversion of liters to kilograms, in fact it was the conversion of liters to pounds, i.e., 1.77 lb/L is the density of the jet fuel. The density of the jet fuel in proper units is 0.803 kg/L and the amount of fuel which ought to have been added was 20,163 L. Using 1.77 without any units led to a near disaster.

FURTHER READING

W. M. Carey, 1985, *Out of Fuel at 26,000 Feet*, Readers' Digest, 126 (May), p. 213.
W. Hoffer and Marilyn Mona, *Freefall, A True Story*, St. Martin's Paperback, New York (1989).
The Gimli Glider, http://www.acs.org/VC2/2my/my2—143.html
International System of Units, Constants, Units, Uncertainty, http://www.physics.nist.gov/

Appendix B

Viscosity

B.1. INTRODUCTION

Viscosity is that property of a fluid that opposes the relative motion of adjacent portions of the fluid and can consequently be regarded as a type of internal friction. Viscosity can be defined as the force required to move a layer of fluid of unit area with a velocity 1 cm/sec greater than the velocity of another layer 1 cm away (see Fig. B.1). Since force is proportional to the velocity difference between the layers and inversely proportional to the distance between the layers, then

$$F \propto \frac{VA}{x} \tag{B.1}$$

where F is force, V is velocity difference, x is distance between layers, and A is area of layer. Therefore,

$$F = \frac{\eta VA}{x} \tag{B.2}$$

where η, the proportional constant, is the viscosity.

The units of viscosity are dyne second per square centimeter or gram per second per centimeter (1 dyne second per square centimeter is simply called 1 poise, after Poiseuille). The reciprocal of viscosity is called the fluidity and is often represented by ϕ; it is a measure of the ease with which a liquid flows.

In a gas the viscosity increases as the temperature increases, whereas in a liquid the converse is true. The interpretation of viscosity in a gas utilizes the high kinetic energy of the molecules and involves the transfer of momentum for one layer of the gas to another, leading to a relationship of the form $\eta \propto v$, where v is the average speed of the gas molecules. In liquids, a completely different interpretation is required, since the molecules are closely packed except for the presence of holes. Over half a century ago it was found that the fluidity of a substance at its melting point is proportional to $V - b$, where V is the volume of liquid and b is the van der Waals constant. This is the effective space occupied by the molecules. $V - b$ is, therefore, the free volume of the liquid. Since most solids expand approximately equal at this temperature the free volumes of liquids are approximately the same; therefore, we could expect the viscosities of most liquids at their melting points to be approximately equal. This is correct within

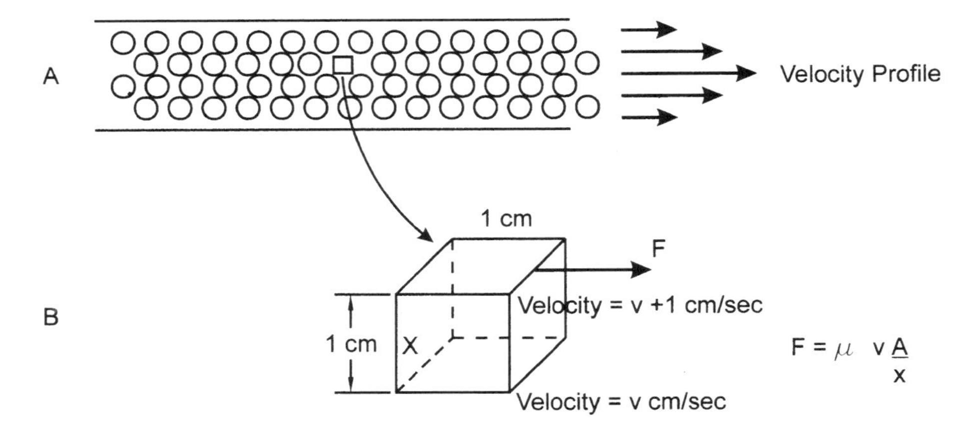

FIGURE B.1. Schematic representation of the viscosity of a liquid. (A) Liquid near the walls of a tube moves slower than liquid near the center or furthest from the walls. Velocity profile depends on the viscosity of the liquid. (B) Viscosity is the force required to maintain a unit velocity gradient for 1 cm^2 plates 1 cm apart.

an order of magnitude. The theory of the significant structures of liquids has been able to relate the viscosity mechanism to fluidized vacancies and to show that the effect of temperature on viscosity is associated with the energy required to create a hole in the liquid. The influence of temperature on the viscosity of a liquid is often represented by the following equation:

$$\eta = Ae^{E_{\text{vis}}/RT} \quad \textbf{(B.3)}$$

where E_{vis} is the energy of activation of viscosity, R is the gas constant, T is the absolute temperature, and A is a constant depending on the substance.

It is interesting that the average heat of vaporization of a liquid is approximately three times the activation energy of viscosity. This means that three times as much energy is required to remove a surface molecule as to move a bulk molecule past a neighbor. The ratio $n = E_{\text{vap}}/E_{\text{vis}}$ was shown by Eyring to be equal to the ratio of the size of a molecule to the size of a hole needed for viscous flow. It has been found that, since a hole of molecular dimensions is not required if, for example, two molecules rotate about their point of contact, the value of n is about 3 for a spherically symmetric nonpolar molecule and increases to 5 as the deviation from spherical symmetry increases.

B.2. MEASUREMENT OF VISCOSITY

The viscosity of a liquid can be measured by several methods. The most convenient for laboratory work are the Ostwald viscometer and the falling ball methods.

B.2.1. Ostwald Viscometer

In 1846 the viscosity of a liquid was related by Poiseuille to the rate of flow of a liquid through a tube under a pressure differential. The Poiseuille equation is as follows:

$$Q = \frac{\pi P r^4}{8\eta l} \quad \text{(B.4)}$$

where Q is the volume of liquid flowing per unit time through a tube of radius r and length l, across which there is pressure drop P. If the pressure drop is due to gravity, then $P = hdg$ where h is the height of the liquid, d is the density, and g is the acceleration due to gravity. By substituting hdg and P in Eq. (B.4), we obtain

$$\eta = \frac{\pi r^4 hdg}{8Ql} \quad \text{(B.5)}$$

The measurement of absolute viscosity by this method requires an exceptional amount of care and patience. For most purposes, it is sufficient if relative viscosities are known. Hence, if the time for a fixed volume of liquid to flow through a capillary is measured, then the comparison of its time of flow with that of another liquid enables us to calculate a relative viscosity. Since

$$Q = V/t \quad \text{(B.6)}$$

where V is the volume of liquid, and t is the time of flow, then

$$\eta = \frac{\pi r^4 hdgt}{8Vl} \quad \text{(B.7)}$$

For a fixed apparatus as shown in Fig. B.2, V, h, g, r, l are constant. Therefore, for one liquid,

$$\eta_1 \propto d_1 t_1 \quad \text{(B.8)}$$

Hence, for two liquids,

$$\frac{\eta_1}{\eta_2} = \frac{d_1 t_1}{d_2 t_2} \quad \text{(B.9)}$$

The viscosities of several substances at different temperatures, which can conveniently be used as references, are given in Table B.1.

B.2.2. Falling Ball Method

The viscosity of a liquid may be determined from Stokes' law. In 1850 Stokes showed that a sphere of radius r under a constant force F will move with constant

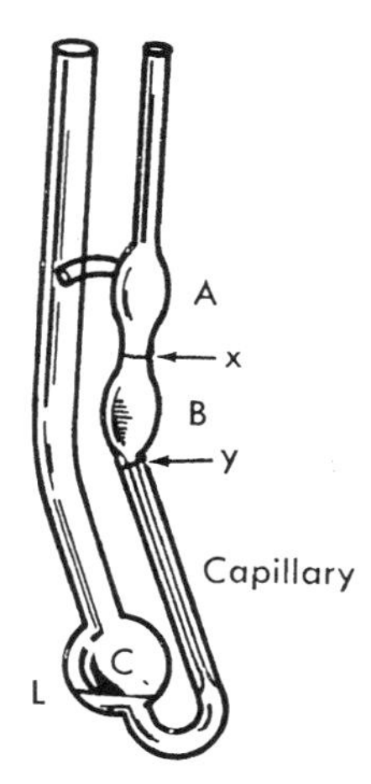

FIGURE B.2. *Ostwald viscometer.* The quaantity of liquid (volume V) required is such that when the liquid is drawn into bulb A, the level in reservoir C is not below level L. The time of flow t, from level x to y, is recorded and compared with other liquids of identical volume V and of known density d.

$$\frac{\eta}{\eta_2} = \frac{d_1 t_1}{d_2 t_2}$$

velocity v in a viscous liquid of viscosity η according to the following relation:

$$F = 6\pi r \eta v \tag{B.10}$$

If the sphere is acted on by gravity alone, then

$$F = \tfrac{4}{3}\pi r^3(d - d')g \tag{B.11}$$

where d is the density of sphere, d' is density of fluid, and g is acceleration due to

TABLE B.1
Viscosities of Several Substances at Various Temperatures, in Units of Centipoise

		Temperature (°C)						
Name	Formula	0	10	20	30	40	50	60
Water	H_2O	1.79	1.31	1.00†	0.801	0.656	0.549	0.469
Ethanol	C_2H_5OH	1.77	1.47	1.20	1.00	0.834	0.702	0.592
n-Propanol	*n*-C_3H_7OH	3.88		2.25	1.72	1.41	1.13	
Isopropanol	*iso*-C_3H_7OH	4.56		2.37		1.33		
Benzene	C_6H_6		0.758	0.652	0.564	0.503	0.442	0.392
Toluene	$C_6H_5CH_3$	0.772		0.590	0.526	0.471		
Chlorobenzene	C_6H_5Cl	1.03		0.799		0.631		
Ethyleneglycol	$C_2H_4(OH)_2$			19.9		9.13		4.95

†Viscosity of H_2O at 20.20°C is 1.0000 centipoise.

gravity. Therefore

$$\eta = \frac{2r^2(d - d')g}{9v} \quad \textbf{(B.12)}$$

Since $v = l/t$, then

$$\eta = \frac{2r^2g}{9l}(d - d')t \quad \textbf{(B.13)}$$

Since r, d, g, l are constant, then the relative viscosity can be easily determined from the following equation:

$$\frac{\eta_1}{\eta_2} = \frac{(d - d_1)t_1}{(d - d_2)t_2} \quad \textbf{(B.14)}$$

For very viscous liquids the falling ball viscometer is generally more suitable than the Ostwald viscometer.

The viscosity and the activation energy of viscosity of a pure liquid substance can tell us very little about the substance. However, the viscosity of a solution is much more informative. Thus, if acetone and chloroform are mixed and the viscosities of the various solutions are measured, it can be determined whether or not "compound" formation can exist. These two substances form a hydrogen bond about 11.3 kJ/mol, and the appearance of a maximum about 1:1 molecular ratio supports this view. Many such compounds, first detected by viscosity anomalies, are now being characterized by more elegant methods. It is interesting to note that substances that show this viscosity effect generally show a volume change when mixed. This suggests a change in the free volume of the solution compared with the free volume available for each of the individual substances.

It must be pointed out that a viscosity effect in the mixture of two or more substances does not necessarily prove the formation of a compound, since other effects, such as steric hindrance* and the association of one of the components may give rise to a viscosity anomaly.

B.3. APPLICATIONS OF VISCOSITY

B.3.1. Molecular Weight of Polymers

From the measurements of viscosity of solutions of polymeric substances, it is possible to determine the average molecular weight of the solute. The empirical relation is as follows:

$$[\eta] = KM^a \quad \textbf{(B.15)}$$

*Steric hindrance is the spatial interference experienced in molecules because of the specific location of bulky chemical groups.

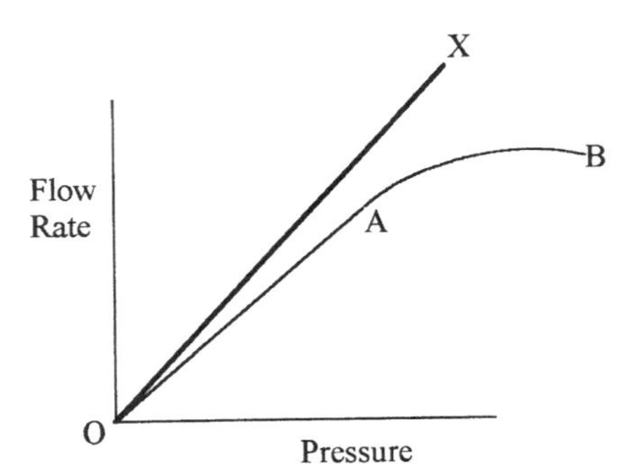

FIGURE B.3. Reduction of turbulence by the addition of a drag reducer (DR) to a fluid OAB, fluid only; OX, with DR.

where K and a are constants that depend on the solute, solvent, and temperature, M the average molecular weight, and $[\eta]$ the intrinsic viscosity and fractional change in the viscosity of a solution per unit concentration of solute at infinite dilution. This is represented as

$$[\eta] = \lim_{C \to 0} \left[\frac{1}{C} \left(\frac{\eta - \eta_0}{\eta_0} \right) \right] \tag{B.16}$$

where η_0 and η are the viscosity of the solvent and the solution, respectively, and C is the concentration of the polymer, usually in wt.%. Although other methods for determination of molecular weight of polymers are more accurate and more reliable, the intrinsic viscosity can give relative molecular weights with reasonable accuracy and facility. It is often the first determination made for the molecular weight of a polymer.

B.4. DRAG REDUCERS

As the flow of a fluid in a tube increases as a result of increasing applied pressures, the Reynolds number (Re) increases to the onset of turbulence where

$$\mathrm{Re} = ud/v \tag{B.17}$$

u is the velocity of flow, d is the diameter of the tube, and v is the kinematic viscosity* and equal to η/ρ, where ρ is the density of the fluid.

This is shown in Fig. B.3 where at point A, turbulence occurs, i.e., $\mathrm{Re} > 2000$. When a low concentration of a long chain soluble polymer is dissolved in the fluid it prevents the onset of turbulence and the flow rate–pressure line is extended from O to

*$v = 0.0100\ \mathrm{cm^2/sec}$ (Stokes) at 20°C for water; $v = 6.8\ \mathrm{cm^2/sec}$ (Stokes) at 20°C for glycerol.

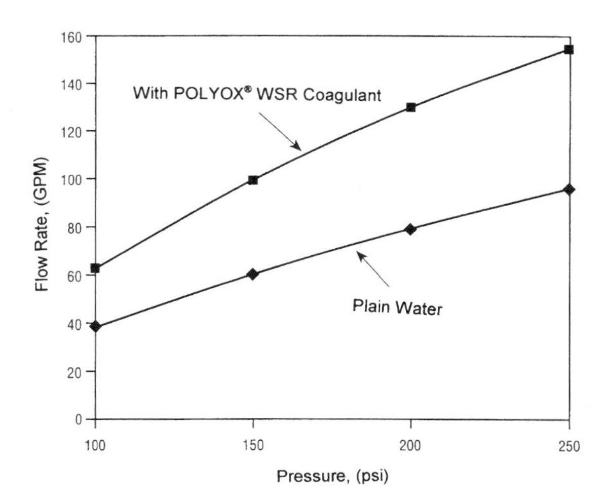

FIGURE B.4. Effect of low concentrations of polyethylene oxide on friction (viscosity) reduction.

X along the straight line plot. This effect is called *drag reduction* and has been applied to the pumping of oil in pipelines, blood in arteries, and free flow of fluids. Figure B.4 shows the reduction in friction (viscosity) as the concentration of polyethylene oxide (PEO, mol. wt. = 5 MD) in water is increased. Figure B.5 shows the increased flow of water through a 1.5 in. hose as a function of pressure when PEO is added to the water. This is illustrated in Fig. B.6 where a 1.5 in. fire hose can deliver water at a rate equal to that of a 2.5 in. hose if polyethylene oxide has been added to the water.

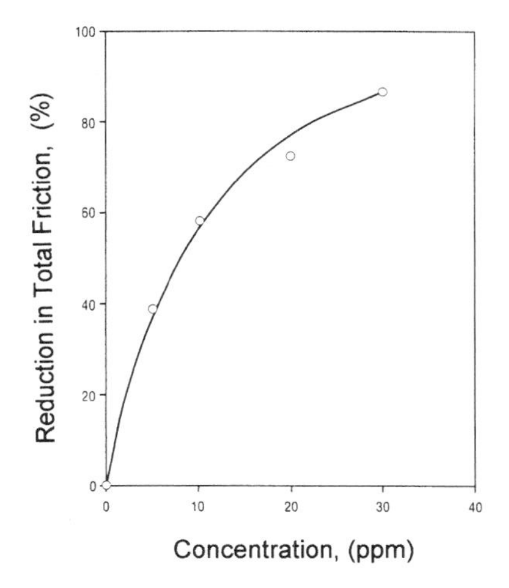

FIGURE B.5. Effect of polyethylene oxide on the flow rates of water through 1000 ft of a 1.5 in. hose at various pressures.

FIGURE B.6. Effect of high molecular weight poly(ethylene oxide) in increasing water flow through fire hoses. Hoses had been adjusted to throw water the same distance and then 30 ppm of poly(ethylene oxide) were injected into the hose on the left.

B.5. ELECTRORHEOLOGICAL FLUIDS

Some liquids with a suspension of solids or colloids can, under the influence of an applied potential, show an increase in viscosity by a factor as high as 10^5, i.e., the liquid is converted to what is essentially a solid. This is called the *Winslow effect* and the liquid is an electrorheological (ER) fluid.

Winslow (1949) reported that silica gel in a low-viscosity oil showed this effect under an electric field of 3 kV/mm. The fluid can be sheared with a force proportional to the square of the electrical field. For example, a 25% by volume of hydrophobic colloidal silica spheres of 0.75 μm diameter in 4-methylcyclohexanol showed ER responses at 40–4000 Hz, although dc fields are also viable. Dispersants are often added to the suspension in order to prevent the settling of the solids.

The ER fluids can have applications as clutches, speed controllers, and valves. Other applications can be expected as work on the subject continues.

EXERCISES

1. From the data shown in Table B.1 convert the value of the viscosity of ethylene glycol at 20°C to SI units, i.e., mN sec/m^{-2}.
2. The viscosity of hexadecane is 3.6 mN sec/m^{-2} at 22°C. The flow of water (20 mL) in an Ostwald viscometer took 47 sec at 22°C. Calculate the time it would take 20 mL of hexadecane (density = 0.7751 g/mL) to flow through the same viscometer at the same temperature.
3. Discuss the differences in viscosity expected for H_2O, H_2O_2, and D_2O.
4. Of the two propanols which one would you expect to have the higher viscosity? Give reasons for your answer.
5. (a) Calculate E_{vis} for trinitroglycerol (TNG) which has the value of $\eta = 360$ mN sec/m^{-2} at 20°C and $\eta = 13.6\,\text{mN sec/m}^{-2}$ at 40°C, (b) calculate η for TNG at 60°C.
6. The fluidity of a liquid explosive doubles between 10° and 20°C. Determine (a) E_{vis} and (b) η_{60}/η_{10}.

FURTHER READING

C. L. Yaws, *Handbook of Viscosity*, 1–3 vol, Gulf, Huston, Texas (1994).
D. S. Viswanath and G. Natarajan, *Data Book on the Viscosity of Liquids*, Hemisphere, Bristol, Pennsylvania (1989).
H. A. Barns et al., *An Introduction to Rheology*, Elsevier, New York (1989).
Viscosity and Stokes' Law, www.math.mcmaster.ca
Norcross Ltd, http://www.viscosity.com/
Viscosity and Surface Tension, http://www.physics.bu.edu/py105/notes/Viscosity.html
http://www.pe.utexas.edu
Drag reducers, http://www.liquidpower.com/about/awhatis.htm

Appendix C

Surface Chemistry

C.1. SURFACE TENSION

A surface is the boundary between two phases. The chemistry of this interface is of great importance to a variety of subjects such as adhesion, corrosion, surface coatings, and many others. The differential attraction of surface molecules in a liquid results in a surface energy that is also called *surface tension*, and accounts for the tendency of a free or suspended liquid to assume a spherical shape in droplets or to expose as small a surface area as possible. Thus the surface of a liquid in contact with air or another liquid phase in which it is immiscible may be considered similar to a rubber elastic membrane or balloon that assumes a spherical shape. To distort this to any other shape would require the expenditure of energy, since the surface area would increase; i.e., the rubber membrane would stretch. A surface energy can therefore be associated with a liquid–gas interface and can be defined as the energy or work required to increase the surface area of a liquid by 1 cm^2 by bringing bulk molecules to the surface; i.e., work per square centimeter. Since the surface is under a tension, a force called the surface tension γ can be defined as the force applied to increase the surface area of a liquid when acting on 1 cm of surface; i.e., force per centimeter. The units of surface energy are

$$\frac{ml^2t^{-2}}{l^2}$$

and those of surface tension as force are

$$\frac{mlt^{-2}}{l}$$

The surface energy and surface tension both have units of mt^{-2} and are equivalent (see Fig. C.1).

When a drop of liquid is placed on a solid, a definite angle of contact exists at the point where the liquid meets the solid. This is shown in Fig. C.2 where the angle θ is called the *contact angle*.

The contact angle has one of two values, advancing or receding, depending on whether the liquid–solid interface area is increasing or decreasing. This is commonly called a *hysteresis* effect. Until recently, it was believed that the ratio of these two angles depended only on the roughness of the surface. It is now reasonably well established

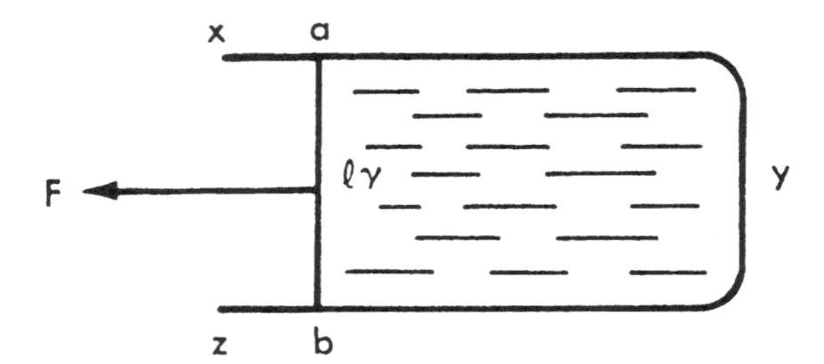

FIGURE C.1. Thin film of liquid (*ayb*) is stretched on an inert wire frame (*xyz*). A wire barrier (*ab*) of length l is pulled toward xz with force F. Since the liquid increases in area on both sides of the frame, then at balance, $F = 2l\gamma$, where γ is the surface tension of the liquid.

that the hysteresis is due to the penetration of the liquid molecules into surface discontinuities and, therefore, depends on the size of the molecules relative to the intermolecular pores of the solid surface. The contact angle thus depends on the type of solid surface, liquid, and to a lesser degree, temperature. Table C.1 lists the values of the contact angles for various substances as well as some surface tension values.

The contact angle depends on the relative bonding between liquid and solid molecules (B_{ls}), compared with the bonding between liquid molecules (B_{u}). Thus, if B_{ls} is greater than B_{u}, θ is usually less than 90° and the solid is considered to be wetted by the liquid. However, if B_{ls} is less than B_{u}, then θ is usually greater than 90° and the solid is not wetted by the liquid. It must be pointed out that the angle of 90° chosen for the demarcation between wetting and nonwetting is quite arbitrary.

The contact angle can be measured by direct observation with the use of magnification of the liquid drop on a flat surface of the solid material, or by inclination of a slide of the solid in the liquid until the meniscus flattens out. The latter method is shown in Fig. C.3.

The wetting of a solid by a liquid is exceedingly important for several applications, including soldering, welding, adhesion and gluing, painting, and dyeing. The presence of fluxes to remove oxide layers in soldering, and the need of special wetting agents in epoxy glues, adhesives, and paints are related to the contact angle and surface tension.

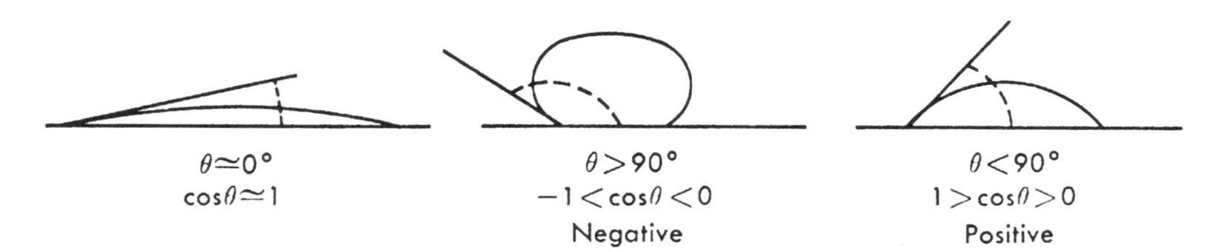

FIGURE C.2. A drop liquid is placed on a solid surface. The angle formed at the interface is called the contact angle. If the volume of the drop is increased, the advancing angle is measured. If the volume of the drop is decreased, the receding angle is measured.

TABLE C.1
Contact Angles and Surface Tensions for Various Substances

Contact angle θ^a

Water on siliconized glass:

T (°C)	γ(dyne cm^{-1})	θ_A	θ_R
4	75	104°	75°
22	72	106°	76°
75	63	104.5°	76.5°

Water on:

Glass	0°
Paraffin	108°
Polyethylene	94°
Teflon	110°
Graphite	86°
Kel F	90°

Standard surface tension in dynes per centimeter, vs. air at 20°C

Water	72.8
Acetone	23.7
Benzene	28.9
Toluene	28.4
Chloroform	27.1
Carbon tetrachloride	26.8
n-Hexane	18.4
n-Octane	21.8
n-Octanol	27.5

[a] θ_A is advancing angle and θ_R is receding angle. When one value of θ is given, it is usually the arithmetic average of θ_A and θ_R.

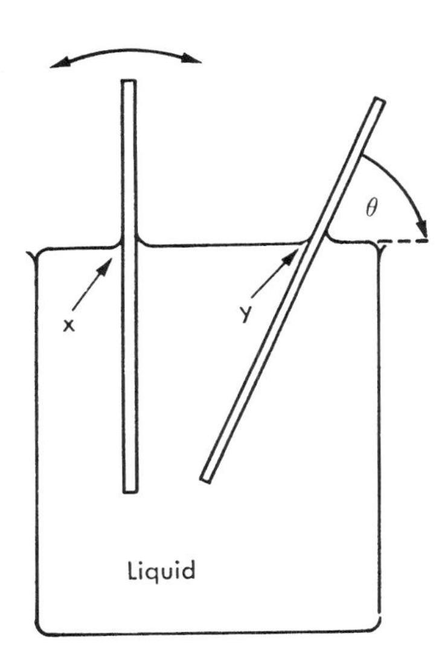

FIGURE C.3. Measurement of contact angle. A solid slide of the material is immersed in the liquid and the meniscus x at the air interface is noted. The slide is then rotated about the point of contact with the liquid surface until the meniscus flattens (y). The advancing and receding angles are associated with decreasing and increasing contact angles, respectively.

C.2. MEASUREMENT OF SURFACE TENSION

The surface tension of a liquid can be measured in a variety of ways, including capillary tube rise, du Noüy ring method, bubble pressure, and drop weight. These methods vary greatly in their applications.

C.2.1. Capillary Tube Rise Methods

The simplest and most common method of determining the surface tension of a liquid is the *capillary tube rise method.* If a glass capillary tube is immersed in a liquid such as water, the liquid in the capillary tube will rise above the outside level of the liquid. This is due to the greater liquid–solid force than the liquid–liquid intermolecular forces, and the liquid tends to wet as much of the solid as possible until the gravitational pull of the column of liquid is equal to the surface tension force (see Fig. C.4).

If the radius of the capillary tube is r and the density of the liquid is d, then the force down, $F\downarrow$, which is due to the column of liquid, is defined as follows:

$$F\downarrow = \pi r^2 hdg \tag{C.1}$$

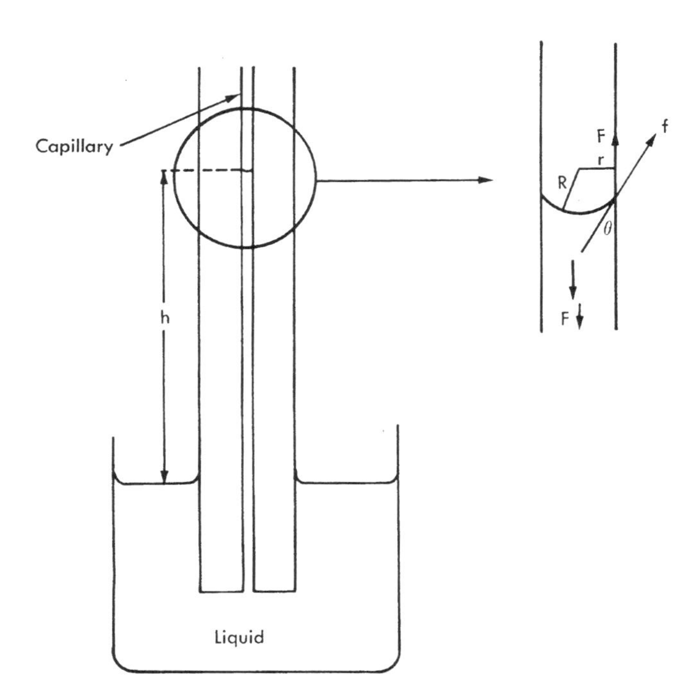

FIGURE C.4. Measurement of surface tension by capillary rise. The liquid rises in the capillary until equilibrium is reached where the gravitational force is balanced by the upward surface tension force. The radius of the meniscus R is assumed to be approximately equal to the radius of the capillary, r.

The force f along the contact angle θ is equal to the surface tension $\times$ the length of the liquid–solid contact, l; i.e., $f = \gamma l$. Since l is equal to $2\pi r$, then f is defined as follows:

$$f = 2\pi r\gamma \tag{C.2}$$

However, the vertical force $F\uparrow$ is defined as follows:

$$F\uparrow = f\cos\theta \tag{C.3}$$

Therefore,

$$F\uparrow = 2\pi r\gamma\cos\theta \tag{C.4}$$

When the column of liquid has reached the same equilibrium position from either a lower or a higher height. (The effect of contact angle hysteresis may make these two heights slightly different.)

$$F\downarrow = F\uparrow \tag{C.5}$$

$$\pi r^2 hdg = 2\pi r\gamma\cos\theta \tag{C.6}$$

or

$$\gamma = \frac{rhdg}{2\cos\theta} \tag{C.7}$$

If $\theta = 0$, then $\cos\theta = 1$ and

$$\gamma = \frac{rhdg}{2} \tag{C.8}$$

If $\theta = 90^\circ$, $\cos\theta$ is negative, and h becomes a negative value; i.e., instead of a rise in the capillary tube, there is a depression. This is observed for the mercury-in-glass system.

C.2.2. Ring or du Noüy Method

Just as it is possible to float objects heavier than water on the ssurface of water, it is possible to pull the surface upward (increase the surface area) by the application of a suitable force, and thereby calculate the surface tension. The *du Noüy* method makes use of a clean platinum ring of radius r that is placed on the liquid surface. The liquid, which wets platinum, tends to adhere to the ring, which is slowly raised by the application of a force, which is previously calibrated, until the net force pulling the ring upward exceeds the surface tension and the ring breaks from the surface. At that point the surface tension force, F, is $2l\gamma$, where l is the circumference of the ring, and since $F = mg$,

$$2l\gamma = mg \tag{C.9}$$

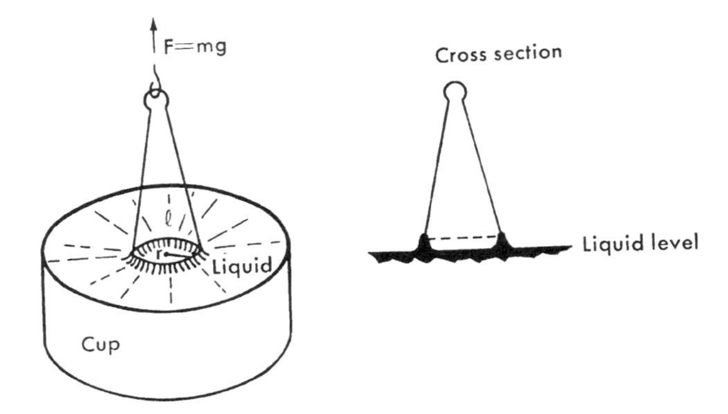

FIGURE C.5. Measurement of surface tension by du Noüy method. The platinum ring of radius r is allowed to touch and be wetted by the liquid surface. It is then raised by a force F (effected by a torsion type of balance) until the ring breaks away from the liquid.

where m is the weight calibration for the system and g is the acceleration due to gravity. Since $l = 2\pi r$, then

$$\gamma = \frac{mg}{4\pi r} \tag{C.10}$$

The factor 2 appears in Eq. (C.9) because two liquid surfaces (one on the inside and one on the outside of the ring) are formed as the ring is raised.

In practice, the absolute value is in error because of the diameter of the wire, the density of the liquid, as well as other terms, and it is often calibrated by a variety of substances to minimize such errors. The most important application of this method is the determination of interfacial tension between two liquids, where other methods do not apply very readily. The ring method is shown in Fig. C.5.

C.2.3. Bubble Pressure Method

If a tube is immersed in a liquid, and the gas pressure in the tube is increased slowly, the liquid level in the tube will drop until the end of the tube is reached, then a further increase in pressure will create bubbles. If P is the maximum pressure measured and P_h the hydrostatic pressure hdg, then

$$P - P_h = P_s \tag{C.11}$$

$P_s \Delta V$ is the work done to increase the volume of the bubbles. However, if the volume increases, the area of the bubble increases, and the energy required to increase the surface area of the bubble is $\gamma \Delta A$ (ΔA is the area increase) (see Fig. C.6). Therefore, at equilibrium,

$$\gamma \Delta A = P_s \Delta V \qquad \text{or} \qquad \gamma dA = P_s dV \tag{C.12}$$

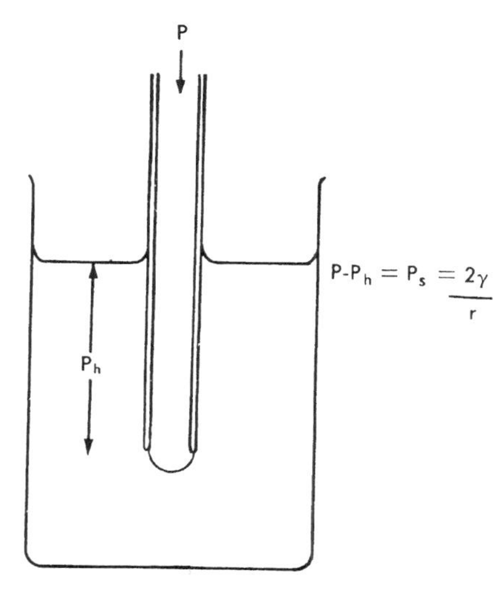

FIGURE C.6. Measurement of surface tension by the bubble pressure method. The maximum pressure is the pressure above which the bubble breaks away from the tube. This value is corrected for the hydrostatic pressure P_h and gives $P - P_h = P_s = 2\gamma/r$.

If r is the radius of curvature of the bubble, then

$$A = 4\pi r^2,\ V = 4\pi r^3/3,\ dA = 8\pi r dr,\ dV = 4\pi r^2 dr$$

By substituting these values in Eq. (C.12), we obtain

$$\gamma 8\pi r dr = P_s 4\pi r^2 dr \qquad P_s = 2\gamma/r \tag{C.13}$$

The value of r is equal to the radius of the tube when the bubble just breaks away from the tube; i.e., when the radius of the bubble is at a minimum, and the pressure is at a maximum. Since $P_h = hdg$, then if h, d, r, and p are measured, it is possible to calculate γ. It should be noted that γ, so evaluated, is independent of contact angle.

C.2.4. Drop Weight Method

When a liquid is allowed to flow slowly from a vertical capillary tube, it forms drops that grow and finally become detached from the end of the capillary tube. The size of the drops will depend on the outside radius of the capillary tube and the surface tension of the liquid. At the moment when a drop falls, the gravitational pull mg, or Vdg, is equal to the surface tension $2\pi r\gamma$, where m is the mass of the drop, V is the volume of liquid, d is the density of the liquid, r is the radius of the drop, and γ is the surface tension. Thus, if we weigh and count the drops of a liquid, we can determine the surface tension from the following equation if the density is known.

$$2\pi r\gamma = mg \tag{C.14}$$

In general, a correction is required, since not all the liquid forming the drop leaves the tip of the capillary tube and the surface tension does not act exactly vertical. Thus

$$\gamma = \frac{mg}{2\pi r \phi} \tag{C.15}$$

where ϕ depends on the ratio $r/V^{1/3}$ and has the value of about 1.7 when $r/V^{1/3}$ is between 0.75 and 0.95.

For a given tip, it is possible, with reasonable accuracy, to determine the relative surface tension from the following relation:

$$\frac{\gamma_1}{\gamma_2} = \frac{m_1}{m_2} \tag{C.16}$$

The *drop weight method* is independent of contact angle and is suitable for the determination of interfacial tension if the tip is immersed into a second liquid immiscible with the heavier dropping fluid. Thus, if the dropping liquid can be collected free from the second liquid and weighed, the liquid–liquid interfacial tension can be determined.

Surfactants reduce the interfacial tension of water either between the two liquids or between liquid and solid. The important properties include:

1 The solubility of the surfactant in at least one of the phases.
2 The surfactant is composed of both hydrophilic and hydrophobic groups on the molecule.
3 The surfactant molecules tend to orientate on the surface of the liquid.
4 The surfactant tends to concentrate at the interface.
5 The surfactant usually lowers the surface tension of the liquid.

The study of surface tension is really a branch of surface chemistry, and its development has been exceedingly rapid in the last decade. Thus, adhesion can be considered to be partially an exercise in wetting and spreading of liquids on solid surfaces. The flotation of ores is accomplished by gravity differences as well as by the adhesion of the solid particle to an air bubble, and it involves solid–liquid–gas interfaces. It is possible to reduce the vaporization of water from bodies of water with large surfaces such as reservoirs and lakes, by adding a monolayer of a substance such as hexadecanol or other surface-active agents. The action of soaps produces a decrease in surface tension on water. Many other applications in our modern environment can be readily identified.

C.3. THE SPREADING COEFFICIENT

Consider a drop of an organic liquid (O) on a water surface (W) as shown in Fig. C.7. At equilibrium the forces exerted at the three interfaces (O-W, O-A, and W-A), where A represents the air phase will be given by

$$\gamma_{W/A} = \gamma_{O/A} \cos \theta_2 \tag{C.17}$$

when spreading occurs θ_1 and θ_2 must approach zero ($\cos \theta = 1$) and

$$\gamma_{W/A} \geqslant \gamma_{O/A} + \gamma_{O/W} \tag{C.18}$$

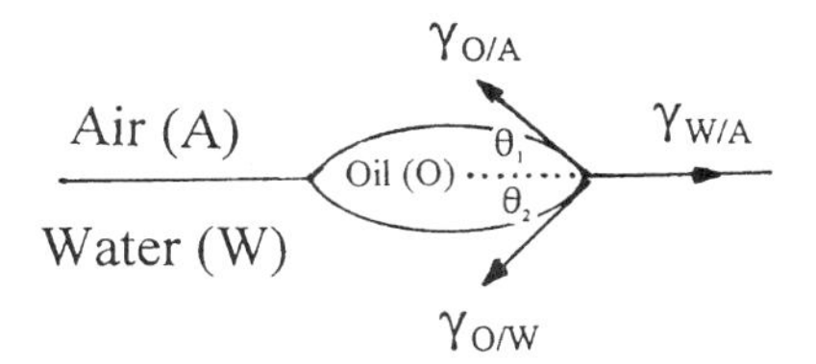

FIGURE C.7. Cross section of a drop of an organic liquid such as oil (O) on the surface of water (W) at equilibrium.

A spreading coefficient is defined as

$$S = \gamma_{W/A} - (\gamma_{O/A} + \gamma_{O/W}) \tag{C.19}$$

and when $S \geqslant 0$ spreading occurs. When $S < 0$ a lens will form on the water surface.

C.4. THE SOLID–GAS INTERFACE

There are two types of gas–solid interaction: (a) physical adsorption, which is due to van der Waals bonds and is reversible, (b) the chemisorption, where the gas forms chemical bonds with the solid surface and results in irreversible adsorption.

If the gas A forms a monomolecular adsorption layer on the solid surface it can be treated as a dynamic process

$$A(g) \underset{k_2}{\overset{k_1}{\rightleftharpoons}} A \quad \text{(adsorbed layer)} \tag{C.20}$$

If P = pressure of gas (A) above the surface and θ = fraction of surface covered by A, then at equilibrium, the rate at which the molecules leave the surface R_1 is equal to the rate at which they condense onto the surface R_2, i.e.,

$$R_1 = R_2 \tag{C.21}$$

$$R_1 \propto \theta \quad \text{or} \quad R = k_1\theta$$

$$R_2 \propto P(1-\theta) \quad \text{or} \quad R_2 = k_2P(1-\theta)$$

$$k_1\theta = k_2P(1-\theta) \tag{C.22}$$

Hence,

$$\theta = \frac{k_2P}{k_1} + k_2P = \frac{P}{a+P} \qquad \text{where} \quad a = \frac{k_1}{k_2} \tag{C.23}$$

If y = amount of adsorbed gas, ym = a monolayer of adsorbed gas (the maximum amount adsorbed), then $y/ym = \theta$. From Eq. (C.23) when P is small, $y \propto P$ and when P is large, y = constant. A plot of y vs. P is shown in Fig. C.8.

$$y = \frac{ymP}{a+P} \tag{C.24}$$

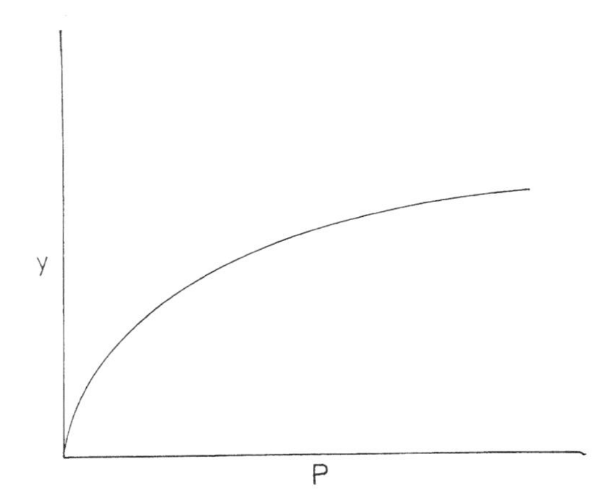

FIGURE C.8. A Langmuir plot of y, the amount of gas adsorbed (per unit mass of solid) on a solid as a function of equilibrium pressure, p of the gas at constant temperature.

and

$$P/y = a/ym + P/ym \tag{C.25}$$

Thus a plot of P/y vs. P is a straight line with slope = $1/ym$ and intercept = a/ym. This is called the *Langmuir adsorption isotherm.* For multilayer adsorption, the more complicated treatment developed by Brunauer, Emmet, and Teller (BET) allows for the determination of surface areas.

If the solid is composed of narrow capillaries then, at high pressures, the gas will condense in the capillaries. This results in adsorption–desorption hysteresis.

C.4. THE SOLID–LIQUID INTERFACE

It is possible to treat the adsorption of a liquid onto a solid surface in a manner analogous to that given for the solid–gas system. However, in the liquid state it is usually a solute which is removed by adsorption onto the solid surface. The Langmuir equation is applicable in most cases. In other cases where the surface is heterogeneous the Freundlich adsorption isotherm

$$\frac{x}{m} = kC^n \tag{C.26}$$

will usually fit the data better than the Langmuir equation, where x is the mass of material adsorbed on the solid, m is the mass of solid, C is the concentration (at equilibrium) of the solute being adsorbed, n is a constant with a value usually between 0.1 and 0.5, and k is a constant which depends on the system.

The empirical constants n and k can be determined from a plot of $\log x/m$ vs. $\log C$, where the resulting straight line has a slope equal to n and an intercept equal to $\log k$.

Adsorption of impurities from solution is used to purify water, decolorize sugar in solution, and many other systems.

EXERCISES

1. It has been suggested by Bikerman that a solid can have no surface energy and that all phenomena attributed to the surface energy of a solid is due to impurities which are adsorbed on such large surfaces. Comment on Bikerman's viewpoint.
2. The traction of an automobile tire in snow can be increased by changing the wetting properties of the surface of the tire. Explain!
3. The contact angle of mercury on glass is about 120° ($\cos\theta = -0.5$). What is the significance of the -0.5 for the $\cos\theta$ in terms of the capillary method of determining surface tension.
4. The motion of a ship, boat, or torpedo through water is significantly influenced by the surface coating. Would friction be affected by changing a hydrophobic surface to hydrophilic and give reasons for your opinion.
5. The solubility of benzene in water changes the surface tension of water, $\gamma_{W/A}$, from 72.8 to 62.2 dynes/cm for a saturated solution. The solubility of water in benzene has only a small effect on the $\gamma_{O/A}$ (28.8 dynes/cm when $\gamma_{O/W} = 35.0$ dynes/cm). If the value of $S = 9.0$ dynes/cm for benzene on water as an initial value—what will eventually occur as the water becomes saturated with the benzene?
6. The dissipation of oil slicks on water was initially effected by adding detergents. What does this do and why has this method ceased to be employed?
7. Methane can be encapsulated in molecular sieves 3A (zeolite) under high pressure and high temperature. The experimental data is given in Table C.A and can be shown to follow the Langmuir adsorption isotherm [Eq. (C.24)]. Plot the data in Table C.A and determine the maximum storage capacity by a plot according to Eq. (C.25).

TABLE C.A
Methane Uptake in a 3A Molecular Sieve as a Function of Pressure

Pressure (Pa) $\times 10^{-8}$	% CH_4 (w/w)[a,b]
0.12	1.5
0.69	3.7
0.69	3.8
1.38	4.7
2.07	5.7
2.76	5.8
3.79	6.6
4.14	6.4
4.14	6.9

[a] Based on weight of zeolite after activation under vacuum at 350°C.
[b] 2 hr encapsulation period at 350°C.

TABLE C.B
Results for the Adsorption of Acetic Acid by Active Charcoal

A[a]	B[b]	C[c]
50.0	10.0	42.23
25.0	10.0	20.3
10.0	25.0	17.8
5.00	50.0	14.3
2.50	50.0	4.70
1.00	50.0	0.22

[a]The volume of acetic acid (1.06 M) diluted to 100 mL for adsorption by 1.00 g of active charcoal.
[b]The volume of solution at equilibrium taken for titration with standardized NaOH (0.1189 M) to determine the residual acetic not adsorbed by the active charcoal.
[c]The volume of base used to neutralize the acetic not adsorbed by the carbon.

8. Determine the parameters of the Freundlich adsorption isotherm [Eq. (C.26)] for the adsorption of acetic acid by active charcoal. The charcoal is used to adsorb the acid from aqueous solutions of different concentrations. When equilibrium is reached the amount of acetic acid was determined by titration with standardized NaOH (0.1189 M). The data is given in Table C.B.

   Plot the mass of acetic acid adsorbed per gram of carbon (x/m) vs. the equilibrium concentration of acetic acid.

   Determine the Freundlich parameters from a plot of $\log x/m$ vs. $\log C$.

FURTHER READING

K. S. Birdi, *Surface and Colloid Chemistry Handbook*, CRC Press, Boca Raton, Florida (1999).
P. V. Brady, Editor, *Physics and Chemistry of Mineral Surfaces*, CRC Press, Boca Raton, Florida (1995).
C. Noguera, *Physics and Chemistry at Oxide Surfaces*, Cambridge University Press, New York (1995).
D. J. Shaw, *Introduction to Colloids and Surface Chemistry*, Butterworth-Heinemann, Woburn, Massachusetts (1992).
K. Christmanmn, *Introduction to Surface Physical Chemistry*, Springer Verlag, New York (1991).
J. B. Hudson, *Surface Science*, Butterworth-Heinemann, Woburn, Massachusetts (1991).
A. Adamson, *Physical Chemistry of Surfaces*, 5th Ed., Wiley, New York (1990).
G. R. Castro and M. Cardona, *Lectures on Surface Science*, Springer Verlag, New York (1987).
R. Aveyard and D. A. Haydon, *An Introduction to the Principles of Surface Chemistry*, Cambridge University Press, Cambridge (1973).
G. G. Parfett, Surface and colloid chemistry, in *Chemical Kinetics*, A. F. Trotman-Dickerson, Editor, Pergamon Press, Oxford (1996).
www.jools.com/
Introduction to Surface Chemistry, http://www.Chem.qmw.ac.uk/surfaces/scc/
Dynamic surface tension, http://www.firsttenangstroms.com/
Aspects of surface tension, http://hyperphysics.phy-astr.gsu.edu/hbase/surten.html
DemiLab, www.ilpi.com/genchem/demo/tension/

Appendix D

Patents

D.1. INTRODUCTION

It is easy to obtain a patent of your invention. It is more difficult to licence, sell, and profit from its implementation and use. Anyone can file a patent application in any country provided it is completed in the required language. Lawyers or patent agents are useful but expensive and should be engaged at a later stage when the application is ready to be filed or appealed.

With the availability of the provisional patent (in USA, Australia, New Zealand, UK) it is possible to have a full year of protection before a formal patent has to be filed. During the one year period it would be possible for the inventor to discuss the invention (preferably under a nondisclosure agreement that ensures confidentiality and does not constitute a public disclosure, and therefore, does not invalidate the filing of patents in foreign countries) with potential buyers without fear of losing rights or control of the invention unless, of course, the patent is not filed. Any public disclosure or sale of the product or operation of the invention for profit could invalidate the filing of *patents* in foreign countries. A typical confidential nondisclosure agreement is shown in Fig. D.1.

If a patent is not filed in a specific country it is possible for the invention to be made and sold there without fear of infringement. The product or process cannot be marketed in a country where it is covered by a patent except by the patent holder or a licensee.

D.2. FILING FEES

The cost of obtaining a patent (Table D.1) varies from one country to another, and the judicious choice of countries in which to file depends on the returns to be expected relative to the *filing fees* and subsequent maintenance fees.

Foreign patents can be filed under the Patent Cooperation Treaty (PCT). The cost depends on the number of pages (30 pages at \$455 + \$10 for each additional page), a designation fee (about \$1000), and some additional fees including transmittal fee, \$240, and search fee, \$1002, making a total of over US\$2700, excluding the patent agent's fee (about \$500). More fees (\$2500) are required within 19 months for the examination of the PCT filing. Costs continue to mount as the countries are selected for specific attention, especially if translations are required.

CONFIDENTIALITY AGREEMENT

You, ____________ (XYZ) (of ____________) the Recipient, produces, manufactures, sells or is interested

in ____________________

______________ and

John D Student (JDS) of ____________ the Discloser, has successfully performed preliminary experiments or has tested or invented a new method or a new application of the Product which information will be disclosed on the following terms and conditions:-

1. The Discloser (JDS) maintains his rights to patent the application (or has applied for a patent) or has Know-How* in relation to the Product which will be disclosed.
   2.1. The Recipient shall limit dissemination of Confidential Information within its Organization to those of its employees who need to receive it for the purposes specified and shall ensure that such employees are made aware of the Recipient's obligations hereunder and are bound to uphold them.
   2.2. A detailed description of any tests which are performed to assess the application or process and the results obtained shall be air mailed, couriered or faxed to JDS as they are obtained.
3. Recipient's obligations hereunder shall not apply or shall cease to apply to any information which:-
   (a) Recipient can demonstrate by written records was known to it prior to disclosure hereunder otherwise than as a result of a previous confidential disclosure by the Discloser
   (b) is in the public domain or come into the public domain through no fault of Recipient; *(except in the case of Know-How which may already be in the public domain)
   (c) is disclosed to Recipient with restriction on disclosure by a third party under no obligation of confidentiality to the Discloser with respect thereto.
4. No right or licence is granted hereby under an intellectual property to which a party is entitled or to use any Confidential Information except as specified herein.
5. The Effective Date of this Agreement is the date by which it has been signed by both parties as specified below.
6. All obligations of the parties under this Agreement expire after a period of seven (7) years following the Effective Date.
7. This Agreement is subject to the Law of the (country, state or province)

Agreed for John D. Student (JDS) Agreed for (XYZ)

______________ ____________________

Date ____________

* The agreement with respect to Know-How may be more complicated because the subject matter may be buried in the literature and is therefore in the public domain.

FIGURE D.1

The invention usually starts with an idea that the inventor wishes to protect and that he or she often would like to discuss on a confidential basis with colleagues, friends, and even potential buyers without fear of being robbed of the idea. Though in most countries the invention belongs to the first person to file a patent, in the USA the person who can prove to have invented first can have priority if diligence is pursued* The "first to invent" can be established for $10 by filing two copies of a description of

*This is expected to change soon to "first to file" so as to be consistent with the rest of the world.

TABLE D.1
Small Entity Fee Schedule[a] for USA and Canada (1999)

	USA ($)	Canada ($)
Basic filing fee	380	150
Independent claims in excess of three, each	39	
Claims in excess of 20, each	11	
Examination fee before 20 months after filing		200
Patent issuance fee	625	150
Provisional application filing fee	75	

[a]Complete fee schedules are available from the respective patent offices and from the web.

the invention with the US Commissioner of Patents under their Disclosure Document Program (see Fig. D.2). The Patent Office in Washington, DC stamps and dates the disclosure and returns one copy to the inventor, the other copy is held on file for two years as proof of date of the invention.

A search of the literature and patents would be necessary to determine if the idea is new and not already patented or published in the open literature. This can be done by a review of patent abstracts or by a computer search of appropriate data banks. Patent agents can do this at a cost ranging from $100 to $500.

Having established the feasibility of the idea, the inventor must now decide if a patent will be filed and where. For $75 a provisional patent can be filed in the USA although other countries may be preferred if the formal patent is to be filed there. This provisional patent can be in the form of a publication to be submitted, a report or a preliminary draft of the patent. There is no requirement to include claims, although it is important to present the supporting data and results as well as the object of the invention and its novelty. The improvement of the invention over previous versions should be stressed.

After filing the provisional patent, the inventor should be trying to licence, sell, or otherwise exploit the invention on a confidential nondisclosure agreement with the view that an interested company would, within the one year period of grace, file the necessary patents in various countries on behalf of the inventor and, of course, pay all the application fees and maintenance fees. If this does not transpire, then it is up to the inventor to pay the application fees which would be classed as a small entity status and is therefore usually half of the regular filing fee.

A license is a means by which an owner (the licensor) grants to the user (the licensee) a license (which can be exclusive, nonexclusive, or limited by time or district) under the patent to use the product or technology in exchange for annual (or semiannual or quarterly) royalties, plus or including, a minimum fee. Minimum royalty fees guarantee the inventor a return on the license and some diligence on the part of the licensee to market the invention. The royalty is usually based on a percentage of gross sales (e.g., 5%) or profits. Profits, however, can be manipulated and reduced by paying high consulting fees to sister companies, thereby lowering or eliminating royalties. Other topics covered by the license include the duration, notice of cancellation, improvements, infringement suits, and settlement of disputes by arbitration, etc.

U.S. DEPARTMENT OF COMMERCE
Patent and Trademark Office
Address: COMMISSIONER OF PATENTS AND TRADEMARKS
Washington, D.C. 20231

DISCLOSURE DOCUMENT RECEIPT NOTICE

Receipt of your Disclosure Document and Government fee of $10 is acknowledged. The date of receipt and the Disclosure Document identification number have been stamped on the attached dupicate copy of your request. This date and number should be referred to in all communications related to this disclosure Document.

WARNING

It should be clearly understood that a Disclosure Document is not a patent application, nor will its receipt date in any way become the effective filing date of a later filed patent application. A Disclosure Document may be relied upon only as evidence of conception of an invention and a patent application should be diligently filed if patent protection is desired.

Your Disclosure Document will be destroyed two years after the date it was received by the Patent and Trademark Office unless it is referred to in a related patent application filed within the two-year period. The Disclosure Document may be referred to by way of a letter of transmittal in a new patent application or by a separate letter filed in a pending application. Unless it is desired to have the Patent and Trademark Office retain the Disclosure Document beyond the two-year period, it is not required that it be referred to in the patent application.

The two-year retention period should not be considered to be a "grade period" during which the inventor can wait to file his patent application without possible loss of benefits. It must be recognized that in establishing priority of invention an affidavit or testimony referring to a Disclosure Document must usually also establish diligence in completing the invention or in filing the patent application since the filing of the Disclosure Document.

You are also reminded that any public use or sale in the United States or publication of your invention anywhere in the world more than one year prior to the filing of a patent application on that invention will prohibit the granting of a patent on it.

Disclosures of invention which have been understood and witnessed by persons and/or notarized are other examples of evidence which may also be used to establish priority.

If you are not familiar with what is considered to be "diligence in completing the invention" or "reduction to practice" under the patent law or if you have other questions about patent matters, you are advised to consult with an attorney or agent registered to practice before the Patent and Trademark Office. The publication, *Attorneys and Agents Registered to Practice Before the United States Patent and Trademark Office 1975*, is available for $3.70 from the *Superintendent of Documents, Washington, D.C. 20402.* Patent attorneys and agents are also listed in the telephone directories of most major cities. Also, many large cities have associations of patent attorneys which may be consulted.

Applicants are cautioned with respect to using attorneys or agents who solicit patent business, directly or indirectly, through advertising. Attention is called to Patent and Trademark Office Rule of Practice 345(a) which reads:

> "The use of advertising, circulars, letters, cards, and similar material to solicit patent business, directly or indirectly, is forbidden as unprofessional conduct, and any person engaging in such solicitation, or associated with or employed by others who so solicit, shall be refused recognition to practice before the Patent and Trademark Office or may be suspended, excluded or disbarred from further practice."

FIGURE D.2

D.3. COMPONENTS OF A PATENT

A patent consists of the following main sections:

Title of invention with names of inventors
Background of the invention
(a) Field of the invention
(b) Description of prior (or related) art
Summary of the invention
Brief description of the drawings
Description of the preferred embodiment
Claims
Abstract

Although these topics will be discussed in sequence they usually appear in different order.

D.3.1. Title

A snappy title is worth some thought and can tell the reader exactly what the invention is all about. The home address(es) of the inventor(s) must be supplied.

D.3.2. Background of Invention

The contents appear under two headings.

(a) Field of the Invention

The field may be broad as well as narrow and may differ from what the patent office may select. Other patents of the subject can be of great help and are essential for this, and the general jargon to be used.

(b) Description of Prior (or Related) Art

This outlines the need for the invention and the problems to be solved. This means that it is important to describe the technology as it is at present and how the invention can change and improve the world. This is analogous to the review of the literature with references and examples wherever possible.

D.3.3. Summary of the Invention

This section describes how the invention solves the problems mentioned in Section D.3.2 and how the invention improves on previously available units. It explains how the state of the art will be advanced and how it will be of benefit to mankind. The Summary is not a rewording of the Abstract but an anticipation of the Claims of the patent. This section is often referred to as Summary of Disclosure.

D.3.4. Brief Description of the Drawings

This section lists the captions or legends to the figures used to illustrate various aspects of the invention. The components of the drawings are usually clearly numbered and referred to by number in Section D.3.5.

If the drawings are inadequate then the examiner might ask for additional figures or clearer versions for some of the drawings. The patent office will supply detailed requirements and specifications for the drawings if asked.

D.3.5. Description of the Preferred Embodiment

This is the heart of the patent where the inventor must now give all the details and basis of the invention. In principle, the details should be sufficient for anyone "skilled in the art" to duplicate the invention. This is not always adhered to and occasionally an important and essential step is omitted or a further improvement is kept secret and not included in another patent. This description makes use of the drawings and figures, with examples, which represent tests and experiments, with tables of results where possible.

The section is usually ended with a general statement which is meant to indicate that, to one skilled in the art, other obvious applications and uses need not be described, e.g., "Since various modifications can be made in my invention as herein above described, and many apparently widely different embodiments of same made within the spirit and scope, it is intended that all matter contained in the accompanying specifications shall be interpreted as illustrative only and not in a limiting sense."

D.3.6. Claims

This section starts with: "What I claim as my invention is":

1. The claims, which are numbered, are the main goal of the patent and the only parts which can be changed once the patent has been filed. The claims are classified as independent claims (limit of 3 — more cost extra) which stand alone, and dependent claims which refer to a previous claim. Extra claims (over 20 in all) add to the cost of the filing fee. Each country has different requirements concerning dependent and independent claims. If one or more of your claims are allowed by the examiner then the patent will be granted.

The first claim should cover the complete invention being broad and encompassing. Two or three independent claims are usually sufficient if supplemented by several dependent claims. These allow for variation in one or more of the parameters or components of the invention to be claimed.

D.3.7. Abstract

This is best written after the application is completed. It is a short paragraph which describes the invention, its purpose, operation, and use. This abstract appears on the front page of the patent and in most collections and lists of patents and abstracts describing new issuances. It is this abstract that attracts potential buyers searching for new products and ideas to market and sell. Hence, considerable care should be taken in its preparation.

D.4. THE PROVISIONAL PATENT

The best approach to writing a patent on your own is to obtain some previously issued patents on the subject and to acquire the jargon and style of the subject. Such patents can be ordered by number from the patent offices.

The response from the patent examiner is usually to deny the validity of all the claims based on previous patents which are included with his reply. The inventor would reply by changing the nature of the claims or point out to the examiner that he/she is mistaken. A telephone call can do much to clarify the changes required to make the claims acceptable.

When there is no possibility of obtaining at least one claim, the patent will be finally denied. At this point it may be possible for the inventor to file a continuation-in-part in which new evidence and results are presented in the revised embodiment to justify the new and altered claims. A patent agent at this time would be a great help in getting the patent approved.

FURTHER READING

D. Pressman, *Patent It Yourself*, 8th Ed., Nolo Press, Berkeley, California (2000).
A. L. Durham, *Patent Law Essentials*, Greenwood, Westport, Connecticut (1999).
J. L. Bryant, *Protecting Your Ideas: The Inventor's Guide to Patents*, Academic Press, San Diego, California (1999).
J. R. Flanagan, *How to Prepare Patent Application*, Patent Educational, Troy, Ohio (1983).
R. O. Richardson, *How to Get Your Own Patent*, Sterling, New York (1981).
A. M. Hale, *Patenting Manual*, 2nd Ed., S.P.I. Inc., New York (1981).
R. A. Buckles, *Ideas, Inventions and Patents*, Wiley, New York (1957).
US Patent Office, (patent search, citation) http://www.uspto.gov/patft/index.html
Canadian Patent Database, http://patents1.ic.gc.ca/intro-e.html
Canadian Intellectual Property Office, Http://www.cipo.gc.ca/
IBM Patent Search, http://www.delphion.com
Introduction to Patents, http://www.derwent.com/patent-intro/
Patent delivery service, http://www.patentgopher.com/
http://www.freepatents.org
Wacky patents, http://www.patent.freserve.co.uk
More on patents, http://patents.cos.com/

Index